THE LIFE CYCLE OF COPPER, ITS CO-PRODUCTS AND BYPRODUCTS

ECO-EFFICIENCY IN INDUSTRY AND SCIENCE

VOLUME 13

The titles published in this series are listed at the end of this volume.

The Life Cycle of Copper, Its Co-Products and Byproducts

by

Robert U. Ayres
Center for the Management of Environmental Resources INSEAD, Fontainebleau Cedex, France

Leslie W. Ayres
Center for the Management of Environmental Resources INSEAD, Fontainebleau Cedex, France

and

Ingrid Råde
Physical Resource Development, School of Physics, Chalmers University of Technology, Gothenburg, Sweden

With contributions from

Roland Geyer
Donald Rogich
Benjamin Warr

Foreword by Luke Danielson
Project Director: Mining, Minerals and Sustainable Development (MMSD)
International Institute of Environment and Development (IIED)
London

KLUWER ACADEMIC PUBLISHERS
DORDRECHT / BOSTON / LONDON

A C.I.P. Catalogue record for this book is available from the Library of Congress.

ISBN 1-4020-1552-6

Published by Kluwer Academic Publishers,
P.O. Box 17, 3300 AA Dordrecht, The Netherlands.

Sold and distributed in North, Central and South America
by Kluwer Academic Publishers,
101 Philip Drive, Norwell, MA 02061, U.S.A.

In all other countries, sold and distributed
by Kluwer Academic Publishers,
P.O. Box 322, 3300 AH Dordrecht, The Netherlands.

Cover photo of Bronze statue of Amita-Buddha, "Diabusu" of Great Buddha, cast in 1252 A.D. by Ono-Goroemon and Tanji-Hisatomo, is used with permission of Ms. Michiko Sato, manager of the Koutoku-in-temple, Japan.

Printed on acid-free paper

Printed in the Netherlands.

TABLE OF CONTENTS

FOREWORD ix

ACKNOWLEDGEMENTS xiii

CHAPTER 1. INTRODUCTION 1
1.1. The life cycle perspective 1
1.2. Historical background 6
1.3. Geology of copper 14
1.4. Geology of lead and zinc 18

CHAPTER 2. COPPER: SOURCES AND SUPPLY 21
2.1. Physical properties and chemistry 21
2.2. Copper production 22
2.3. Process technology 23
2.3.1. Mining 27
2.3.2. Beneficiation 30
2.3.3. Leaching 33
2.3.4. Cementation and solvent extraction (SX) 34
2.3.5. Roasting, and smelting 35
2.3.6. Converting 37
2.3.7. Electrowinning 38
2.3.8. Fire refining and electrolytic refining 38
2.3.9. Future trends in primary processing 39
2.4. Exergy and exergy flows 40
2.5. Sulfur recovery 44
2.6. Production-related wastes and emissions 46
2.6.1. Mining wastes 46
2.6.2. Beneficiation wastes 48
2.6.3. Leaching (acid) wastes 50
2.6.4. Smelting wastes 51
2.6.5. Wastes from finishing operations 54
2.6.6. Recycling (secondary recovery) wastes 54
2.6.7. Toxic releases 54
2.6.8. Global estimates of airborne emissions 54
2.7. Optimal extraction/production 56

CHAPTER 3. COPPER: DEMAND AND DISPOSITION 59
3.1. Consumption patterns and trends 59
3.2. Accumulation of copper stocks in the anthroposphere 70

3.3. Dissipative uses and losses of copper 80
3.4. The future of demand for copper 83

CHAPTER 4: LEAD, ZINC AND OTHER BYPRODUCT METALS 101
4.1. Context 101
4.2. Physical properties and chemistry of lead and zinc 101
4.3. Lead process technology 103
4.3.1. Ore mining and beneficiation 103
4.3.2. Sintering 105
4.3.3. Smelting 105
4.3.4. Drossing and final refining 106
4.3.5. Exergy and exergy flows 106
4.4. Lead sources and uses 109
4.5. Zinc processing 112
4.5.1. Ore mining and beneficiation 112
4.5.2. Roasting and sintering 112
4.5.3. Smelting and refining 114
4.5.4. Exergy and exergy flows 114
4.5.5. Recycling old zinc scrap 117
4.6. Zinc sources and uses 117
4.7. Lead and zinc wastes and emissions 120
4.8. Other byproduct metals 127
4.8.1. Antimony 130
4.8.2. Arsenic 131
4.8.3. Bismuth 135
4.8.4. Cadmium 135
4.8.5. Germanium 140
4.8.6. Gold 141
4.8.7. Indium 142
4.8.8. Rhenium 142
4.8.9. Selenium 143
4.8.10. Silver 143
4.8.11. Sulfur 143
4.8.12. Tellurium 147
4.8.13. Thallium 147

CHAPTER 5. THE FUTURE OF RECYCLING 149
5.1. Background 149
5.2. Recovery and recycling of copper from old scrap 153
5.3. Recovery and recycling of electronic scrap 157
5.4. Copper as a contaminant of recycled steel 159
5.5. Copper recycling wastes and emissions 161
5.6. Recovery and recycling of lead 162
5.7. Recovery and recycling of zinc 164

5.8. Recovery and recycling of byproduct metals 165
5.8.1. Antimony 165
5.8.2. Arsenic 165
5.8.3. Cadmium 165
5.8.4. Germanium 166
5.8.5. Gold 166
5.8.6. Indium 166
5.8.7. Selenium 167
5.8.8. Silver 167
5.8.9. Tellurium 167
5.9. Further comments on losses and potential recoverability 167

CHAPTER 6. CONCLUSIONS AND QUESTIONS 169
6.1. Introduction 169
6.2. Copper availability 169
6.3. Copper demand: the continuing electrification of the global energy system 171
6.4. Lead, zinc and byproduct metals availability and uses 173
6.5. Concentration, reduction and refining technology 174
6.6. Sulfur recovery and acidification of the environment 174
6.7. Copper, lead and zinc recycling 175
6.8. Emissions and accumulation of copper and zinc in agricultural soils; probably a non-problem 177
6.9. Accumulation of arsenic, cadmium and other toxic metals in the terrestrial environment: A real problem 177
6.10. The threat of 'toxic time bombs' 180
6.11. The long-term prospect 180

REFERENCES 183

APPENDIX A: THE EXERGY CONCEPT 193
A1. Definition and description of exergy calculations 193
A2. Exergy as a tool for resource and waste accounting 195
A3. Composition of mixtures, including fuels 197

APPENDIX B: THE BEHAVIOR OF COPPER, LEAD AND ZINC IN SOIL 201
B1. Metals in soils 201
B2. Aqueous phase speciation 203
B3. Solid phase constituents and complex formation 204
B4. Summary 210

APPENDIX C: GLOBAL COPPER MODEL 211
C1. Introduction 211
C2. A model of the global copper system 211

C3. Calibration of the model 219
C4. Copper consumption scenarios 227
C5. Copper system scenarios 232

APPENDIX D: GLOSSARY 253

INDEX 257

FOREWORD

Achieving the goals and objectives of sustainable development requires better information about the consequences of proposed actions. Partial information accounts for many failed efforts in the past. The financial implications for the proponent of the projects have often been more thoroughly analyzed than the implications for other actors. The impacts on biological diversity, or on the social fabric of local communities, have often been ignored. Decision-makers may also focus more on the short-term consequences instead of longer-term impacts, creating negative unintended consequences.

It is clear that better decision-making processes are needed. Making better decisions requires identifying, obtaining, synthesizing and acting on larger and more diverse data sets, including information that has previously been overlooked in development decisions. The good news is that better processes are being developed and are becoming available.

If the goal is to reach decisions that are broadly understood and accepted, affected communities need to be consulted. Early public participation in defining problems is a prerequisite to effective decision-making. There is no universal formula or checklist of information applicable to every proposed project. The scope of information required should not be determined from the start by small cadres of experts. It is unlikely that any individual or small group processes all of the expertise to achieve the kind of profound inter-disciplinary synthesis that is needed.

Determining the kind of information needed for decision-making is an important start. However, the practicalities of gathering the information are often daunting. Existing processes have serious limitations. The standards of reporting vary. Often the information gathered is based more on past necessities, or on what is inexpensive to gather, than on the demands of a holistic decision-making process for sustainable development. Information from different sources is often not comparable. Companies, governments or others frequently hold useful information in confidence. In some cases it is deliberately distorted or concealed. And it is not surprising that high quality information is more frequently gathered about the concerns of the rich and powerful than those of the poor and dispossesed.

Synthesizing available information also presents an enormous challenge. Access to information alone is insufficient unless there are effective ways of sorting and synthesizing the data. Sustainable development will be hindered rather than helped by compilation of masses of undifferentiated and poorly digested information.

Life cycle approaches to the materials used in national or world economies show considerable promise in helping to elaborate the architecture of decision-making for sustainable development. As practitioners better understand and work with life cycle analysis as a practical decision-making tool, they help define the kinds of information that is needed, learn to tap into sources for obtaining it, refine methods of synthesizing it more effectively, and how it can be used to make better decisions.

This book, and the research and workshop on which it was based, identifies some of the key issues in the use of life cycle approaches in support of sustainable development.

Life cycle approaches to better decision-making require access to reliable data at a level of detail not generally publicly available outside companies. Many companies are reluctant to share this information, particularly where their competitors refuse to folow suit. Those who are given access to this information are often asked to disguise the origins, or to present it in the aggregate. This makes the processes of peer review and quality assurance difficult.

As a result, life cycle approaches have made more headway inside companies as an aid to company decision-making than in public bodies to help public policy formulation. A key to gaining greater acceptance of life cycle tools in the public policy arena is greater confidence in underlying data, which can only gained through broader access to the data and more rigorous use of peer review.

The pursuit of greater eco-efficiency is a necessary but not a sufficient condition for sustainable development. Life cycle analysis has in general focused on the material: consumption of water and other resources, levels of emissions, energy consumption and the like. While this is very useful, decision-making for sustainable development requires balancing these with other factors such as employment and livelihoods, equity, and the needs of developing countries. Bluntly, the most modern and eco-efficient industrial technology is concentrated in the richest countries. Criteria and policies that make decisions only on eco-efficiency grounds without integrating these other factors might well wind up creating yet more barriers to the products of the global South, and yet more concentration of wealth in the countries that already have the most. As just one example, the wealthier countries consume more of most materials. The post-consumer material available for recycling therefore tends to be concentrated in the countries that do the most consuming; a drive for higher levels of recycling could, without some compensating mechanisms, create empoyment principally in the richest countries.

It may be possible to develop life cycle approaches that better reflect values such as poverty alleviation, global equity and livelihoods, though these will require considerable thought. And they are likely to be hampered by the fact that data on the factors most important to the developing world may not be as available or usable as data relevant to the physical eco-efficiency factors.

There are some areas very important to public policy, such as current rates of recycling, where data are often sketchy and hard to obtain. Better data on recycling should be a high priority for better public policy in the metal sector.

The tradeoffs are complex. While for example eco-efficiency in general might suggest economizing in the use of materials, greater efficiency in the use of energy may well suggest greater use of some materials such as copper.

This work was undertaken to support the Mining, Minerals and Sustainable Development Project (MMSD), a two year process of research and consultation designed to examine the potential contribution of the minerals sector to sustainble development. The MMSD Project, of which I had the privilege of serving as Director, was managed by the International Institute for Environment and Development in London in collaboration with many other institutions and research teams worldwide. This book is one of a substantial number of research projects supported by MMSD, designed to test the state of knowledge in a variety of fields relevant to that goal. The authors' research was subject to intensive discussion, debate and criticism in a workship in New York on August 9–10, 2001. The proceedings of that workshop and the overall report of the MMSD Project, *Breaking New Ground*, are available at www.iied.org/mmsd.

This book has been a valuable and much appreciated contribution to the MMSD Project, and a major contribution to the public literature on the life cycles of copper, zinc and other metals. It would not have been possible to produce this important work without the authors' expertise and years of experience in life cycle analysis, and their intimate knowledge of data sources.

I would like to express my personal appreciation as well as the thanks of IIED and the entire project team for the diligence, patience and insight of Professor Robert and Leslie Ayres and their colleagues for this excellent contribution to our work.

Luke Danielson
Director: Mining, Minerals and Sustainable Development Project (MMSD)
International Institute for Environmenta and Development (IIED)
London

ACKNOWLEDGEMENTS

First and foremost we acknowledge the direct contributions of PhD student (now Dr.) Roland Geyer, who translated a number of German documents for us, Donald Rogich, formerly Director of the Mineral Economics Division of the US Bureau of Mines (ret.), who provided some valuable archival data, and PhD student (now Dr.) Benjamin Warr, who contributed Appendix B.

Next, we acknowledge the support of the Mining, Minerals and Sustainable Development (MMSD) Project at the International Institute for Environment and Development (IIED), London, which financed this research, during the 12 month period October 2000 through October 2001. In particular, we thank Luke Danielson, Director of the project, and Caroline Digby, Director of Research, and Bruce Howard, staffer. The MMSD Project organized a two day workshop, August 9–10, 2001 in New York City. Mary Stewart, of the Center for Risk, Environment and Systems Technology and Analysis (CRESTA), Department of Chemical Engineering, University of Sydney, Australia prepared the workshop report, which was of great assistance. It would be excessive to list all the attendees here, but we cannot overlook the contributions of Mary Ann Curran (EPA), Janice Jolly, USBM (ret.), Gustavo Lagos (Catholic University of Chile), Eric Rodenburg (USGS), John Tilton (Colorado School of Mines). Industry representatives included Scott Baker and Jan Smolders of the International Copper Association (ICA), Martin Ruhrburg of the International Copper Study Group (ICSG), Craig Boreiko of the International Lead Zinc Research Organization (ILZRO) and Don Smale of the International Lead/Zinc Study Group (ILZSG), Leonard Surges of Noranda, and Louis Webberly of BHP Minerals, all of whom provided useful documents and advice.

We also acknowledge a considerable debt to the Department of Physical Resource Theory, Chalmers Institute of Technology, Gothenborg, Sweden, which hosted one of us (RUA) for three months as a visiting Jubilee Professor, during that period. In particular, we thank Christian Azar (Professor), John Holmberg, and Sten Karlsson, for many productive conversations, ideas and suggestions. Undergraduate students of the department, Julia Hansson and Johan Rootzen, undertook a research project on the history of copper mining in Sweden. Although we decided not to include their report directly in this book, it was nevertheless very illuminating. Their report has been included in our report to the MMSD and we understand that it is is accessible through the MMSD website www.iied.org/mmsd.

Finally, we acknowledge valuable help from Prof. Arnim von Gleich, of Fachhochschule, University of Applied Sciences, Hamburg, who provided helpful advice and documents from German sources about which we might not have known.

RUA
LWA
IR

CHAPTER 1

INTRODUCTION

1.1. The life cycle perspective

The life cycle analogy originates from biology. The life of an individual (higher) animal begins with conception – the fertilization of the egg – and proceeds through a series of stages including growth of the foetus in the womb, birth, infancy, adolescence, maturity, senescence and death. For most purposes the briefer characterization 'from cradle to grave' suffices (in biology), although in the present study we allow also for reincarnation (recycling). In the case of a metal such as copper one automatically thinks of the 'cradle' as the mine and the 'grave' as the ultimate disposal site, whether in a landfill or a sediment. Recycling is analogous to reincarnation, in the sense that it is the beginning of a second life.

However in biology it is said that 'ontogeny recapitulates phylogeny', meaning that the development process of an individual organism recapitulates the evolution of that organism. For example, the developing human foetus grows a vestigial but recognizable tail, which subsequently disappears before the baby is born. We also have other vestigial organs such as the appendix and the tonsils. Some of our evolutionary history remains in our bodies. So it is with industrial systems. There is an important historical component to a life cycle. This is true also for technological systems. The mining process itself has evolved considerably since its crude beginnings, as has the pattern of utilization and disposal. While the future is not a straightforward extension of the past, there are some useful lessons for the future to be learned from the historical record.

It is important to emphasize at the outset that, despite its title, this study is not a formal life cycle assessment (LCA) approach. It is, at best, a sort of 'kissing cousin'. LCA methodology has now been standardized by the International Organization for Standardization (ISO 14040), where it is defined as follows: 'LCA is a technique for assessing the environmental aspects and potential impacts associated with a product by compiling an inventory of relevant inputs and outputs of a system; evaluating the potential environmental impacts associated with those inputs and outputs; and interpreting the results of the inventory and impact phases in relation to the objectives of the study' (Udo de Haes 2002).

In fact, formal LCAs are essentially 'cradle-to-grave' comparisons between two or more products performing the same function and serving the same market niche or (less frequently) two or more processes for producing the same product or material. The comparisons are made at three levels. The simplest and quantitatively most reliable is known as *inventory analysis*. This analysis consists of compiling a detailed list of all of the direct, as well as indirect, inputs to the system (e.g. to mining, smelting, refining, energy use, etc.). The direct and indirect outputs are also compiled. These compilations can be in mass terms or in exergy terms, or preferably both.

Most LCA studies provide figures in mass terms, except for fuels, which are typically given in terms of heat content (enthalpy). This approach has serious drawbacks, among which is the impossibility of utilizing mass balances for data verification since energy sources and other inputs are given in different units (Ayres 1995a). The term exergy is unfamiliar to most non-specialists, which is a problem that cannot be avoided. However exergy is defined and discussed in *Appendix A*, which also includes some relevant data for later use in this report. For a published account of the use of exergy for environmental analysis and LCA see (Ayres 1995b; Ayres and Martinas 1995; Ayres et al. 1998).

The second level of comparison in conventional LCAs is known as *characterization*. In this step the inputs to and outputs from the product system are grouped into a number of impact categories. Examples of such categories are climate change, stratospheric ozone depletion, acidification, eutrophication, eco-toxicity, human toxicity, and abiotic resource depletion. The different impacts of a product system in each of these categories are aggregated in terms of indicators for these categories, resulting in the environmental profile of the given product system.

The third level develops 'weight factors' in order to aggregate the characterization results of the second level into one single score. In LCA terms this step is called 'weighting'. There is a very large literature on this topic, which cannot (and need not) be summarized here.

Most economists would prefer to use 'shadow prices' for this purpose. The methodology, therefore, is known as 'contingent valuation' (Johansson 1987; Freeman 1993). In the absence of competitive markets for determining the prices of environmental services (or human health) the usual approach is to do surveys of consumers or other interested parties to find our 'willingness to pay (WTP) or 'willingness to accept' (WTA). Despite serious problems (e.g. the lack of built-in budget constraints and the well-known problem that experienced pollsters can obtain almost any result by choosing the way in which the question is asked) the WTP methodology has been used in practice in some cases, with mixed results. But it is clearly limited by the fact that it undervalues environmental effects that people do not understand or cannot imagine, or that are outside their daily experience, while over-valuing other impacts that do directly concern their living conditions or have somehow been exaggerated or caught the public's imagination.[1]

---

[1] Examples might include public fears of radiation poisoning (due to Three-Mile Island and Chemobyl), dioxin (Times Beach, Seveso), and asteroid collisions (Hollywood). Every risk analyst knows that tobacco and drunken driving are hundreds or thousands of times riskier than any of the three examples. Yet people smoke and drink.

The limitations of the WTP methodology has led to the increasing use of non-monetary approaches by various researchers, since the 1970s. The first idea was to use a single physical indicator of impact. The most popular was 'net energy analysis' which compared products and processes in terms of the aggregate energy inputs and outputs associated with all stages of the life cycle. Others were total materials input, total materials input per unit service (MIPS) and total solid waste output (Hunt et al. 1974; Schmidt-Bleek 1993a; Schmidt-Bleek 1993b; Udo de Haes 2002).

The above stepwise procedure of life cycle impact assessment, distinguishing between characterization and weighting, has been introduced by researchers at Leiden University (Heijungs et al. 1992), although at that time under different terms, in order to separate the scientific and normative components of the analysis as much as possible (See ISO 14042). A discrete number (typically about 10) of impact categories are identified, overlapping as little as possible. As stated above, most lists include categories like climate change, ozone depletion, acidification, eutrophication, eco-toxicity (itself hard to define) human toxicity, and resource depletion. However some lists are longer, and some of these items are recombined into other groups. The categories are often subdivided into two groups, namely those related to extractions and those related to emissions.

Within a category, it is generally possible to model the impact of the different pollutants more precisely. For instance, a given greenhouse gas can be assessed in relation to others in terms of its share in the total greenhouse forcing effect at a given time. An important point concerns the level in the impact chain which is chosen for the aggregation of the effects. One option is that the aggregation takes place early in the chain, for instance using Global Warming Potentials as a measure for the impacts on climate forcing. This was the original 'environmental themes approach' proposed by Heijungs et al. (1992). Another option is that the aggregation takes place late in the chain, for instance describing the effects in terms of damage to human health or to biodiversity. Steen and Ryding (Steen and Ryding 1993) were the first to propose such an 'environmental damage approach'. They argued that each emission can be allocated to one or more environmental damage categories, to which it makes a quantifiable contribution At present attempts are being made to integrate the environmental themes approach and the environmental damage approach into one encompassing framework (Bare et al. 2002).

One of the main areas where substantial progress has been made is in refining the models for the characterization of eco-toxicity and human toxicity impacts.

When the USEPA began publishing its Toxic Release Inventory (TRI) data in the late 1980s many analysts simply added up the release data in mass terms, on the basis of the toxicity of the compounds, to construct comparisons of toxic emissions by industries, states or even counties (Newell 1998, p. 110). This implicitly gave each chemical on the list an equal hazard-factor, per unit of mass. An influential study of the packaging sector tried to improve on this by weighting chemical hazards based on known toxicological factors, finally arriving at dollar values by comparison with lead (Tellus 1992b; Tellus 1992a; Zuckerman and Ackerman 1995). However, this approach fails to take into account the environmental processes that intervene between emissions and effects, especially transport and immobilization mechanisms. These can spell differences of many orders of magnitude in hazard (Hertwich et al. 1997).

The next step was to model the emissions to concentrations step, using a simplified 'box model' and including exposure parameters for human toxicity. The idea is to divide the world into interacting compartments (air, fresh water, sediment, topsoil, seawater, etc.) with empirically derived equations describing phase changes, immobilization and remobilization processes and flows to other compartments. Such models, assuming steady state conditions, have been applied most successfully to organic chemicals, using the fugacity concept to model equilibrium concentrations, based on volatility and solubility (Mackay and Patterson 1991; Mackay et al. 1993). Unfortunately heavy metal concentrations are more difficult to model this way, due to inadequate empirical knowledge of many environmental processes such as suspended particle formation, deposition, erosion and sedimentation (Hertwich et al. 1997).

The state of the art at present appears to be multimedia modeling, e.g. based on the 'Uniform System for the Evaluation of Substances' or USES developed at RIVM and adapted for LCA at Leiden University, Center of Environmental Science (CML) (Huijbregts 1999; Guinée 2001); or based on CALTOX (Hertwich et al. 1997). The original USES model included 100 chemical species, including some metals. At present the USES-based model gives characterization factors for some 180 substances. It includes a large variety of intermediate processes (e.g. transport to soil and to the ocean, plant uptake and bio-concentration) as well as assumptions about human exposure mechanisms via ingestion and inhalation.

Much more is known about carcinogens than other toxics. For carcinogens there is some data from which estimates of damage functions can be derived. Nowadays it is usual to assume that dose-effect functions for carcinogens are linear, and that there is no lower threshold. This is not a valid assumption for many metals, however, both because some toxic metals (including chromium, copper and zinc) are actually needed in trace amounts and because there are detoxification mechanisms for low doses that determine their availability.

At present, these aspects are not yet taken into account but probably will be in the future. This is not very easy however, because LCA is fundamentally a linear tool, assuming linear dose-response functions and using a concept of 'unit risk', based on such measures as the 'no observable effect' level (known as NOEL). The results of life cycle impact assessment are expressed in units which compare the emission of a given substance with that of a reference substance.

For eco-toxicity impacts, the results are based on changes in the species composition (known as PNEC, or Predicted No-Effect Concentration for water or soil ecosystems). Here generally some 'safety factor' is used, taking into account the number of species for which toxicity data for water and soil ecosystems are available. Unfortunately, these factors are still rather arbitrary.

For human toxicity there is an increasing tendency to express the impacts in terms of damage to human health. In order to arrive at one score for human toxicity, this requires a weighting between different types of disablement. Standardization takes place in terms of the so-called DALYs, or Disability Adjusted Life Years (Murray and Lopez 1996). The advantage of this procedure is that it enables a comparison with other types of impact, also influencing human health, such as effects of fine particles or accidents.

The present study is not about any particular metal product, and comparisons between products made from copper or copper alloys *vis à vis* similar products made from other materials are necessarily crude and simplistic and focused mainly on the reasons why one is substituting for the other (or not). Furthermore, the present study does not (and cannot) compile complete inventories of all inputs and outputs associated with a given copper product, or even a given stage in the production process. We aim for a somewhat more aggregated assessment. Similarly, quantitative impact analysis at the process level of detail cannot be attempted, except at the highest level of aggregation. The mining, concentration and smelting/refining processes in use are quite diverse, partly due to significant differences in the grade and concentration of ores being exploited in different locations and partly due to local factors such as energy costs and environmental regulations. In any case the relevant input/output (and cost) data are mostly considered proprietary.

On the other hand, there is a plethora of data, of variable quality, on emissions, concentrations, health impacts and ecosystem impacts. Because of extremely diverse and sometimes inconsistent data, there is no possibility of presenting a comprehensive yet comprehensible inventory of inputs and outputs from each stage of the life cycle process. Such inventories would have to be compiled for particular mines or smelter/refiners.

To summarize, we do not mean to say that life cycle analysis has no value. Very much to the contrary, there are enormous benefits to thinking in life cycle terms. Indeed, it can be argued that a life cycle perspective is now essential for decision-making in the mining and minerals sector. LCA is a powerful tool in its place. However, for several reasons, we do not feel competent to apply the full panoply of LCA tools and models, as defined by SETAC, to the task of attempting to assess the environmental or health impacts of heavy metal production, use or emissions, in general terms.

This book focuses on one metal, on copper. In this respect our approach can perhaps be more accurately characterized as substance flow analysis (SFA), in a life cycle perspective. SFA focuses on the different processes in the economy which are related to a given substance or group of substances. This tool aims at the analysis of a substance from cradle to grave, rather than at a product from cradle to grave as in LCA. But in other respects there is a strong resemblance to LCA. For instance, it makes no sense to compare copper, or lead, or zinc, *as such* with each other or any other element. Each element is unique, with unique attributes. One can compare copper with aluminum for high voltage transmission, or copper with glass fiber for information transmission, or one can compare brass with other materials for particular purposes. Or one can compare different smelting-refining processes provided it makes sense to assume similar conditions (ore grades, scale of production, availability of water, cost of electric power, markets for sulfuric acid, etc.).[2] In this respect the present approach follows the pattern of LCA. Also in LCA it is not the material as such that counts, but the specific contribution it provides to the defined function of the system in question.

What we have tried to do in this book is to present briefly what is known about the past present and future of copper (and also the closely related metals with which it is commonly associated) as a major industrial commodity in its larger social, economic and environmental context, with particular emphasis on downstream uses, disposal and recycling. Our approach might be characterized as 'cradle to cradle' (rather than 'cradle to grave'). It presents the relevant facts and conclusions primarily from an industrial-ecology perspective, rather than a purely economic perspective. It is not intended as a book about the applicability of life cycle methodology, either in the form of LCA or SFA, to the copper sector. *It is a state-of-the-art report about the copper, lead and zinc sectors from a life cycle perspective.*

This book is divided into six chapters including this introduction. The remainder of this chapter concerns itself with historical and geological background, for context. The next *chapter (2)* is devoted to copper sources and production, as well as production-related emissions. *Chapter 3* is about copper uses, flows, consumption, and accumulation. *Chapter 4* is devoted to lead, zinc and the most important byproduct metals (excluding molybdenum, nickel and cobalt, which are mostly mined for themselves). *Chapter 5* considers the problem of recycling in general and in the long term. *Chapter 6* gives the summary and main conclusions. *Appendix A* provides background information on the concept of exergy. *Appendix B* discusses copper, lead and zinc in soil. *Appendix C* gives a detailed description of the global copper model used in this book. *Appendix D* is a glossary of specialized terminology.

1.2. Historical background

Metals and other minerals are extracted from the earth because they are wanted or needed for various purposes. The French economist Auguste Say famously said that 'supply creates its own demand'. Copper was the second metal used by man, after gold. Availability, historical availability of the metal in native

---

[2] This has been done and published in the case of nickel processes (Anne Landfield, Ecobalance/Price Waterhouse Coopers, personal communication, August, 2001).

form certainly encouraged early applications, mainly for decorative purposes (c. 7000 BCE in Turkey). Hammered native copper has been found among Chaldean remains c. 4500 BCE (McMahon 1964). Copper-arsenic was probably the first alloy (c. 3400–3200 BCE),[3] probably created by hammering native copper at red heat with small amounts of arsenic. (Arsenic is a byproduct of some copper ores.) This discovery was followed shortly (c. 3000 BCE) by tin-copper (bronze), ushering in the 'Bronze Age'. The most important source in the ancient world was from deposits in the Sinai desert (originally exploited as sources of turquoise). Other early mines were in Syria and Cyprus, The discovery of bronze was undoubtedly facilitated by the discovery of complex copper-tin ores in a number of places, including Turkey, Cornwall and Wales, Indochina, Spain and a few other places (Thornton 1995). Copper-tin ores are virtually unknown nowadays, presumably having been exhausted in the distant past.[4] Yet, if there had never been any such ores, the bronze age might never have occurred. Ore minerals and materials related to copper are shown in *Table 1.1*.

Bronze has a much lower melting point than pure copper, which makes it suitable for casting, while it is hard enough for tools, swords and spears, and cannons. Smelting of copper oxides and carbonates began between 5000 BCE and 3000 BCE, both in Turkey and what is now the Israel-Egypt border. Annual production in the early Bronze Age was of the order of one or two hundred metric tons. Smelting of sulfide ores followed about 2500 BCE. Thereafter, pure copper was restricted largely to decorative uses (beads, bracelets), and coinage.[5] Bronze dominated utility goods until the iron age, but retained a place in several markets up to modern times. In fact, the Ottoman Turks conquered Constantinople in 1453 using wrought iron cannons that fired stone balls and made a terrifying noise. Yet cast bronze cannons were more compact (hence more mobile) and equally powerful. The French army that (temporarily) conquered northern Italy in 1494 used bronze cannon, and inaugurated the period of European dominance of artillery technology (Kennedy 1989, pp. 22–23). Today bronze is mainly used for ships propellers and castings of sculptures. In the Middle Ages pure copper sheet found a new use for roofing in Sweden and Alpine countries (where its strength was important in supporting the weight of snow accumulations).

---

3 A small admixture of arsenic makes copper harder. Pure copper is too soft and malleable for cutting tools or other implements, but the alloy with arsenic is much harder. The so-called 'iceman' (c. 3300 BC), whose frozen corpse was uncovered in the Austrian Alps a few years ago, carried a copper arsenic axe and had traces of both copper and arsenic in his hair (Landner and Lindeström 1999).

4 No such composite ores are mentioned in recent surveys of the mineral industries, such as (Kesler 1994).

5 Another use that has been important at some times and places is for casting statues, particularly of Buddha. The large copper Buddha at Kamakura in Japan (cover photo) may be the biggest single copper casting in the world.

Table 1.1: Ore minerals and materials

Element Ore mineral	Composition	Maximum content of desired element
Antimony Sb		
Stibnite	Sb_2S_3	72%
Tetrahedrite	$Cu_8Sb_2S_7$	25%
Arsenic As		
Arsenopyrite	FeAsS	46%
Realgar	AsS	70%
Orpiment	As_2S_3	61%
Löllingite	$FeAs_2$	73%
Smaltite	$CoAs_2$	72%
Enargite	Cu_3AsS_4	19%
Bismuth Bi		
Native bismuth	Bi	100%
Bismuthinite	Bi_2S_3	81%
Cadmium Cd		
Greenockite	CdS	78%
Sphalerite	(Zn,Cd)S	<1%
Copper Cu		
Native copper	Cu	100%
Azurite	$2CuCo_3.Cu(OH)_2$	56%
Bornite	Cu_5FeS_4	63%
Chalcopyrite	$CuFeS_2$	35%
Chalcolite	Cu_2S	80%
Chrysacola	$CuSiO_3.2H_2O$	39%
Cuprite	Cu_2O	89%
Enargite	Cu_3AsS_4	48%
Malachite	$CuCo_3.Cu(OH)_2$	58%
Gold Au		
Native gold	Au	100%
Electrum	(Au,Ag)	variable
Calaverite	$AuTe_2$	44%
Lead Pb		
Galena	PbS	87%
Anglesite	$PbSO_4$	66%
Cerrusite	$PbCO_3$	76%
Rhenium Re		
Molybdenite	$(Mo,Re)S_2$	<1%
Silver Ag		
Native Silver	Ag	100%
Argentite	Ag_2S	87%
Argentiferous galena	(pb,Ag,Bi,Sb)S	variable
Ag-ferrous tennantite-tetrahedrite	$(Cu,Fe,Ag)_{12}(As,Sb)_4S_{13}$	variable
Tellurium Te		
Calaverite	$AuTe_2$	56%
Copper ores	–	traces
Zinc Zn		
Sphalerite	ZnS	67%
Willamite	Zn_2SiO_4	56%

Source: Extracted from Kesler (1994; Appendix III).

Brass was introduced by the Romans during the reign of the Emperor Augustus.

Cumulative copper production between 2000 BCE and 700 BCE has been estimated at 500,000 metric tons, based on Greenland ice core data (Hong et al. 1994). Cumulative production during Roman times (350 BCE to 350 CE) has been estimated at 5 million metric tons, with annual production peaking at 15,000 metric tons (Ibid). Spain accounted for about half of the Roman total. World production declined to around 2000 metric tons per year (tpy) until the 9th century when new ores were discovered in Europe and China. The next peak in output (13,000 tpy) occurred during the Sung dynasty in China (10th–12th centuries) (ibid.). The history of global copper production is shown graphically in *Figure 1.1*.

The Falun mine in Sweden opened in the 13th century and reached peak production of the order of 2000 tpy in the 17th century. It continued to operate until the end of the 19th century. Sweden continued as the dominant copper producer in the 16th and most of the 17th centuries. *Figure 1.2* shows an outline of production history at Falun.

Global production at the beginning of the industrial revolution was around 10,000 tpy, mostly from Britain (Wales and Cornwall). Uses at the time were mostly in the form of brass or bronze, plus some kitchenware and coinage. World production in 1800 was 18,000 tpy.The modem industrial era of copper use may be said to have begun with the advent of the telegraph and the use of copper wire for electrical purposes of all kinds. The first practical telegraph line was demonstrated by Samuel F.B. Morse in 1844 (between Washington DC and Baltimore); a company was formed by him to extend the line to Philadelphia and New York City in 1845. By 1850 there were 50 telegraph companies registered in the U.S. In 1850 the first undersea telegraph cable was laid, under the English Channel. Telegraphy was soon followed in the 1880s by telephone lines and Edison's electrical generating and transmission systems.

Britain was the leading copper producer in the early 19th century, with 45 percent of world output in 1836 (Lovins 1973, p. xv). In those days Cornish ores averaged 8.3 percent in grade (ibid.). Britain was still the second largest producer in 1869, with 15,698 tons from native ores (*Encyclopedia Britannica* 1911). In that year the Calumet and Hecla Company, in Northern Michigan became the world's largest single producer (less than 6200 tons) (Coppa 1984, p. 134). By then the US was also the world's largest consumer of copper.

In 1877 the Rio Tinto mines of Spain[6] led the world with an output of 27,000 tons, and maintained leadership until 1892 when the Anaconda Company of Montana achieved an output of 37,500 tons (ibid.), shortly increased to 50,000 tons. From 572 tons in 1850 and 12,600 tons in 1860 US

---

[6] The river (which gave its name to the Rio Tinto Zinc Co. (RIZ), now Rio Tinto Ltd, was colored by mine wastes.

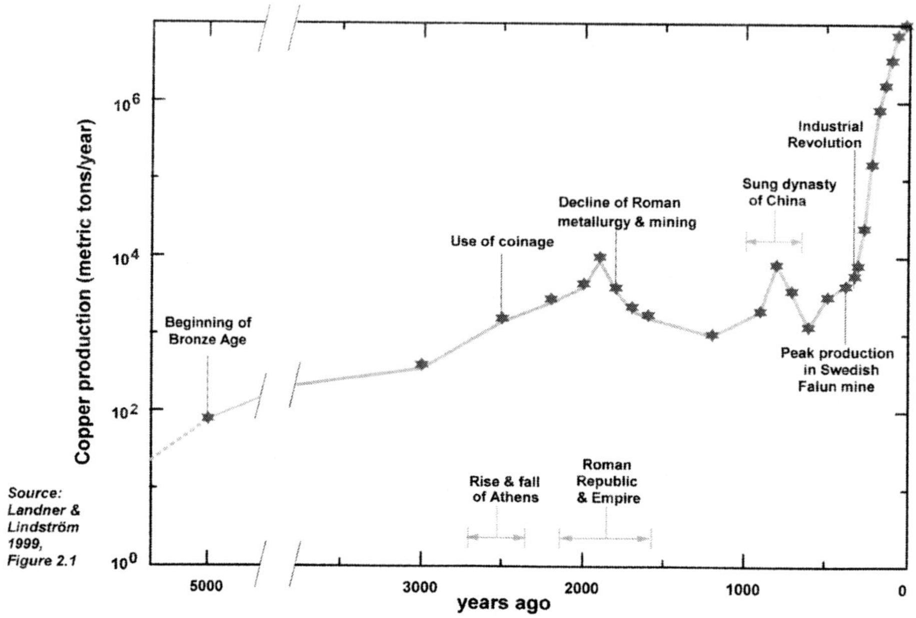

Figure 1.1: Variations in the rate of copper extraction, past 5000 years.

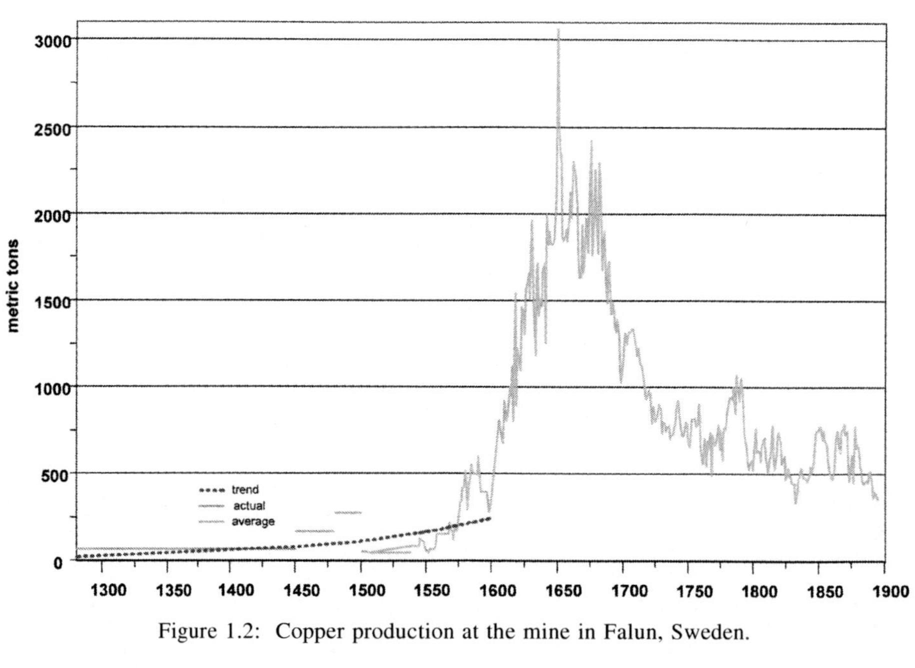

Figure 1.2: Copper production at the mine in Falun, Sweden.

production soared to 175,294 tons in 1895 and 397,003 tons a decade later (*Encyclopedia Britannica* 1911). By the end of the 19th century the US already produced 60 percent of the world's total production, which had already increased to 496,819 tons.[7] Seven years later, global output had increased to 723,807 tons (ibid.). In 1925 the US still produced over half of the world's copper. Production history since 1800 in the 'Western World' is shown in *Figure 1.3*; *Figure 1.4* gives data on consumption from 1950 on for the same region. A more detailed analysis of global production *vis à vis* prices and industrial production is given later in *Section 2.8*.

Lead and zinc were known in ancient times. Lead was first mined 9000 years ago. It was extensively used in metallic form because of its low melting point (327.5 deg. Celsius) and the ease with which it can be smelted from the main ore (lead sulfide or galena). It was used by the Romans for water pipes and wine receptacles, among other purposes (with probable adverse effects on the health of well-to-do Roman citizens) those who could afford to drink piped water and wine. Global production history is shown in *Figure 1.5*. and *Figure 1.6*. shows the lead concentration in Greenland ice by the age of the ice.

Zinc melts at a slightly higher temperature than lead (419.45 deg. C) but still much lower than the melting point of copper (1068 deg. C) and well below the 600 deg. C temperature achievable by a charcoal hearth (without bellows). Zinc was not widely used in metallic form, but zinc compounds were used for healing wounds and sore eyes, in ancient times. Zinc was the basis for an extremely useful alloy of copper, namely brass, first introduced by the Romans in the time of Augustus Caesar. Zinc metal was apparently produced in India from the 12th to the 16th centuries CE. China began smelting zinc in the 17th century. It was first recognized as a separate metal in Europe in 1546. The first European zinc smelter was located in Bristol, UK, in 1743.

Gold and silver were valued in prehistoric times because of their beauty and rarity – and both were (and still are) occasionally found as pure metallic nuggets. Their resistance to corrosion by air or water, and their malleability (which made them easy to hammer into desired shapes, especially decorative objects and coins) contributed to their utility. Silver and gold both have relatively high melting points (960.5 deg. C for silver, 1063.6 deg. C for gold), but silversmiths and goldsmiths did not need to melt the metal for most purposes.

Native sulfur was known in ancient times (as brimstone) and was used mainly for 'medicinal' or military purposes. Sulfur was a major constituent

---

[7] From the 1890s through the 1920s New York City and the Hudson-Raritan basin was the world's leading center of copper refining, due in large part to the nearby location of copper and brass consuming industries, especially the burgeoning electrical goods manufacturing sector (Ayres et at. 1988).

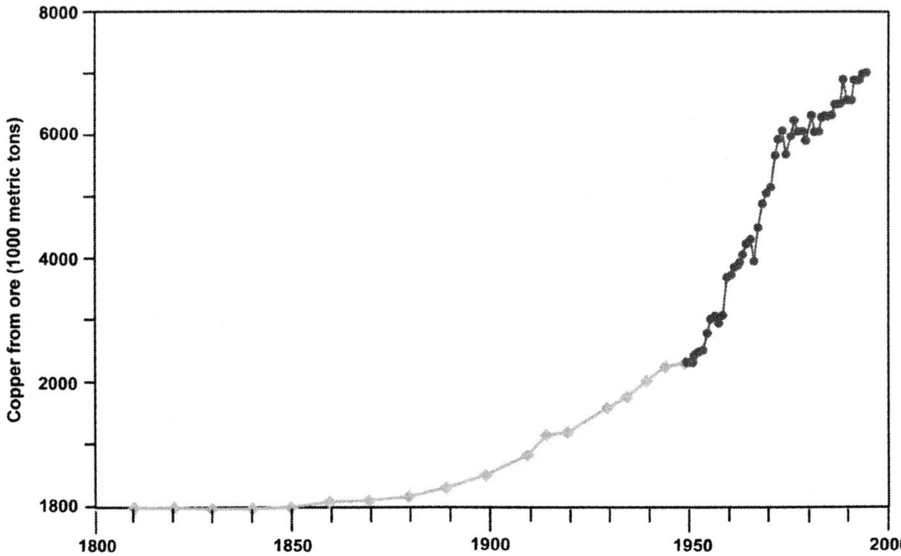

Source: Landner and Lindström 1999, Figure 4.1

Figure 1.3: Total production of copper from ores in the 'Western World', 1810–1995.

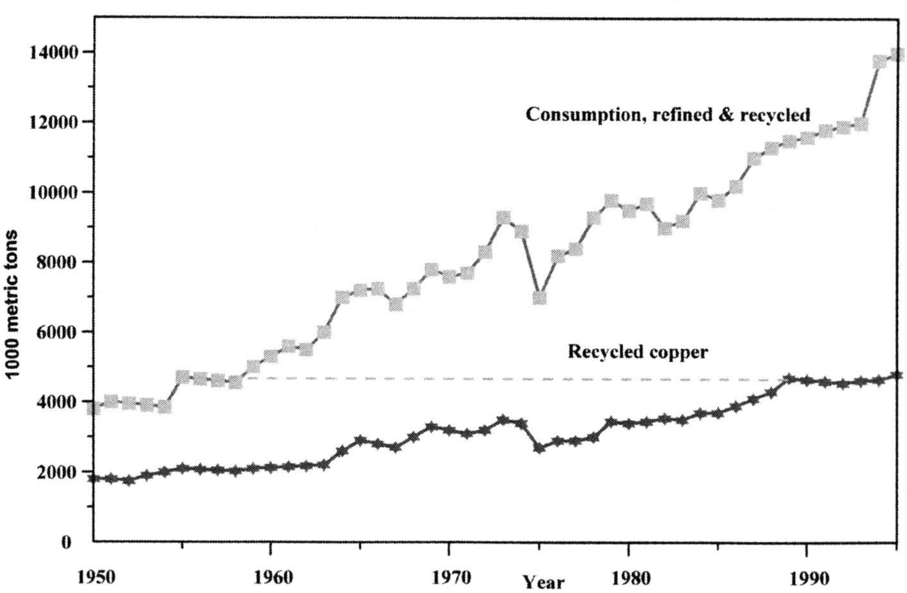

Source: Landner and Lindström 1999, Figure 4.2

Figure 1.4: Total consumption of copper in the 'Western World', 1950–1995.

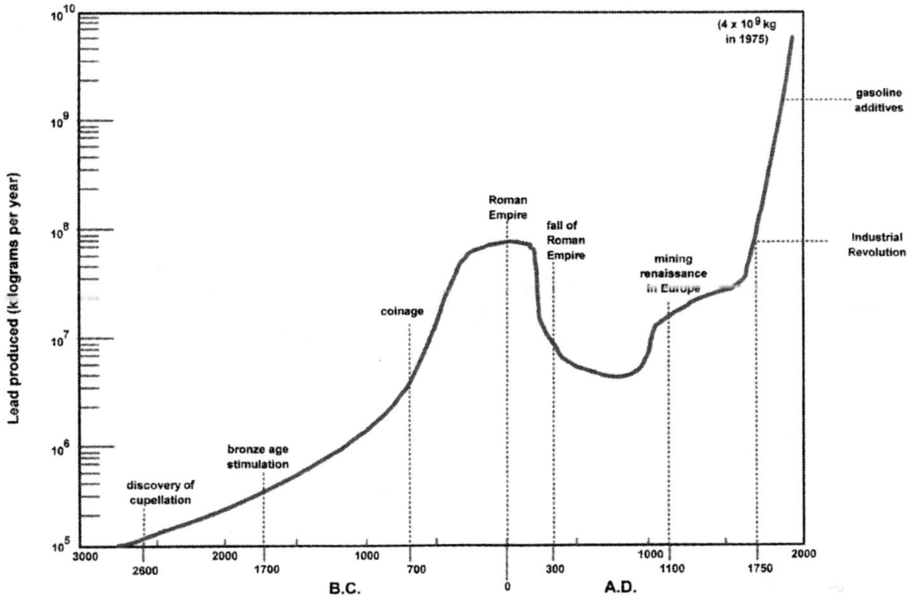

Source: Adapted from Patterson et al. 1970

Figure 1.5: Historical global lead production.

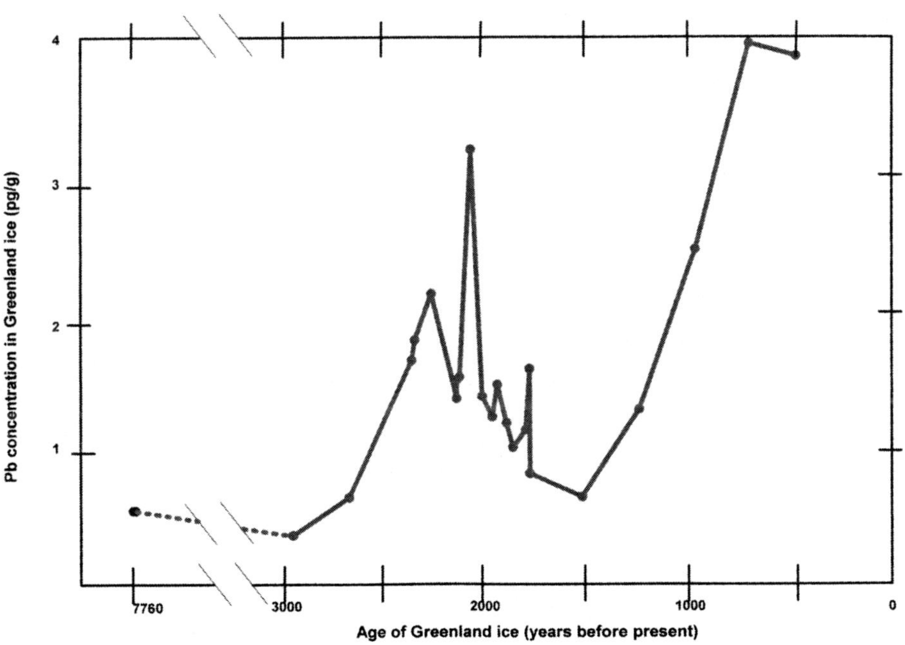

Source: Boutron 1998, p. 158

Figure 1.6: Lead concentration levels in Greenland ice.

of 'fireworks' and gunpowder, for instance. Sulfuric acid was known to the alchemists in medieval times. The industrial era began when the acid was first produced commercially in the early 18th century (by the 'chamber process' invented by Samuel Roebuck), and has since maintained its place as one of the most important industrial chemicals. Sulfuric acid has literally thousands of uses. In Europe, where sulfur in native form was unavailable, it was obtained from iron pyrites. Sulfur (and sulfuric acid) recovery from non-ferrous metal smelters is a modern ohenomenon.

Arsenic was also known in ancient times, and was the first copper alloying agent. Most of its uses subsequently have been for 'medicinal' or biocidal purposes, or for cosmetics. Antimony was occasionally used in ancient times as a cosmetic and medicinal. The other byproduct metals were not discovered until the 19th century.

1.3. Geology of copper

Geologists classify elements as geochemically abundant or geochemically scarce. The first group consists of 12 elements, of which 4 are widely used metals (aluminum, iron, magnesium and manganese) that accounts for 99.23 percent of the mass of the earth's continental crust. Silicon, calcium, sodium, potassium, titanium and phosphorus are also in the top 12, though rarely used in metallic form. Only iron, among them, has an atomic weight greater than 40. The other 90 or so known elements (including all other metals) altogether account for just 0.77 percent of the crustal mass. (A number of the trans-uranic elements, including plutonium, are not found in nature at all, but can only be synthesized in nuclear reactors or high energy physics laboratories.) The point is that all common rocks are composed of compounds of one or several of the abundant light elements, especially silicates. Hence it has appeared safe to assume (e.g. McKelvey 1960) that there is a standard (quasi-Gaussian) grade-abundance distribution, such that larger quantities of every element will be available at progressively lower grades, down to the crustal average.

For the geochemically scarce elements there are surprisingly different distributional patterns. Beryllium is an interesting example. It is not one of the scarcest elements, but although it is a component of something like 40 minerals, and 50 other minerals containing beryllium are known, yet only one (beryl) is found in (a very few) mineable deposits. Rubidium is another peculiar case: it is not a primary constituent of any known mineral. Yet rubidium is the 9th most common metal and constitutes close to 0.31 percent of crustal mass.

On the other hand, some of the scarcer elements are found in sizeable deposits at concentrations hundreds or even thousands of times greater than the crustal average. Examples in this group include chromium (0.2 percent of crustal mass); zinc (0.132 percent of crustal mass); nickel (0.08 percent), copper (0.058 percent), tin (0.040 percent) and lead (0.016 percent).[8] These metals, especially copper, are all mined in large tonnages today. Nevertheless, copper is a scarce element, number 28 in order of occurrence in the earth's crust, according to the most quoted sources.

The peculiarities of distribution are consequences of the earth's structure and the chemical characteristics of the elements. In the first place, heavier elements are naturally more concentrated in the core near the center of the earth, where they remain in a solid state under very high pressure. The solid (nickel-iron) core of the earth is surrounded by a semi-liquid mantle, on top of which floats a lighter solid crust made up of the lighter elements and their compounds. The geological system resembles a blast furnace in which molten metals fall to the bottom and the lighter slag floats on top. The earth's crust is essentially slag.

The presence of heavy metals in the crust ('slag'), as distinguishable minerals – rather than atomic substitutes – usually reflects a magmatic intrusion from the molten mantle, resulting from some tectonic or volcanic disturbance. Mineral deposits are formed following such intrusions by differential cooling and crystallization, dissolution in super-heated brines which later penetrate and crystalize in cracks in the rock, and by chemical reactions with the crustal rock (Bachmann 1960). Other mechanisms for mineral formation and segregation on the earth's surface include differential weathering, differential deposition of weathered minerals, reactions with atmospheric gases, and biological activity. Mineable copper, zinc and lead ores are all highly enriched as compared to the crustal average. The same is typically true of byproduct metals associated with those ores. But the opportunities for scarce elements to mineralize near the crustal surface are themselves very scarce.

Only the naturally enriched minerals in the crust can be concentrated by conventional physical-chemical means. Mineral compounds of the scarcest elements do not exist, as such, in common crustal rocks such as granite or feldspar. Atoms of the very scarce elements are almost entirely distributed as atomic substitutes in minerals of the more abundant elements, especially silica (SiO_2) in various combinations. These substitutions are not entirely random, of course, because they depend to some extent on the crystal structure of the 'parent' rock and the size and shape of the gaps that occur therein. For instance, lead is mostly an atomic substitute for potassium, while zinc is mainly an atomic substitute for magnesium. Extraction of metals that are present only as atomic substitutes in other minerals is feasible only in a few cases, mainly where they are contaminants that must be removed for other reasons. Antimony, arsenic, bismuth, cadmium, selenium, silver, tellurium and

[8] The foregoing figures, from standard USGS publications are widely quoted. However a recent publication gives much lower estimates for the crustal abundance of copper and zinc, namely 25 ppm for copper and 63 ppm for zinc (Wedepohl 1995).

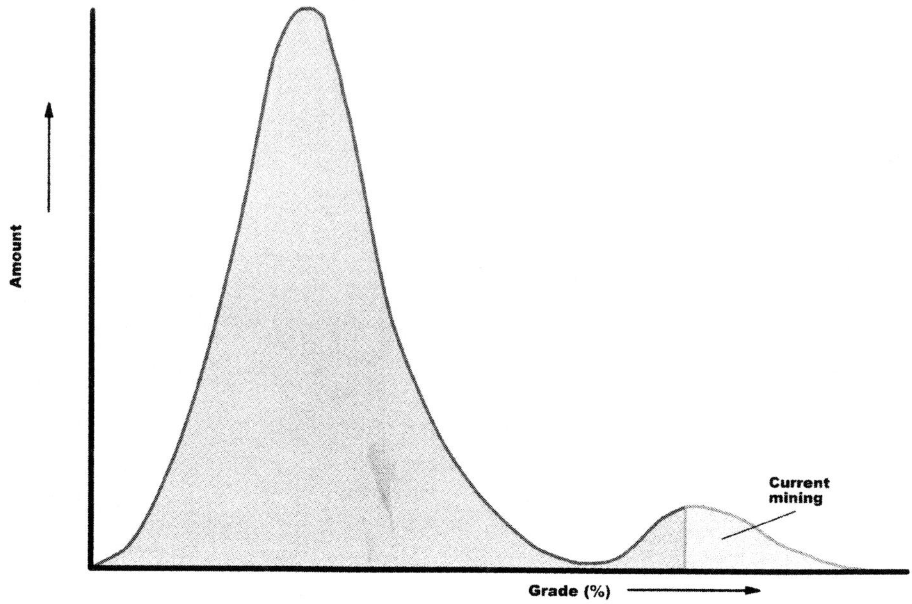

Source: Skinner 1976, Figure 4

Figure 1.7: Probable distribution of a geochemically scarce metal in the Earth's crust.

gold are all associated with lead, zinc and copper ores. Once the 'parent' minerals have been concentrated and smelted, it is feasible to separate and extract these minor contaminants in secondary processes.

Taking the foregoing facts into consideration, it seems likely (though as yet scientifically unproven) that certain geochemically scarce elements tend to have a bimodal distribution (*Figure 1.7*), in which the smaller peak – corresponding to relatively high concentrations – reflects geochemical mineralization, while the main peak reflects atomic substitution in more common minerals (Skinner 1976; Skinner 1987).[9] The implication is that certain metals – including some of the ones that have been industrially important for a very long time – may be effectively exhausted (i.e. for practical purposes) within a few decades.

Copper and lead exemplify the problem. Copper from currently mined ores averages around 0.8 percent in grade globally (0.5 percent in the US, and as low as 0.32 percent in some Swedish mines). Grades have been declining, as

[9] Apparently the bimodal distribution is still regarded by most geologists as speculation. It is true that there is no direct evidence of bimodality. However, this is not a serious objection, inasmuch as there is equally no direct evidence for the McKelvey hypothesis of single modality. On the other hand, if copper, lead and zinc were distributed according to the McKelvey rule, the total quantities in the earth's crust would have to be hundreds or thousands of times greater than the accepted crustal abundance.

higher grade ores tend to be exploited first. However, it is not possible to extrapolate current technologies from current practice (c. 0.5 percent or so) much below 0.1 percent grade. Certainly the same techniques will not be applicable to recovering copper from industrial or municipal wastes (which average 0.02–0.05 percent in grade) or to the crustal average (0.0058 percent), which is 100 times lower than the current average and 20 times lower than the lowest grade now regarded as exploitable.[10]

In crustal rock copper will not be found in grains of sulfides or oxides such as malachite or chalcopyrite, but only as atomic substitutions in common minerals. This is known as 'the mineralogical barrier'. To separate the copper chemically from the silicate matrix (as opposed to physically separating grains of copper minerals from other minerals, as is done today) would involve much higher energy investments, of the order of 100 to 1000 times greater per ton than current costs (Gordon et al. 1987; Skinner 1987). To mine copper from average crustal rock in quantities comparable to current US consumption would also require an enormous quantity of water – something like 5 times the annual flow of the Mississippi River (Gordon et al. 1987).

It is worth noting that refined copper is currently much less energy-intensive than aluminum (c. 60 GJ/t for copper from average ore (OECD countries) and as little as half of that for the most efficient refiners (see *Chapter 2, Section 2.3*) vs. 175 GJ/t for aluminum). Since about half of this is for materials handling and concentration, this energy advantage will disappear as the grade of copper ore mined approaches the mineralogical barrier (around 0.1 percent grade). At that point energy costs will sharply increase the economic incentives for copper recovery and recycling *vis à vis* mining virgin ore. On the other hand, we do not have enough data (or geochemical theory) to make a robust estimate of the amount of recoverable copper as a function of ore grade.

Recoverable copper deposits now are mostly (90–95 percent) in the form of disseminated sulfide minerals, such as chalcopyrite ($CuFeS_2$), chalcocite (CU_2S), bornite (Cu_5FeS_4) and enargite (Cu_3AsS_4. On average, a ton of sulfur is associated with each ton of copper. By far the most important deposits are of the disseminated porphyry copper type, in which the dominant mineral is chalcopyrite. Globally, 90–95 percent of copper is derived from sulfide ores. One sees different estimates in the literature. Arsenic is associated with some porphyry copper deposits, such as Butte, Montana, and EI Indio (Chile).

---

[10] It is not yet clear how the undersea manganese nodules fit into the overall geological picture. They do contain copper (0.53 percent on average) but both the availability and recoverability of the nodules are highly uncertain. It is now known that much of the oceanographic 'research' on the subject that was done in the 1970s was financed by the US Navy to disguise its real interest in trying to recover a sunken Soviet submarine. We know little about activity in this field at present.

Another type of porphyry deposit is the copper-molybdenum type, in which the mineral molybdenite (MoS_2) is interspersed with chalcopyrite, as in Henderson Colorado, Quartz Hill Canada and a few other locations in Chile and the Philippines. These ore bodies are mainly mined for molybdenum, with copper as a byproduct. We do not discuss them further.

So-called oxide ores, which account for 5–10 percent of global output, are weathered products of sulfide minerals, notably malachite ($CuCO_3.Cu(OH)_2$), azurite ($2CuCO_3.Cu\ (OH)_2$), cuprite (CU_2O) and chrysacolla ($CuSiO_3{\cdot}2H_2O$).

Exploitable reserves and 'total' resources of copper have been estimated at intervals by the US Bureau of Mines and (recently) by the US Geological Survey. Exploitable reserves of recoverable copper were estimated at about 100 million tons in 1935; new discoveries raised this to 212 million tons in 1960 (McMahon 1964). Reserves grew again sharply to 340 million metric tons (MMT) in 1984, but since then they have declined slowly to 321 MMT in 1990 and 310 MMT in 1994 (Landner and Lindeström 1999, p. 37). Estimates vary according to prices and assumptions.[11] Total potential resources have increased somewhat over the same period, from 500 MMT to around 590 MMT.[12] On this basis, current demand could be satisfied for about 60 years. As a matter of interest, cumulative global production of copper between 1970 and 1996 was 216 MMT.

A summary of the resource-reserve situation for copper as compared to other important metals, along with accumulated usage to date, is shown in *Table 1.2*.

1.4 Geology of lead and zinc

Lead and zinc are generally found in so-called hydrothermal deposits. The main lead mineral is the sulfide PbS (galena); the main zinc mineral is also the sulfide, ZnS (sphalerite). There are three major categories. The first, so-called Mississippi Valley Type or MVT deposits (because they were first discovered in Missouri) are found in porous limestone, dolomite or sandstone. The minerals were precipitated from brines which followed aquifers until conditions were right for precipitation and replacement of pre-existing minerals in the sedimentary strata. The lead and zinc minerals in MVT deposits are usually combined with other minerals such as barite and fluorite.

The second category is known as sedimentary exhalative (SedEx) deposits.

---

11 Known recoverable reserves as of the early 1980s were estimated by another source as 544 million metric tons (DKI 1984).

12 It was estimated in 1975 by the Committee on Natural Resources and the Environment of the US National Academy of Sciences (COMRATE) that the mineralogical barrier for copper occurs at 0.1 percent grade. Below that grade, copper would be found only as atomic substitutes in other rocks. The committee also calculated that the maximum amount of copper embodied in all ores would be 1000 MMT (COMRATE 1975).

Table 1.2: Accumulated use of metals compared to various measures of future availabilty (MMT)

Metal	Accumulated metal use	Reserves	Resources[a]	Resource maximum[b]
Abundant metals				
Al	488	4300	8000	very large
Fe	18400	66000	23000	very large
Scarce metals				
Zn	280	150	4400	15400
Cr	95	420	3400	15100
Cu	344	320	2300	9600
Ni	28	50	130	11600
Pb	180	70	1400	2500
Cd	0.71	0.5	6	21
Hg	0.9	0.1	0.6	15

[a] Crowson (1992).
[a] Maximum resources have been estimated as 10^{-5}–10^{-4} of all materials in the Earth's crust to a depth of 4.6 km Skinner (1987). The maximum for each metal is used here.
Source: Karlsson et al. (1997) Table 7.3.

These were precipitated from submarine hot springs. A related group is known as volcanogenic massive sulfide (VMS) deposits, many of which are also sources of copper. These were also formed by brines from submarine hot springs. The third category consists of vein, chimney-manto or skarn deposits, which are found in many locations along the Pacific Rim. They resulted from the injections of very hot brines, of magmatic origin, into a crack or chimney in a sedimentary rock such as shale or limestone, where precipitation occurred. Silver is frequently found in such deposits.

The double peak distribution and the mineralogical barrier for copper was discussed above. The case of zinc is comparable to that of copper, while for lead the situation is even more extreme. Lead sulfide ores with grades ranging from 2 percent to 7 percent are being mined now, and less than 10 GJ/t is needed to produce a ton of the pure metal. However to obtain lead from crustal rock, where it has a grade of around 0.002 percent would require something like 400,000 GJ/t, or 40 thousand times more than the current energy cost of lead. Even the most optimistic assessment of future energy availability would necessarily exclude common rock (or ocean water) as potential resources of these metals.

CHAPTER 2

COPPER: SOURCES AND SUPPLY

2.1. Physical properties and chemistry

Copper has atomic number 29, belongs to the same group (1B) as silver and gold, and shares some of their properties, including color, high electrical and thermal conductivity, high ductility (making it easy to draw into wire) and high malleability. Copper is second only to silver in electrical and thermal conductivity, and is significantly better than the third and fourth highest metals in both categories (gold and aluminum, respectively). Copper is also relatively corrosion resistant, although it does oxidize slowly in air. It has a high melting point (1083 C), with specific heat of 0.39 kJ/kg per degree C, and its melting heat is 343 kJ/kg, whence the theoretical heat requirement for melting pure copper is less than for competing metals such as iron and aluminum. It has a very high boiling point (2595 C) and has a very low tendency to fume, as compared to other non-ferrous metals such as arsenic, cadmium, lead, and zinc.

Pure copper is very ductile, as mentioned, and consequently it is easy to draw into wire. However the wire from pure copper is not very strong. Unfortunately, most alloying elements significantly reduce conductivity. Adding 10 percent aluminum doubles the strength but cuts electrical conductivity by over 80 percent. On the other hand, adding 0.8 percent cadmium significantly increases the strength and hardness of copper, while cutting conductivity only 10 percent (Henstock undated). This use accounts for 1–2 percent of cadmium consumption at present (See also the section on cadmium in *Chapter 4*).

Copper has several valence states, ranging from 0 to +3. Pure copper Cu(0) is very stable. It oxidizes at first to $Cu(1)_2O$ (cuprite), which is black and unstable. On heating, the cuprite oxidizes to normal cupric oxide, Cu(II)O which is quite stable and insoluble. Some other salts of Cu(II), notably the sulfate, $CuSO_4$, are extremely soluble (See *Table 1.1* for a list of copper ores).

The attractive blue-green patina on exposed copper surfaces consists of several compounds, depending on the surroundings. For example in coastal and marine environments where there is salt in the air, the chloride $Cu_2(OH)_3Cl$ is likely to be formed. Where there is sulfur in the air, on the other hand, the sulfate $Cu_4(OH)_6SO_4 \cdot H_2O$ (posjnakite) will be formed after some years the latter tends to evolve to $Cu_4(OH)_6SO_4$ (brochantite). The latter compounds are relatively insoluble, but not totally so. Also, they can be washed off an exposed surface, over time, by wind or flowing water.

Copper is quite biologically active. This is due to its tendency to form strong complexes in water or soil with organic ligands containing nitrogen or sulfur. It also forms complexes with some large macromolecules (proteins, nucleotides, etc.) The tendency of divalent copper ions to form complexes with organic substances accounts for the fact that in natural waters only a very small fraction (typically less than 1 percent) of the copper is found in free ionic or hydrated form (Landner and Lindeström 1999, p. 19]. The rest is normally complexed with hydroxyl (OH) ions or carbonate (CO_2) ions. The distribution of copper in fresh water depends on the pH and alkalinity of the water. Other inorganic ligands such as hydrogen sulfide (HS-), phosphate (PO^{4}_{3-}), chloride (Cl^-) and ammonia (NH_3) can also play a role. However organic ligands such as humic acids, amino acids, polysaccharides and polyphenols tend to dominate in both freshwater and marine environments, with 75–99 percent of all copper bound in organic complexes (Flemming and Trevors 1989). Humic acid-copper complexes, in particular, tend to be very stable.

Another important chemical mechanism governing the fate of copper in natural systems is precipitation, primarily as hydroxides or sulfides. (Two of the major copper ores, malachite and azurite, are precipitated hydroxides; the rest are primarily precipitated sulfides.) Conditions for precipitation depend on dissolved oxygen level, pH, and alkalinity.

The final chemical mechanism involved in determining the fate of copper in the environment is sorption. Copper is adsorbed on both inorganic particles (such as clays) and organic colloidal materials, including living or dead cells. Again, the organic materials tend to be preferred. Moreover, copper tends to out-compete most other divalent metal ions in finding absorption sites on humic particulates (Flemming and Trevors 1989). In soils it has been shown that up to 98 percent of the copper is bound to the organic fraction (Thornton 1983).

Trace levels of copper are essential to all living organisms. However, copper is also quite toxic at high exposure levels. An important consideration is that only soluble copper is bio-available. Copper complexed with organic matter, or copper oxides and other insoluble compounds are not accessible to living organisms. For this reason, the total amount of copper in a water body is not a measure of potential hazard to biological organisms. Even the soluble fraction is only bio-available under restricted conditions as to pH, dissolved organic carbon (DOC), dissolved carbon dioxide, hardness and salinity. A more detailed discussion of this issue can be found in Appendix B.

2.2. Copper production

By 1950 global output of primary copper had reached 2.7 million metric tons (MMT). In 1998 it was 12.3 MMT, up from 8.96 MMT in 1990. Trends in production, imports, exports and consumption are shown in *Figures 2.1–2.6* for major industrialized countries since 1913.[1]

Copper refinery output generally exceeds mine output by 15 percent or so, because of old scrap recycling. Thus in 1990, global refinery output was 10,841 kMT compared to 8956 kMT mine output. By 1998 refinery output was 13,880 kMT compared to mine output of 12,311 kMT. Evidently the fraction attributable to recycling old scrap is declining. This point will be discussed in more detail in Chapter 4. The largest producer of refined copper is the US (2,480 kMT in 1998, followed by Chile (2,335 kMT), Japan (1,277 kMT), China 1,150 kMT), Germany (696 kMT) and Russia (625 kMT). A large portion of the Chinese production was from imported scrap (USGS 2000).

2.3. Process technology

Even though many systems have been carried out in the last decades for copper production – in order to improve both process efficiency and to reduce its environmental impact – the particular uses of copper – such as electrical equipment production – require a high level of purity, which can be achieved only by means of a final stage of electrolytic refining. Thus, the standard process for the production of primary copper in use since the late 19th century involves three major stages, namely: ore mining and beneficiation, smelting and desulfurization and final electrolytic refining. Commercial grades of cathode copper today range from 99.95 percent to 99.97 percent pure.

A newer hydrometallurgical process, known as solvent extraction-electrowinning (SX-EW), discussed later, essentially by-passes the beneficiation and smelting stages, substituting a leaching stage followed by cementation or electro-winning. At present the SX-EW process is in a rapid growth stage and it is to some extent displacing the older pyro-metallurgical approach. However, whereas the pyro-metallurgical approach is a net producer of byproduct sulfuric acid, the hydro-metallurgical process is a net consumer of the acid. Thus, in the long run it is likely that the two will be increasingly combined.

---

[1] These figures are generated from published national statistics. It is evident that the statistical series for France, Germany and Sweden are anomalous during the years of World War II. Wartime production, imports and consumption of copper soared in the UK and the same must have been true for Germany and occupied France. Germany certainly imported large quantities of copper – either as scrap, concentrates or cathodes – from Sweden, but this does not show up in the statistics. Incidentally, quite a lot of copper must have been embodied and lost in ships and their cargos sunk by military action during the war. The total number of ships of all kinds sunk was in the thousands, and the tonnage of copper in each was significant. (For instance, ships propellers are generally made of bronze, a copper alloy, and the electrical wiring in a warship would be comparable to that in a small town.)

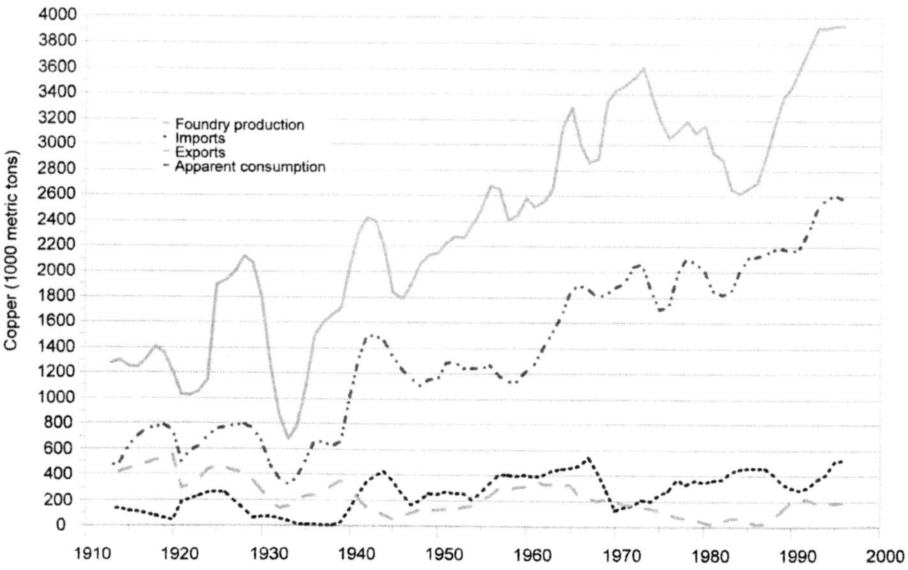

Figure 2.1: USA; copper foundry production, imports, exports & apparent consumption, 1913–1998 (3 year moving averages).

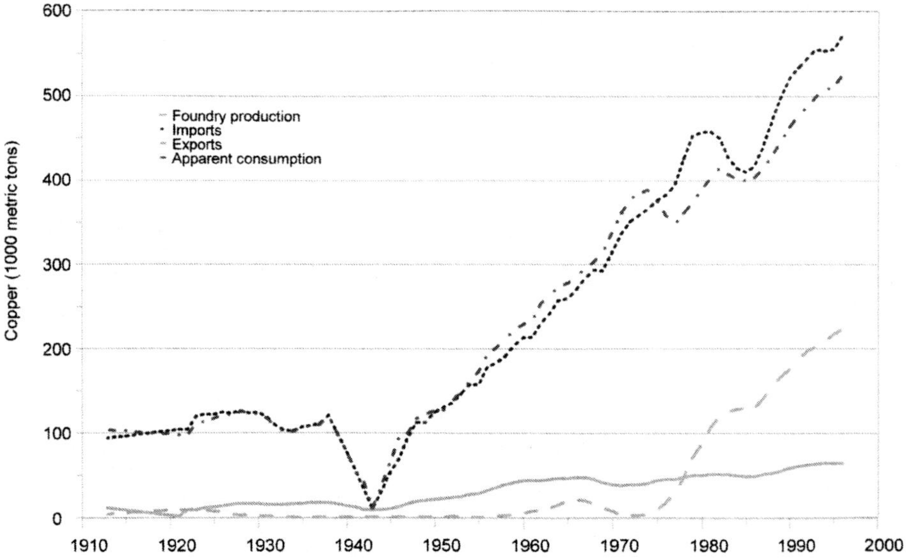

Figure 2.2: France; copper foundry production, imports, exports & apparent consumption, 1913–1998 (3 year moving averages).

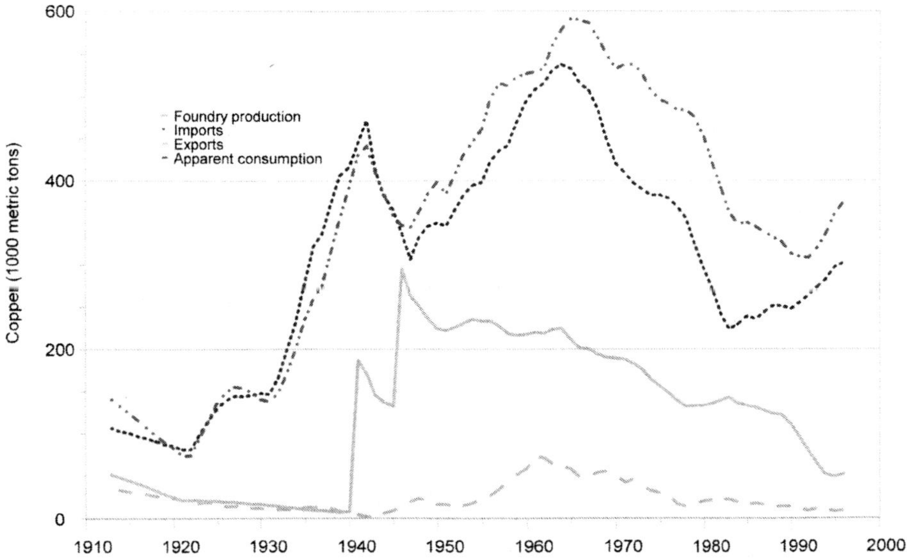

Figure 2.3: United Kingdom; copper foundry production, imports, exports & apparent consumption, 1913–1998 (3 year moving averages).

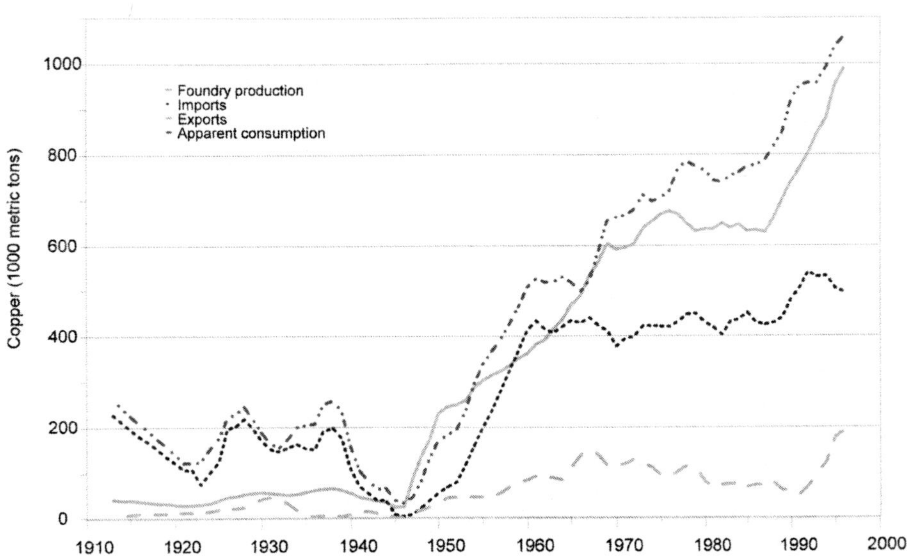

Figure 2.4: Germany; copper foundry production, imports, exports & apparent consumption, 1913–1998 (3 year moving averages).

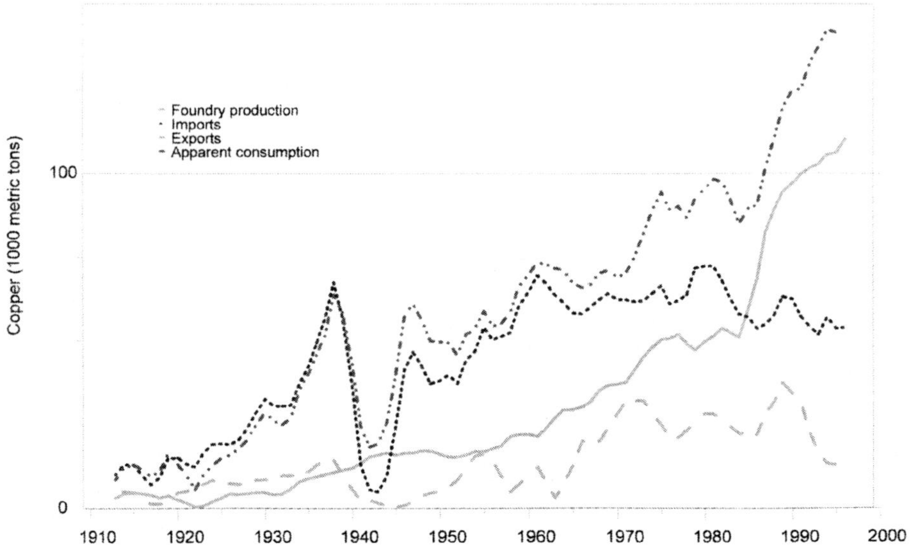

Figure 2.5: Sweden; copper foundry production, imports, exports & apparent consumption, 1913–1998 (3 year moving averages).

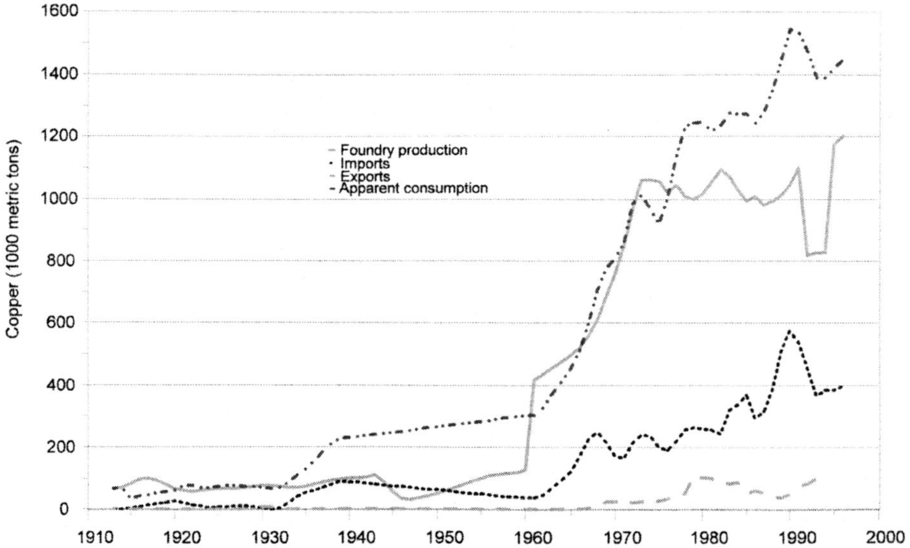

Figure 2.6: Japan; copper foundry production, imports, exports & apparent consumption, 1913–1998 (3 year moving averages).

The composite mass flow diagram in *Figure 2.7* reflects this perspective. It is a composite derived from specific processes described in (Gaines 1980; PEDCoCu 1980; USBuMines 1991a). See also (Ayres and Ayres 1998). Needless to say our analysis depends upon a variety of assumptions regarding the composition of all substances (ores, concentrates, wastes, etc.) involved. *Appendix A Table A1* gives details on the compositions assumed. However it must be emphasized that the numbers are exemplary and approximate; they do not refer to any specific combination of proprietary technologies.

2.3.1. *Mining*

Copper was mined exclusively from underground seams until the 20th century. The copper mined in Cornwall in 1836 had a grade of 8.3 percent according to an official report at the time. Copper mined in the 1860s and 1870s from the Michigan copper district in the Keweenaw peninsula included many nuggets of native metal. They were sorted by hand and primitive equipment.[2]

According to one source, the average grade of copper ore mined in the US in 1900 (entirely underground) was 3.4 percent, At that time the recovery rate was only 61 percent, using gravitic (density gradient based) methods of separation (Coppa 1984). Average copper concentration in US mine tailings at the time was 0.783 percent, which is significantly higher than ores now being mined (ibid.). However, data before 1907 was scarce and unreliable. Bureau of Mines documents assessed the average grade of US copper reserves in 1902 as 2.75 percent, but an estimate in 1905 estimated the grade at only 1 percent – clearly an anomaly.

The next USBM estimate was 1907, when the figure given was 2.1 percent. It was 1907 when the first open pit mine was opened (Bingham Canyon, Utah.) Open pit mining permits the use of very large equipment. Resulting economies of scale enable the exploitation of lower grade disseminated (porphyry) ores – the ores now mainly mined. By 1920 ore grades mined in the US had fallen to 2 percent. Grades actually increased slightly during the early 1920s (due to the discovery of new mines). Meanwhile the introduction of flotation concentration techniques raised the recovery rate (in concentrate) to 91 percent, and to as much as 95 percent in some mines by the end of the decade (Coppa 1984).

Reduced demand due to the great depression of the 1930s, together with more efficient recovery techniques, led to the closure of some mines and a temporary upward blip in the grade of officially recognized copper reserves to above 2 percent in 1933 (Ruth 1995 Figure 2). However, by 1936 the average grade of US ores was again down to 1.5 percent, of which 45 percent was from open pits (NAS/NRC 1969). Three years later it was down to 1.3 percent; by 1950 the grade hade fallen to 0.94 percent and by 1960 it was

---

[2] The largest known 'nugget' of pure metallic copper weighed 420t (DKI 1984).

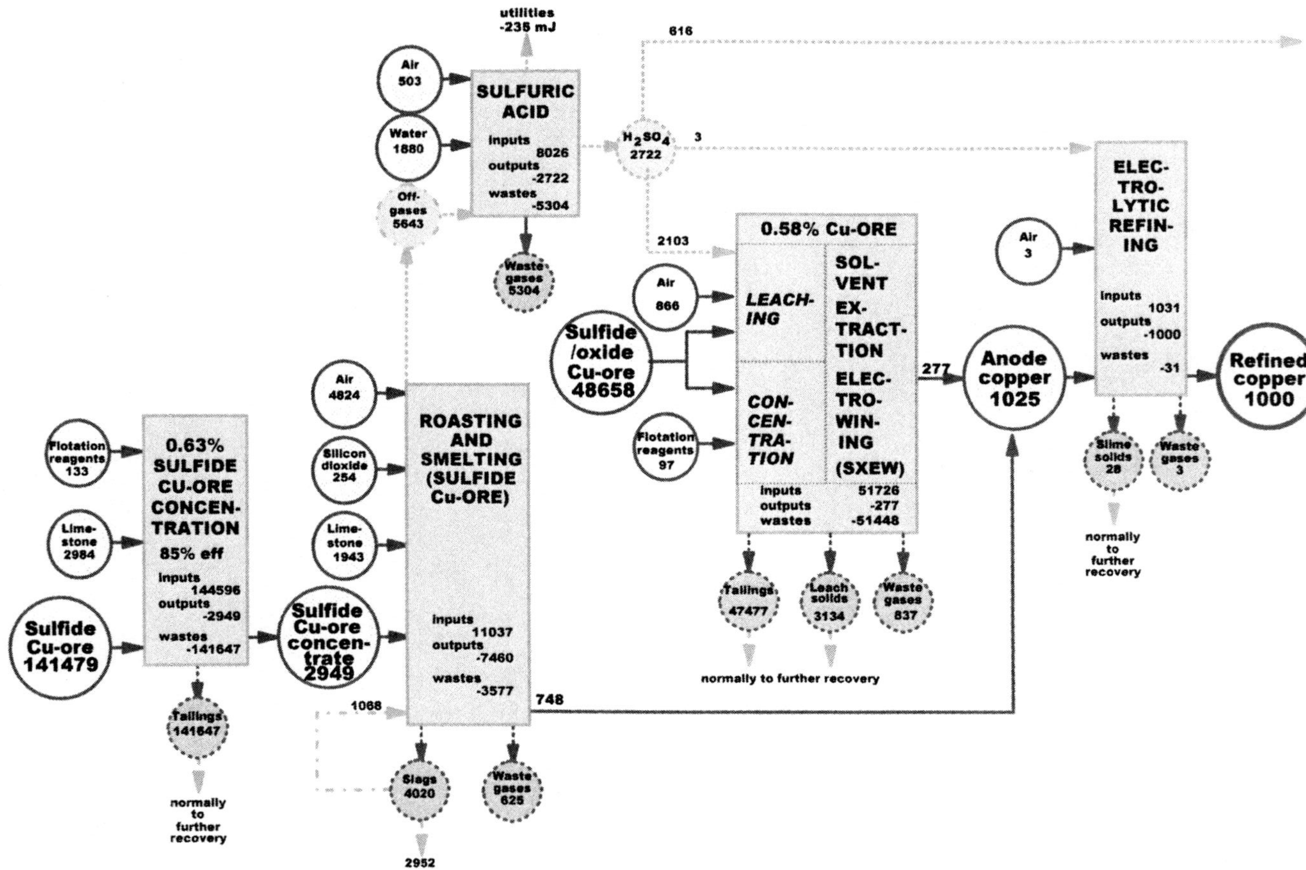

Figure 2.7: Mass flows (kg) in the production of 1 MT copper (simplified processes, typical material mixes).

down to 0.78 percent (McMahon 1964 Table 11). In 1968 the average grade of US ores was down to 0.6 percent, of which 83 percent was from open pit mines (NAS/NRC 1969). By the mid 70s it had fallen to 0.5 percent, but there was a slight apparent increase in the 1980s, probably due to the closure of some older mines. By 1990 open pit mines accounted for 88 percent of US copper mine output.

Today the grade still hovers above 0.5 percent, although the general downward trend is clear. See *Figure 2.8*. In general about 210 tons of material – excluding overburden – are mined in the US, on average, to produce 1 ton of refined metal. The corresponding worldwide figure is 125 tons/ton.

For purposes of modeling, Ruth has fitted the historical series covering the years 1930–1987 to a function of the form

$$\ln g(t) = -0.335 - 0.314 \ln Q(t)$$

where $g(t)$ is the grade of ore and $Q(t)$ is the cumulative production up to time t, where Q is indexed to unity in the year 1930 (Ruth 1995). The fit was done econometrically with an adjustment for first order auto-correlation. The R^2 for this fit was 0.98, which is reasonably good (ibid.). This functional form can be used in a model for projections into the future, as will be noted later.

A material efficiency for measure has also been defined by Ruth to express both the fact that beneficiation mining involves handling much more material than the crude ore itself, and the fact that mining and beneficiation (discussed below) does not actually recover all of the copper in the ore (Ruth 1995). Exploratory drilling and overburden removal roughly double the amount of material that must be handled to yield a ton of copper: altogether 420 tons of

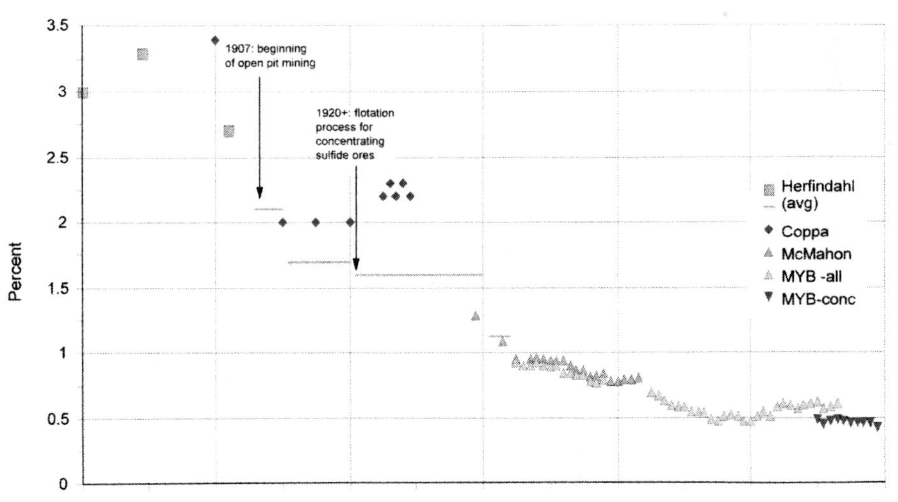

Figure 2.8: US copper ore grade percent, 1880–2000.

material were handled in 1990 per ton of copper produced in the US. The efficiency measure is defined as the ratio of crude ore weight to the total weight of materials handled, multiplied by the efficiency of beneficiation (discussed below). This efficiency measure has been calculated for the US for the years 1958–1987 (Ruth 1995 Figure 2). It has ranged between 0.25 and 0.4, with a mean of 0.3045 and a standard deviation of 0.048. For future reference this efficiency ratio has been denoted as $a(t)$.

Global mine production (copper content) reached 12,600 kMT (Cu content) in 1999, the last year for which we have data (USGS monthly February 2002). The two largest mine producers, by far, were Chile and the USA, followed by Indonesia, Australia and Canada. See *Table 2.1*. Copper mine output has increased roughly 1000-fold since 1800. The rate of annual increase in the 1990s actually accelerated to 3.4 percent p.a., which was considerably faster than population growth or economic growth.

2.3.2. *Beneficiation*

The specific processes appropriate for beneficiation of copper ores are dependent on the combination of minerals in the ores. Individual mining companies have developed a variety of proprietary processes to recover various by products or co-products.

The main technique for concentration of sulfide ores is froth flotation, sometimes preceded by gravity separation.

Organic wetting agents, known as collectors, permit air bubbles to adhere preferentially to some minerals (in finely divided suspensions) carrying them to the surface in froth. Collectors for sulfide minerals are usually xanthates or other chain hydrocarbons. Frothers in sulfide flotation are usually alcohols, pine oil or ring-carbon molecules like cresylic acid.

Table 2.1: World copper production, reserves and reserve base, 1999 (million metric tons Cu content)

Country	Production	Reserves	Reserve base
Chile	4.382	88	160
United States	1.660	45	90
Indonesia	0.740	19	25
Australia	0.735	9	23
Canada	0.614	10	23
Peru	0.536	19	40
Russia	0.530	20	30
China	0.500	18	37
Poland	0.460	20	36
World	12.600	340	650

Source: USGS monthly (February 2001).

Flotation agents used depend on the specific copper minerals in the ore (*Table 2.2*). According to one source, sodium mono-sulfide is used as a precipitation agent for the oxides (azurite, cuprite and malachite). None is needed for the sulfides (bornite, chalcocite, chalcopyrite). Sodium silicate is used as a dispersant for all copper minerals, sulfide and oxide. Sodium cyanide is a depressant for the sulfides; quebracho for azurite and cuprite, tannic acid for malachite. Sulfide minerals require no activator; polysulfide for the oxides. Alkyl or alkyl aryl xanthate aerofloats are collecting agents for the sulfides and azurite; fatty acids are added in the case of oxides. Pine oil is used as a frother in all cases, with vapor oil and cresylic acid added in the case of the oxides. Lime is a pH regulator and pyrite depressant for the sulfides; sodium carbonate for the oxides.

A Canadian copper mine producing sulfide ores, has released the following information about its use of flotation agents (per metric ton of ore): 1.1 kg of lime; 35 g. of x-amyl-xanthate and 30 g. of x-isopropyl xanthate 30 g. plus 14 gm of 'Dowfroth 250' and 20 gm. of pine oil (UNEP/IEPAC 1991 Table 8b).

Copper concentrates from sulfide ores consist mostly of chalcopyrite ($CuFeS_2$) and other copper sulfide minerals, plus iron pyrites with a small admixture of silicates and other ordinary rock. The composition of the concentrate depends entirely on the copper-sulfur ratio of the ore.[3] The average sulfur-copper ratio for all US ores is about 1.2:1 (Towle 1989 Table 4), which suggests that a plausible *average* concentrate composition for sulfide ores only might have a sulfur-copper ratio as high as 1.3:1 with a composition of 23 percent Cu, 30 percent S and 31 percent Fe, the remainder (16 percent) being divided between ordinary rock and byproduct metals. The latter include 0.2–4.0 percent Zn, 0–4 percent As, 0–0.67 percent Pb, and 0.13 percent Ag, plus 31.5 gm Au per metric ton of copper and variable traces of other metals such as cobalt (Co), nickel (Ni), molybdenum (Mo), palladium (Pd), platinum (Pt), selenium (Se), and tellurium (Te) (PEDCoCu 1980 Table 10). In fact most US output of the latter metals, except molybdenum, silver, gold, platinum and palladium are byproducts of copper mining, as discussed later in this report. On the other hand, a lower sulfur-copper ratio (1.2:1) and less waste rock (10 percent) would result in a composition closer to 27 percent Cu, 32.4 percent S and 28.6 percent Fe.

Clearly, for ores with less sulfur the copper fraction in the concentrate can be somewhat higher. Non-US ores on average have a higher grade (around 0.8 percent) and a sulfur copper ratio of around 0.92 (Towle 1989 Table 4), which allows for concentrates with higher copper content (up to 35 percent).

---

[3] The US range is reported to be from 20–50 percent Cu, 30–38 percent S, 20–30 percent Fe (PEDCoCu 1980 Table 10). However based on mass balance considerations, it is unlikely that the copper content can ever exceed 35 percent for sulfide ores. Note that for chalcopyrite, the most common ore mineral, the copper-sulfur ratio is 1:1. Additional sulfur is probably mainly from iron pyrites in the ore.

Table 2.2: Typical flotation reagent consumption in non-ferrous metal mills (g/t ore)

	Pb-Zn (sulfides) Les Malines France	Pb-Zn (oxide + sulfides) Zellidja Morocc	Cu-Pb-Zn Brunswick Mining & Smelting Canada	Ni (sulfide) Falconbridge Canada	Cu (sulfide) Lornex Homestake Canada	Au cyanidation + CIF USA	Cu-Zn (pyrite) Pyhasalai Finland
Acids							
H_2SO_4				500–600			5000[b]
Alkalis:							
Lime	1000		2500	225–400	1100	1200	3150
Sodium carbonate		550	3300				
Sodium hydroxide		246					
Modifier							
Copper sulfate	200	120	815	35–60			330
Sodium cyanide	10	13				550	28
Zinc sulfate	60	91					1450
Sodium sulfate		2800					
Sodium silicate		2700					
Sulfur dioxide			700				
Starch			100				
Collectors							
x-Amylxanthate	45	130	270	60–85	35		220
x-Isopropylxanthate					30		
x-Ethylxanthate	5	20					
Diesel oil		69					
Amine		250					
R-242[a]		60					
Frothers							
Dowfroth 250				20–25	14		
Hexylic acid	40						
Pine oil					20		
HBTA frother		85					
Carbon						30	

Source: UNEP/IEPAC (1991) Table 8b.

[a] R242 = Aniline dicresyl dithiophosphate plus thiocarbonilide.

[b] Sulfuric acid is used for pyrite recovery.

Table 2.3: Chilean copper smelters

Item	Units	Total	Avg.	Min.	Max.
Capacity	kt Conc. per year	5115	731	325	1550
Mix of concentrates	Cu/Fe/S %		27/19/26	20/17/12	22/25/39
Revert/concentrates	Ratio	0.13	0.15	0.03	0.26
Recovery from concentrate	% Cu	98%	97%	96%	98%
Copper production	kt Cu per year	1554	222	72	490
Energy consumption	kWh per t Conc.	305	294	84	704
Water consumption	m^3 per t Conc.	2.4	2.3	0.8	3.7
Acid plants	kt H_2SO_4 per year	3748		50	200
Sulfur capture	% S	70%	83%	39%	96%

Source: Demetrio et al. (1999).

The concentrate used at the only German smelter, in Hamburg, obtained mostly from Chile (40 percent), Portugal (23.5 percent) and Papua New Guinea (22 percent), is probably more typical of Europe. It has a sulfur-copper ratio of 1.06:1, resulting in approximately the following composition: 30 percent Cu, 32 percent S, 26 percent Fe, 1.5 percent Zn, 0.2 percent Pb and 0.1 percent As. The rest (10.2 percent) consists of other minerals, mainly silicon dioxide (7 percent) and other light metal oxides (Bruch et al. 1995). Chilean concentrates average 27 percent Cu, 26 percent S and 19 percent Fe. Data on Chilean smelters are shown in *Table 2.3* (Demetrio et al. 1999).

2.3.3. *Leaching*

Oxide ores and supergene sulfide ores (i.e. ores not containing iron) can be recovered most easily by hydro-metallurgical techniques,[4] notably *in situ* leaching, dump leaching, heap leaching and vat leaching. *In situ* leaching recovers copper from small oxide ore bodies (too small to mine economically) or worked-out underground mines. Dump leaching is used to recover copper from large quantities (millions of tons) of strip mine waste with a very low grade. Dilute sulfuric acid is trickled through the material, alternating with periods during which the waste is exposed to air (to oxidize the sulfides). The acid also reacts with pyrites in the ores, creating additional sulfuric acid and ferric sulfate, which dissolves copper minerals. Once the process starts it continues naturally if water and air are circulated through the heap. The time required is typically measured in years. (Sulfur dioxide is emitted during such operations.) Bacterial action contributes to the process, and ferric salts accelerate it. Soluble copper is then recovered from drainage tunnels and ponds. Copper recovery rates vary from 30 percent to 70 percent.

---

[4] The two major categories are hydrometallurgy and pyrometallurgy. For definitions, see GLOSSARY.

Heap leaching is essentially the same as dump leaching, except that it is applied to oxide ores, which are previously ground or crushed and heaped on a prepared surface. The natural oxidation process requires months or years to go to completion. Heap leaching is only feasible when very large quantities of ore are involved, though it is the principal method of recovery at the Rio Tinto Zinc mines in Spain, where it has been used for hundreds of years. At one time 20 million tons of ore were being treated by this method over an area of 350 acres at the RTZ mine (McMahon 1964, p. 101). Recovery of 70–80 percent is possible.

In vat leaching, oxide ores are crushed and placed in large concrete vats (up to 18,000 metric tons capacity), then flooded by concentrated sulfuric acid, which converts most of the copper to copper sulfate. Extraction of one metric ton of copper from ore with a grade of 1 percent requires 4400 liters of 96 percent sulfuric acid. Vat leaching typically recovers 75 percent of the copper. The process is fairly rapid (hours to days).

There have been experiments with chloride leaching and electro-winning of chalcopyrite ores. A plant in Arizona using Duval's CLEAR process was producing 80 metric tons/day in the early 1980s. However the purity was not high enough to justify the cost. There are other chloride processes being tested (e.g. Intec, Activox, etc.) and research continues. To be successful it will have to be combined with a solvent extraction stage (Davenport 1999). Biological leaching processes are also being tested (e.g. the Billiton-Codelco joint venture based on Billiton's Bio-Cop process; see www.billiton.com).

2.3.4. *Cementation and solvent extraction (SX) (Davenport 1999)*

Solutions of copper sulfate from vat or heap leaching are too dilute for electro-winning (EW). Recovery is either by cementation or solvent extraction (SX). The leaching-cementation process has been practiced for centuries, especially in Europe (Wales, Spain) and later in Chile and Zaire. It is a precipitation process, in which soluble copper sulfate reacts with iron – usually scrap, such as old tin cans – yielding soluble iron compounds (ferrous or ferric sulfate) and precipitating the copper. The 'cemented' precipitate – typically 70 percent Cu, with iron and other impurities – is not pure enough for direct use as metal. Cemented copper must be fed to a smelter-refinery, often along with other feeds (mainly concentrate). The liquor is recycled back to the heap or vat. Overall recovery from solution by cementation is around 95 percent.

The leaching-cementation process has now been almost universally overtaken by solvent extraction. Agitation leaching with solvent extraction is a method to produce a relatively pure solution of copper sulfate suitable for electro-winning. Kerosine, with a 12 percent mixture of LIX hydroxy oximes (a proprietary organic solvent), is used in an agitated vessel. The solvent combines with the copper, forming a complex. When agitation is stopped, the copper-containing organic complex forms a separate dense layer from which the water can be drained off. Sulfuric acid then breaks down the complex and regenerates the solvent for further use. The sulfuric acid, together with dissolved copper sulfate, is then recycled through the electro-winning (EW) cells. Recovery from solution in this case is about 95 percent. The SX-EW process can be continuous and is suitable for large scale operations. Most important, the quality of the metal is fully competitive with smelted/refined copper. The only significant impurity is lead (from Pb-Sn-Ca cathodes).

The SX-EW process is still being improved. Extractant consumption is down from around 3 kg/metric ton of copper in the early days to about 2 kg/metric ton of copper today. Costs are down in proportion. The SX-EW process has almost no economies of scale. Costs per ton of Cu for a very small operation are virtually the same as for a large operation.

In 1990 the SX-EW process accounted for only 6.4 percent of world copper mine production, and 8.2 percent of production in the western world. The SX-EW share has grown rapidly in the 1990s, reaching 14.4 percent and 18 percent respectively by 1998 (Simada et al. 1999 Table 4) and over 30 percent in the US.

2.3.5. *Roasting and smelting (Gaines 1980; Davenport 1999)*

Roasting is a process to convert copper and iron sulfide concentrates (25–30 percent Cu) to oxides, thereby removing most of the sulfur (as sulfur dioxide). In the late 19th century ore was roasted in open air fires to drive off the sulfur and other volatiles (such as arsenic). This practice led to widespread environmental damage, especially in Montana and Tennessee. More recently, two main types of furnace have been used, viz. multiple hearths and fluidized bed roasters. In the most recent systems the roasting stage is combined with the smelting stage. The sulfur dioxide, if sufficiently concentrated, is captured and subsequently converted to sulfuric acid (H_2SO_4) in a contact acid plant.

The next stage is smelting. There are four major types of copper smelter in use, namely reverberatory furnaces, electric furnaces, flash smelters and Noranda smelters. The choice among them depends on the specific characteristics of the concentrate feed. However, the first two are essentially batch-type and consequently inherently less efficient than the latter two types (flash smelters, and Noranda smelters) both of which are suitable for continuous operation. As of 1988 at least 3000 kMT of global capacity was still using reverberatory furnaces, although the figure had dropped to 2000 kMT a year later (M. Stewart, University of Sydney, personal communication, 2001). These furnaces are obsolescent because they do not make much use of the heat value of the sulfur, and their off-gases have such low sulfur dioxide concentration that recovery is uneconomic. Some reverberatory furnaces, mainly in Chile, converted to oxygen-sprinkle flash smelters by retrofitting with feed driers, an oxygen plant and some minor modifications of the furnace itself (Gaines 1980, p. 41). There were a few old smelters still using blast furnaces or submerged arc electric furnaces, mostly being phased out. (There are still several operating in Poland.)

Electric furnaces were always limited to areas with very cheap electric power. Most are now closed, although they are still in use in Sweden and Zambia, where hydroelectric power is available.

There are two types of flash smelters, Outokumpu and INCO. In both cases a finely divided dry concentrate is fed into a hearth-type furnace. Effectively, the smelting is done on suspended particulates. Most of the process heat is derived from oxidation of the sulfides, making the process very energy-efficient. The Outokumpu design, which has been by far the most widely adopted of the two, uses preheated air (possibly oxygen enriched) and yields a gas stream with 10–15 percent SO_2 concentration.

The INCO design, introduced in 1983, utilizes pure oxygen and yields a gas stream that is 80 percent SO_2. The higher the sulfur dioxide concentration in the gas stream, of course, the easier it is to recover sulfuric acid, liquid SO_2 or pure elemental sulfur. Disadvantages are loss of copper in the slag (1 percent) and flue dust, both of which must be retreated. INCO itself is the only recent adaptor of this process, at Sudbury, Ontario.

The Noranda submerged tuyere rotatable furnace design (introduced 1973) comes closest to one-step operation, combining the functions of roaster, smelter and partial convertor. It was designed to yield blister copper, but the blister copper thus produced has higher impurity content than the conventional product. In practice Noranda units in Canada and the US (Kennecott) are used to produce a low sulfur high copper (70 percent) matte, which is subsequently converted conventionally in batches. Unfortunately, the slag has a high copper content (10 percent) which means that it must be reprocessed by froth flotation. It is also the most efficient design in terms of energy conservation. The Noranda process is now in operation at Port Kembla, Australia (since 1990) and two new units are apparently operational in China. The Australian operation is a special case, though, because it utilizes a very unusual ore (copper, gold and uranium) (Mary Stewart, University of Sydney, personal communication, 2001).

The El Teniente smelter, introduced in 1977 at Caletones, Chile, is also of the submerged tuyere type, with a rotatable furnace. It requires 'seed matte' whence it must be used in combination with some other type of smelter. It is mainly used to increase the capacity of existing plants. Eight more units have been installed since 1984), four of them in Chile, one in Peru (Ilo), one in Mexico (La Caridad), one in Zambia (Nkana) and one in Thailand (Rayong).

The only commercially operational continuous process design is the Mitsubishi lance smelter, introduced in 1974. It utilizes three units, a smelting furnace, an electric slag cleaning furnace and a relatively conventional air-blown convertor, all under computer control. The smelter discharges a slag-matte combination to the slag cleaning unit, where they are separated. The slag (0.5 percent Cu) is discharged, while the liquid matte (60–65 percent Cu) is conveyed directly to the convertor. Convertor slag (10 percent Cu) is then granulated and fed back into the smelter unit. Units are now operational at Naoshima (Japan), Timmins (Ontario), Onsan (Korea) and Gresik (Indonesia).

Three other processes are worth mentioning. Contop cyclone smelting was originally an adaptation of the reverberatory furnace, the only stand-alone unit being operated successfully by ASARCO at El Paso during 1993–1999. It has been mothballed because of high energy costs. The Vanyukov smelter is a submerged tuyere non-rotatable furnace. It is operating at the Balkash refinery in Kazakhstan and possibly in Russia.

The only new pyro-metallurgical process that may be gaining new adherents around the world is the Isasmelt lance process, first adopted by Cyprus's Miami Arizona plant in 1992, and subsequently (after some teething troubles) by Sterlite (Tuticorin, India) and Union Miniére (Hoboken, Belgium). It is thermally efficient and particularly good for low sulfur concentrates and mixed charges of moist concentrates, residues and scrap. However this process is unproven and most of the industry seems to be leaning toward hydro-metallurgical processes that are better able to manage low-grade ores and separate out desired products (M. Stewart, University of Sydney, personal communication, 2001).

2.3.6. *Converting (Gaines 1980; Davenport 1999)*

Molten copper-iron matte from the smelting furnace, together with copper scrap, is 'converted' by blowing air through it, to oxidize any remaining sulfur. A flux (silica and lime) is added that combines with the iron yielding an iron rich slag and relatively pure blister copper (98–98.5 percent Cu). There are two main types of batch convertor, principally Pierce-Smith (the most common) and Hoboken. They differ primarily in the method of capturing sulfur dioxide off-gases. The Pierce-Smith convertor is chemically efficient, with a reasonably good sulfur dioxide concentration, but it is still a batch process. The Hoboken design, introduced by Union Miniére (c. 1970) is preferable, because it achieves better sulfur dioxide recovery. There are now 20 units in operation. Most Chilean refineries use the so-called Teniente modified convertor.

There are three types of continuous convertor: flash converting, lance converting and submerged tuyere converting. All three offer better control over oxygen intake than batch convertors and generate more concentrated sulfur dioxide, which is easily captured for acid production. (In fact, the higher concentrations are triggering changes in the catalyst technology for acid plants. Cesium-based catalysts are in the ascendant.) The continuous flash convertor uses finely ground solid matte from an Outukumpu flash furnace. A drawback is that the lime-based slag must be recycled to the smelter. A continuous flash convertor was pioneered by Kennecott at Garfield Utah (1995). The technology is also being adopted by the Southern Peru Copper Co. at Ilo.

The other two continuous convertors utilize liquid matte, which is thermodynamically preferable to batch conversion. The first continuous lance convertor was the Mitsubishi unit at Naoshima, Japan (1974, 1991). Others are incorporated with Mitsubishi smelters at Timmins, Ontario (1979), Onsan, Korea (1998), and Gresik, Indonesia (1998). The first stand-alone unit was installed at Port Kembla, Australia (1999). In combination with a Mitsubishi smelter, the lime-base slag from the convertor need not be recycled, but can be discarded directly. The slag is somewhat corrosive, however. The submerged tuyere (Noranda) convertor was first introduced at the Noranda plant in Quebec (1998) is an extension of the Noranda smelter technology; it feeds liquid matte from the Noranda smelter into a submerged tuyere rotary furnace, where it is stirred and oxidized by a jet of oxygen-enriched air. The slag is silica based, but both smelter and convertor slags must be recycled. The sulfur dioxide passes on to an acid recovery plant.

2.3.7. *Electrowinning*

As noted above, pure copper sulfate dissolved in sulfuric acid is generated by the vat leaching process and the solvent extraction (SX) process. This material is suitable for feeding directly to electrolytic cells. The electricity requirement is quite high because a comparatively high voltage is needed (3 V), compared to the electro-refining discussed later. The combination, accounted for 540 kMT of US primary output in 1995 (and 574 kMT in 1996), against a world total of 1073 kMT in 1995, and 1423 kMT a year later. The adoption of SX-EW technology since 1995 has been very rapid: the world total produced by that method nearly doubled to 1997 kMT (1.98 MMT) in 1998, which represents 18 percent of western world production and 14.4 percent of total world production (ICSG 1997). The new SX-EW capacity is mostly in Chile (where there are essentially no competing uses for smelter byproduct sulfuric acid.) Note that the growth of SX-EW is accompanied by an increase in leaching.

The SX-EW process differs from traditional electrolytic refining (below) in that the anodes are inert (Pb-Sn-Ca). Copper is deposited on the cathode, liberating oxygen and regenerating the sulfate ion as sulfuric acid. Purity of cathode copper today can be as high as 99.99 percent, which is comparable to the best electrolytic copper. However, the fact that the EW cells decompose water (yielding oxygen at the anode) is a disadvantage, in terms of energy consumption.

2.3.8. *Fire refining and electrolytic refining*

Blister copper (about 98 percent or 98.5 percent Cu) is not pure enough for electrical applications, and needs to be further treated. Fire refining is an old technique, similar to the Bessemer convertor, in which air is blown through the melt and flux is added to eliminate the last impurities. First air is blown through the liquid copper to oxidize any residual sulfur. Excess oxygen was originally removed by 'poling' with logs, which decomposed into carbonaceous products that quickly oxidized. Nowadays natural gas, hydrogen or ammonia are used for this purpose. The product is mostly cast into anodes (>99.5 percent Cu) and sent on to an electrolytic refinery; however given a high enough quality blister, fire refined copper can be used for alloys and special castings.

The electrolytic refining process uses fire refined anodes from the fire refinery, or recycled anodes from the electrolytic refinery itself, copper sheets as cathodes and sulfuric acid as an electrolyte. Impurities collect as a 'slime' at the bottom of the cell. This slime, or anode mud contains many of the precious and other byproduct metals and is recycled for additional processing. At the end of the process pure copper is collected on cathodes, together with small amounts of precious metals deposited as anode mud which can be easily recovered.

2.3.9. *Future trends in primary processing*

Although we have disclaimed any attempt to carry out a formal LCA for copper, the three main competing reduction technologies, namely reverberatory furnaces, flash furnaces and heap leach-SX-EW can be compared on the basis of more or less standard assumptions about ore grade, scale and other features. The main differences between the three cases are that the first two achieve total recovery of 88 percent, while the SX-EW process achieves only 62 percent recovery. The electric power requirements are assumed to be 6000 kW/t for the flash smelter, 5700 kW/t for the reverberatory smelter and 9350 kW/t for the SX-EW plant. In all cases the electric power is assumed to be obtained from steam-electric coal-burning power plants. The flash furnace uses additional fossil energy from oil to obtain oxygen, which permits high temperature operation and enables efficient burning of sulfides and 93 percent recovery of sulfur dioxide as sulfuric acid. The reverberatory furnace uses even more extra fuel simply to maintain high temperature reaction conditions in the furnace, but only 5 percent of the sulfur dioxide is recovered. A detailed analysis along these lines, with respect to four main environmental impact categories, has been performed by (Giurco et al. 2001).

We summarize the main results here and refer the reader to the authors for more details.

As regards global warming impact (from CO_2 emissions) the first two technologies are comparable, but the SX-EW technology has a 25 percent greater impact, due to high (coal-based) electricity requirements (Electro-winning is considerably less efficient than electro-refining due to higher voltage requirements, viz. 3V vs 0.2 V).

As regards acidification, of course the reverberatory furnace is by far the worst, generating 2500 kg of SO_2/metric ton of copper, compared to around 150 kg/t for the flash furnace and even less (but not zero) for the SX-EW process, attributable to the use of coal for generating electricity.

Giurco et al. also compare the three processes in terms of eco-toxicity (mainly due to the use of fuel oil) and smog generation (the latter due to diesel engines) These effects are not directly observable or measurable, and are imputed indirectly. We note that Giurco et al. attribute significantly higher eco-toxicity effects to reverberatory furnace than to the others, with SX-EW having the lowest score in that category. Eco-toxicity is related to diesel oil consumption. As regards smog (micro-particulates) the three are comparable, with SX-EW is slightly higher and flash furnace technology slightly lower than the reverberatory technology.

We find these results plausible but not very meaningful, since there is no way to make robust generic comparisons between the four different environmental impact categories, three of which, in any event, would be given different weights in different locations.

2.4. Exergy and exergy flows[5]

As noted previously, exergy is a measure of the maximum work that can be extracted when a subsystem that is not initially in equilibrium with its surroundings equilibrates. One virtue of using exergy rather than energy as a unit of analysis is that the two are nearly equivalent for utility inputs (as in most traditional studies where 'energy' is considered), whereas exergy analysis also takes into account non-fuel materials. For most mineral ores the exergy content of the ore is small or even negligible compared to the exergy consumed in the beneficiation and smelting processes. For instance, iron ore has an exergy content of 0.695 GJ/t of steel, whereas finished steel has an exergy content of 6.750 GJ/t. Other process inputs (coal, scrap iron, utilities) amounted to roughly 22.2 GJ/t (US average, 1993) (Ayres and Ayres 1999 Appendix D).

For aluminum the exergy content of ore (bauxite) is 4.118 GJ/t of refined metal, whereas the metal itself has an exergy content of 32.805 GJ/t and the other process inputs add up to 131 GJ/t (ibid.). In both cases the ore contributes only about 3 percent of the total process inputs.

However for copper, lead and zinc the situation is quite different, because of the significant exergy content of the sulfur in the ores. For copper, in particular, the exergy content of mixed sulfide/oxide ores typical of the US amounts to 61.6 GJ/t of refined cathode copper (2.112 GJ/t). By contrast, other process inputs in 1993 added 6.2 GJ/t (mostly chemicals for the beneficiation stage) plus 45.3 GJ/t fuels and utilities provided externally (ibid.). It is only the latter figure that is comparable with data from most of the studies

[5] An alternative term that is slightly more intuitive is 'available energy'. Available energy is energy that can do work. The important distinction is that exergy is not conserved; it is 'used up' in contrast to energy, which is conserved. Thus exergy is essentially what we mean when we use the term 'energy' for most purposes. However, it is a more precise concept. The exergy concept is discussed in more detail in *Appendix A*.

in the literature. It is extremely important to note that, in contrast with the two previous cases, the exergy embodied in the sulfide ores of copper, lead and zinc accounts for more than half of the total exergy consumed in the process chain. See *Figure 2.9* and *Figures 4.2* and *4.5*.

In this connection it is worthwhile to mention the energy analysis for the copper industry by Ruth (Ruth 1995). Ruth's dynamic optimal extraction model depends upon two dynamic relationships. One is the ore grade vs. cumulative output relationship discussed previously in connection with mining. The other is the efficiency of energy use in the mining-smelting-refining chain, also expressed as a function of cumulative production.[6] The energy learning curve derived by Ruth takes the form

$$\ln(e(t) - e^*(t)) = 10.292 - 0.953 \ln Q(t)$$

where $e(t)$ is the energy consumed by the mining sector, $e^*(t)$ is the minimum energy thermodynamically required to separate atoms of copper from the rest of the materials in the ore, and $Q(t)$ is cumulative production, indexed to unity in 1958 (Ruth 1995). The mathematical formula above was derived by econometric fitting to historical data from the US Census of Mineral Industries (various years from 1958 to 1987), with a correction to account for first-order auto-correlation. The data points were interpolated between census years by using annual value figures converted to quantities by means of derived coefficients. The R^2 value for the fit was 0.849, which is only fair.

The resulting formula suggests that energy savings of 51.64 percent (with an uncertainty of 20 percent or so either way) are achieved for each doubling of cumulative copper production after 1930. This is a very high rate of 'learning' as the author admits. He suggests that the period during which the fitting was done might not be typical, for various reasons (ibid.).

Ruth calculates e^* for mining as a function of ore grade, based on a 'perfect gas' model. The details need not be reproduced here, since the method of calculation is somewhat dubious. Suffice it to say that for an ore grade of 1 percent the minimum energy required to overcome the entropy of mixing (in the ore) is about 4MJ/t, while for an ore grade of 0.5 percent it is 4.25 MJ/t.

Ruth obtained learning curves for copper smelting and refining by a similar procedure, again using historical data from the US Census of Manufactures, between 1958 and 1981, interpolated between census years as before. The resulting formula is as follows:

$$\ln(e(t) - e^*(t)) = 12.591 - 0.208 \ln Q(t)$$

---

[6] This is generally known as the 'learning curve' or 'experience curve' and has been found to be a useful relationship for projecting costs. See (Ayres and Martinás 1992).

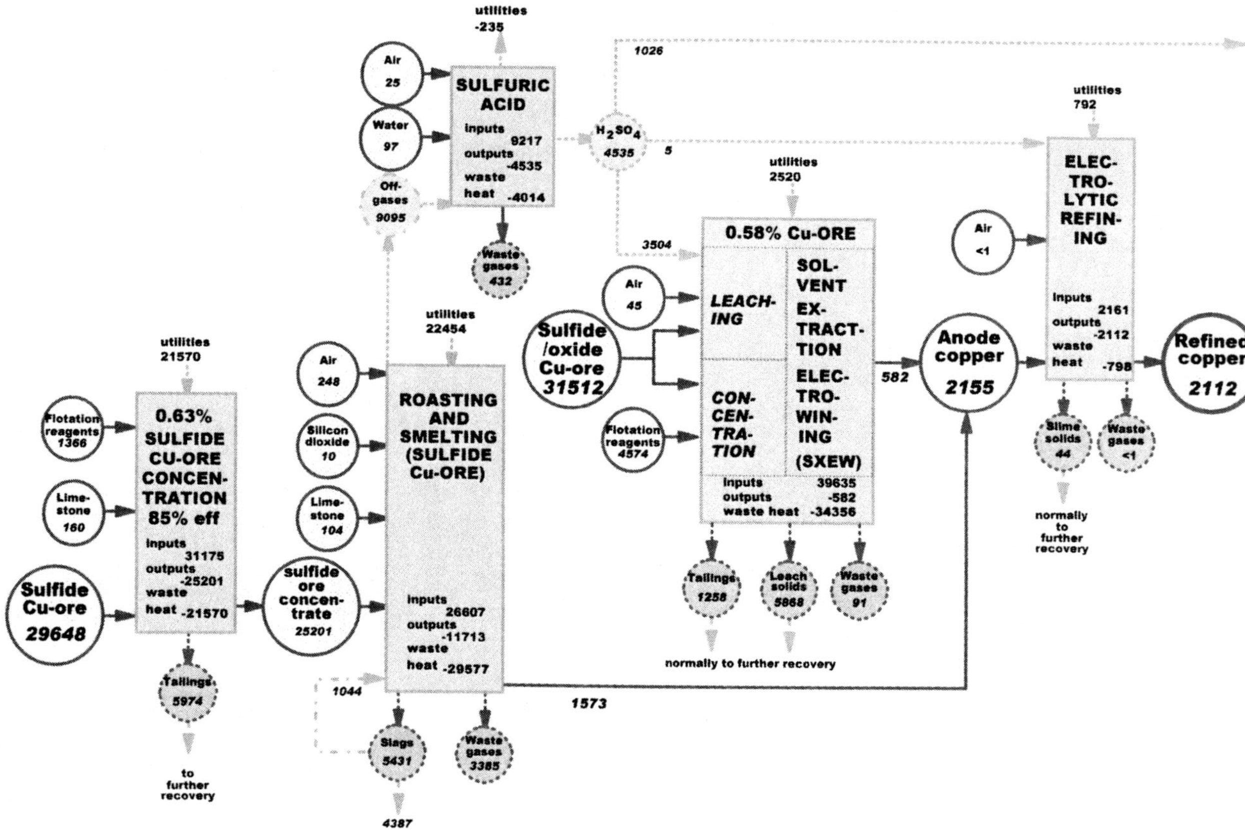

Figure 2.9: Exergy flows (mJ) in the production of 1 MT copper (simplified processes, typical material mixes).

where e^* is the minimum smelting and refining energy requirement and $Q(t)$ is cumulative production, again indexed to 1958 (Ruth 1995). The correlation in this case is rather poor, with an R^2 value of only 0.147. The derived learning rate for energy reduction in copper smelting and refining, based on this experience curve works out to 86.57 percent for each doubling of cumulative output $Q(t)$. This is an implausibly high learning rate as compared to historical experience in other industries (Ayres and Martinás 1992), and we cannot take it too seriously.

Minimum energy requirements for copper smelting and refining were estimated by Ruth to be 1682 MJ/t (1.682 GJ/t), which is related to the energy released by the basic exothermic reaction $CuS + O_2 \rightarrow Cu + SO_2$. However the rationale for this determination is not clear,[7] and other important reactions are involved, notably the converting reaction that separates iron-rich slag from copper in the matte. Again, for future reference Ruth introduces an energy efficiency measure which we cite later.

$$\eta(t) = \frac{e^*(t)}{e(t)}$$

Reverting to real-world experience, detailed German data for production units where energy-saving technologies had been introduced give the weighted average exergy consumption (of fuels and utilities) for the entire sequence of operations from mining and concentration through smelting and refining as 41.8 GJ/t and 57.3 GJ/t for the lowest grade of ore considered, 0.5 percent (Bruch et al. 1995). As of the late 1990s, the International Copper Institute (ICI) estimated that 60 GJ/t as fuels and utilities was required on average to produce refined copper in Europe from ore (Landner and Lindeström 1999 Table 4.2).

The Swedish Boliden Company has reported even better performance at its Ronnskar smelter, 47 GJ/t using ore with a grade of 0.35 percent of which 32 GJ were needed for mining, concentration and transport and only 15 GJ/t for smelting at a production level of 100 kMT/y in 1995 (Landner and Lindeström 1999, p. 60). At a production level of 130 kMT/y in 1996, energy consumption at the smelter had fallen to 10 GJ/t; further reduction to 7.3 GJ/t is expected as a result of the installation of a newly designed Outokumpu flash smelter and additional scale economies.

Technological progress in sulfur recovery/utilization will reduce some of these figures further, over time, although the lower limit calculated by Ruth above seems unrealistically low and the high learning rate he calculated is extremely unlikely to continue for long. In other words, while the

[7] The reaction as described by Ruth is "the reaction of copper sulfates into copper and sulfur dioxide" (Ruth 1995, p. 207). Unfortunately this reaction does not balance and it is not an adequate description of what happens in the smelter.

smelting/refining stage of the process can become more efficient – and could conceivably become an energy exporter – the mining, transport and beneficiation stages will probably consume ever more energy in the future. There is no realistic possibility that technological progress will overcome the mineralogical barrier, discussed at the end of *Chapter 1*.

Note that the recent energy efficiency improvements are in the smelter – primarily in the more efficient utilization of the chemical energy bound in the sulfides – not the mining/concentration stage. Roughly half of the energy consumed – about 20 GJ/t on average – was for mining and concentration, rising to as much as 32 GJ/t in the case of the lowest grade ores currently mined.[8]

The above data refers to pure refined rolled copper sheet. Rolled copper tubes (for water pipe), made from sheet, required an additional energy input of about 11.5 GJ/t as of the mid-90s. Drawn copper wire is somewhat more energy intensive, depending on the gauge (thickness) of the wire. As a matter of interest, brass (63 percent Cu, 37 percent Zn) from average ores averaged 52.2 GJ/t while bronze (6 percent Sn) averaged 43.5 GJ/t. A simplified diagram with exergy content of the main material inputs and outputs for each stage is presented in *Figure 2.10*.

The copper industry shares one of the features of the aluminum industry, being extremely energy (exergy) intensive. This is due partly to the large amounts of waste material that must be processed at the mine and concentration unit and partly to the purification processes of electrolytic refining and electro-winning. The overall mass and exergy flows of the process were shown in *Figure 2.9*. See *Appendix A*, *Table A1* for details on the assumptions used regarding the composition of ores, concentrates and compounds.

2.5. Sulfur recovery

As mentioned previously, approximately 1 ton of sulfur is associated with each ton of primary copper in average ore. Of course, there is no sulfur in oxide ores but most ores (90–95 percent) are sulfides. The sulfur-copper ratio varies slightly from mine to mine. Until the last few decades (i.e. before the introduction of hydro-metallurgical SX-EW processes) this sulfur was invariably emitted to air as sulfur dioxide, along with other volatiles such as arsenic. However, regulation in recent decades has discouraged such emissions. Moreover, most of the sulfur can now be economically recovered. The most common method of recovery is via an acid plant. However a few refineries produce liquefied sulfur dioxide. Sulfuric acid from the smelters is increasingly being used for leaching, to produce feedstock for the SX-EW process.

---

8 It is somewhat surprising that much higher figures are occasionally quoted, as high as 140 GJ/t. Some of these may simply be obsolete. We cannot explain the discrepancies.

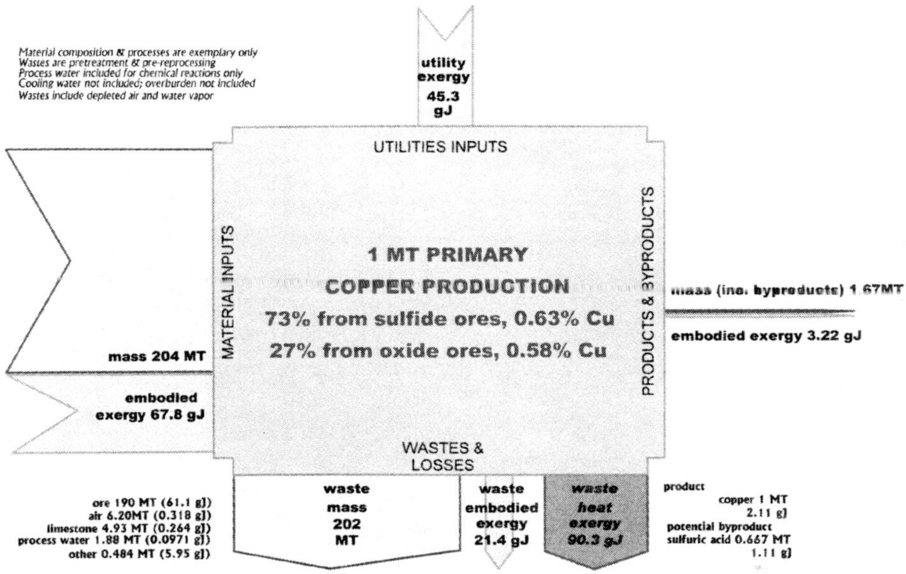

Figure 2.10: Primary copper mass and exergy flows per MT Cu.

In 1988 US smelters produced 1.043 MMT of copper from concentrates containing 1.242 MMT of sulfur, of which 1.124 MMT was captured for use (90.4 percent control) (Towle 1989). The efficiency of sulfur recovery (as opposed to emissions control) from non-ferrous metal smelters outside the US is not easy to determine. One study (Chapman 1989 Table 3) estimated the sulfur control indices for a number of exporting countries for 1985 as follows: Chile (14 percent), Peru (0 percent), South Africa (39 percent), Zaire (76 percent), Zambia (70 percent), Philippines (0 percent), Australia (0 percent) and Canada (26 percent). It estimated the average control level for copper exporters as 32 percent, compared to an average of 89 percent for the major copper importing countries.

Based on the US Bureau of Mines survey, 29 non-US smelters in 1988, accounting for 4.455 MMT of copper and 4.095 MMT of sulfur inputs, captured only 2.263 MMT and achieved just 55.3 percent control, resulting in sulfur dioxide emissions of 3.898 MMT (1.951 MMT S) (Towle 1989). Together, the 37 US and non-US smelters surveyed in 1988 accounted for 77 percent of world primary copper production. Unfortunately, it is likely that the refineries in countries not surveyed (e.g. China, the former USSR, Indonesia, Zambia, Zaire) achieved an even lower level of control than those included in the USBM survey.[9]

[9] Apparently a more recent survey has been carried out by the Finnish Environment Institute, Helsinki. However we have not yet seen it.

Substantial improvements in sulfur recovery have occurred since 1988, both in developed countries with strict regulation and in developing countries. However detailed data at the global level are unavailable. US data on sulfur recovery from non-ferrous smelters are available from the US Geological Survey (USGS). By-product sulfur from US non-ferrous smelters (as a group) amounted to 1.4 MMT in 1995, and increased to 1.61 MMT in 1998, of which 1.43 MMT was from copper smelters. Output fell to 1.32 MMT in 1998 (of which 1.13 MMT was from copper smelters) due to the closure – at least temporarily – of three copper smelters (out of seven). The closures were a direct consequence of declining prices which, by 1999, were the lowest (in real terms) in the 20th century. Because of increasing use of sulfuric acid for the SX-EW process, however, US copper producers as a group switched from net sellers of sulfur (or sulfuric acid) to net importers, in 1999.

Recent data on recovery are available for Chile, the world's biggest copper producer (Demetrio et al. 1999). As of the late 1990s, Chilean smelters captured 70 percent of the sulfur in the ore (see *Table 2.3*). Actual sulfur tonnage was not given, but copper output was 1.554 MMT, and if the sulfur-copper ratio for Chile is 1:1, something like 1.1 MMT of sulfur was captured.

Globally, sulfur recovery from non-ferrous smelters was at least 8.22 MMT in 1995, out of a global total of 54 MMT, with unspecified sources accounting for 3.92 MMT (USGS 2000). The metallurgical share increased to at least 10.3 MMT in 1998 and 10.2 MMT in 1999, out of global totals of 56.7 MMT and 57.1 MMT, respectively. Of this, 4.3 and 4.5 MMT respectively was from unidentified sources, mostly oil and gas refineries or metallurgy. In short the metallurgical share of total sulfur production (and consumption) is now between 19 percent and 20 percent and growing. Meanwhile the US contribution to the global total is declining. The decline accelerated in 1999.

2.6. Production-related wastes and emissions

2.6.1. *Mining wastes*

In mass terms the overburden from mining constitutes by far the largest single solid waste emission from the copper cycle. In addition to the ore itself, a large amount of waste rock (overburden) is removed to get access to the ore, especially in surface mines. The ratio of overburden to concentrator mill tailings in the US is now about 1.9:1. This represents a decline from the 1970s when the ratio was closer to 3:1 (Coppa 1984, p. 144). If the figures are accurate, it would seem that fewer new mines are being opened.

Overburden has no economic value. Any topsoil that was present is likely to be buried or eroded away. Heaps of overburden are also likely to be unstable and subject to occasional landslides. If the overburden is truly inert one could safely assume that overburden is indistinguishable from the (local) earth's crust, which means it contains no exergy, (available energy) by *definition* (See

Appendix A). We have made this assumption, for convenience, in our exergy calculations. However it must be acknowledged that the distinction between ore and overburden is an economic one, which means that some (at least) of the 'overburden' is actually low grade ore that has been set aside for future use when prices rise. Thus the declining ratio of overburden to mill tailings can also be interpreted as delayed utilization (possibly via the SX-EW process) of lower grade ores that were previously stored.

An important point to consider is that the compacted material removed from a mine, whether underground or open pit, cannot, in general, be put back into it. There is an unavoidable 'swell ratio' which means that the volume increases, not only because of air spaces between the particles, but because of water absorption. For hard rock the swell factor ranges from 1.6 to 1.9, or roughly 1.75 on average (40 percent to 45 percent of the volume consists of empty space) (Lovins 1973, pp. 44–45).

The most serious environmental problem associated with copper (as well as gold, lead, zinc, nickel and molybdenum) mining is acid rock drainage (ARD), formerly known as acid mine drainage, and largely associated with coal mining. The problem is, in brief, that sulfide-bearing rock, exposed to the air by mining, generates acid wastes which also mobilize many other heavy metals. Some of the sulfides in question – mostly iron pyrites – are left in piles of overburden and gangue. But, also, the walls of both underground and open pit mines are exposed to moisture and atmospheric oxygen. Abiotic oxidation is slow but the oxidation process is also catalyzed by certain anaerobic bacteria, especially *Thiobacillus ferrooxidans*, often present in the rock. This can increase the oxidation rate by a factor of as much as one million (Mitchell 1999).

As oxidation proceeds, the exposed rock surface cracks (spalls) exposing still more sulfides to the process. The oxidation process, in turn, acidifies the environment and mobilizes many other metals, including lead, arsenic and cadmium. In fact, due to feedback processes, the ARD process is regarded as autocatalytic and, once begun, can be very difficult to control. On the other hand, there is usually some time lapse between the closure of a mine and the onset of a full-scale ARD problem. This fact implies that prediction and prevention are essential tools for minimizing ARD (ibid.). However, up to now, prediction methodologies have not been notably effective.

Several treatment strategies are being explored. Active treatment, where prevention has failed, involves collection of the ARD and neutralization of the acid by liming (Mitchell 1999). (Some other alkaline reagents are also being tested.) However, liming is unlikely to constitute an adequate long-term solution, if only because it is likely to continue indefinitely. Thus, passive treatments are also being explored, notably constructed wetlands that are engineered to duplicate natural ecosystems insofar as possible. The idea of using artificial wetlands is to reduce the oxidation rate. Such systems typically involve several processes, including anoxic ponds (to reduce the dissolved oxygen in the water), anoxic limestone drains (to add alkalinity), aerobic cells (to precipitate iron and aluminum hydroxides and arsenic by absorption onto the precipitates), anaerobic cells to regenerate organic matter and filter suspended solids and rock filters on which algae and manganese oxidizing bacteria grow, to increase pH (alkalinity) and remove manganese from the water (ibid.).

It is important to note that artificial wetlands are probably impractical in many of the arid areas where copper is now being mined, especially the southwestern US and northern Chile.

Other mine-related emissions include dust, grinding aids (usually steel rods or ceramic balls), explosive residues, and combustion wastes from heavy earth-moving equipment. Airborne wastes from these sources are listed later. Explosives used (mainly ammonium nitrate) range from 6 kg/t for low grade ore from underground mines (9 kg/t for high grade ore from underground mines) to 80 kg/t for high grade ore from open pit mines (e.g. Chile) and 138 kg/t for low grade ore from open pit mines (e.g. US) (Bruch et al. 1995). Sulfuric acid is used in the SX-EW process, e.g. in the case of an underground mine producing low grade ore, which is leached (273 kg/t) (ibid.).

2.6.2. *Beneficiation wastes*

The second major category of wastes consists of mill tailings or *gangue*, which is unmineralized rock separated from the mineralized concentrate during the beneficiation process. Mine waste (tailings) greatly exceeds the quantity of usable ore concentrate. For example, a ton of concentrate delivered to Germany corresponds to 37.65 tons of tailings left at the mine or mill, not including overburden (FfE 1999). The global average is 37 tons of waste per ton of concentrate.

Tailings from the concentrators constitute a major solid waste. They are generally stored in ponds which also contain the various chemicals used for froth flotation, or their degradation products. Moreover, such ponds contain all of the chemicals added during the froth flotation process (See *Table 2.2*). Flotation chemicals added range from 6 kg/t for low grade underground ore to 14 kg/t for low grade ore from open pit mines. These chemicals are not especially toxic (except for the cyanide), but many of them are potentially reactive. Concentrator wastes are typically acidic (after weathering) and phytotoxic. They contain no plant nutrients. Other emissions, based on German data, are summarized in *Tables 2.4* and *2.5* (Bruch et al. 1995 Tables 7, 8).

The solids in the tailings consist mostly of common rock minerals (silicates, etc.) But they also contain a significant fraction of the pyrites and 10 percent to 15 percent of the heavy metals, including the copper. Any pyrites in the rock or hydrocarbons (kerogen) – normally present in sulfide gangues – will weather and oxidize on exposure to air, thus causing acidification. One significant component of these wastes is arsenic, which is discussed in more detail in a later chapter. However, US copper ore is estimated to contain 6.5 kg or arsenic per metric ton of copper, or 9.5 kMT in copper ore processed

Table 2.4: Emissions and wastes from four German copper mines in 1992

	Underground		Open pit	
	High-grade ore	Low-grade ore	High-grade ore	Low-grade ore
Emissions per metric ton of Cu concentrate				
Fumes (Nm^3)	*10300*	*3550*	*17360*	*26940*
Dust (kg)	18.56	65.52	95.56	230.34
SO_2 (kg)	3.3	0.34	4.0	6.4
NO_X (kg)	7.5	5.5	18.3	27.6
CO (kg)	2.5	3.4	9.2	13.7
HC (kg)	1.1	1.6	4.2	6.3
CO_2 (kg)	2330	500	3350	5260
Solid wastes: metric tons per metric ton of Cu in dry concentrate				
Rubble		6	292	76
Residue	28.8	114	92	140
Ash	0.082	0.008	0.099	0.158
Gypsum	0.031	0.003	0.038	0.060

These mines account for 15% of total German copper ore production.
Waste allocation per metric ton of Cu is difficult because other metal concentrates are produced as well.
Source: Adapted from Bruch et al. (1995) Table 7, p. 6.

in 1989, of which 2.4 kMT was left in mill tailings in that year (Loebenstein 1994).[10]

Even though the concentrate is removed for further processing, the tailings occupy from 2.2 to 2.4 times more volume than the original ore, because of the 'swell factor' noted above, plus the fact that the tailings from a flotation process are wet – usually 50 percent water by volume (ibid.). For this reason, tailings are normally stored in ponds, created on or near the mine site, usually behind an artificial dike. In the US the pondwater is generally recycled and these dikes are theoretically designed to last indefinitely. There are stabilization programs to prevent wind or water erosion by planting grass or some other hardy plants. Dust is the major problem, at present.

Most international mining companies follow good practices with respect to gangue and overburden everywhere they operate, although there are some disputes about this, particularly in connection with projects in Papua New Guinea and Indonesia. In fact, virtually every year there is a report of some dike failure in a remote area. Some of these have resulted in significant amounts of toxic mine waste suddenly being released into a river. Even if the

[10] Arsenic content of ores varies considerably. The ore processed at the ASARCO smelter in Tacoma Washington (closed in 1985) was as high as 4 percent As. On the other hand, the ore processed at the Tennessee Chemical Co. Copperhill plant had an arsenic content of only 0.0004 percent (Loebenstein 1994, p. 7).

Table 2.5: Energy consumption, emissions, wastes and byproducts from the production of 1 metric of copper from concentrate[a] and scrap, Germany 1992

	Total	Production only	Electricity & other
Primary energy (GJ/t)[b]	*21.8*	*6.3*	*15.5*
Emissions			
Fumes (*Nm³*)	16000	12100	3900
Dust (kg)	0.425	0.043	0.382
SO_2 (kg)	4.75	4.06	0.69
NO_X (kg)	3.93	0.34	3.59
CO (kg)	1.08	0.09	0.99
HC (kg)	0.71	0.01	0.70
CO_2 (kg)	1390	460	920
Solid wastes (kg/t)			
Acid mud	<1	<1	
Arsenic deposit	2	2	
Ash	30		30
Gypsum	<10		<10
Waste water (*m³*)	*~1*	*~1*	
Marketable byproducts (kg/t)			
Slag	1120	1120	
H_2SO_4[c]	1880	1880	
Anode mud	10	10	
$NiSO_4$	10	10	

[a] Major inputs are 2130 kg concentrate and 570 kg scrap. The concentrate is assumed to be Cu(30%), Fe(26%), S(32%), Zn(1.5%), SiO_2(0.2%), As(0.1%), Al_2O_3, CaO and MgO (3%), other heavy metals (<0.2%).
[b] Overseas transport of concentrate accounts for 10% of the energy consumption.
[c] H_2SO_4 production accounts for 13% of total energy consumption.
Source: Bruch et al. (1995) Table 8, p. 7.

tailings are not toxic, the resulting inundations may cause other problems, such as damage to wildlife. Highly publicized incidents recently occurred in Spain and Romania, for instance, and there was a case in the western US a few years ago. Admittedly the most recent cases were not specifically copper related, but the problems are similar.

2.6.3. *Leaching (acid) wastes*

The heap leaching (SX) process is another source of waste, seldom discussed. Large amounts of sulfuric acid (a smelter byproduct) are used for this purpose. A crude mass balance for the leach process is interesting. To obtain 1 t of copper in the form of copper sulfate (as feedstock for the EW process) approximately 0.5 t of sulfur is required as acid, all of which is presumably recovered during the electro-winning (EW) stage. In principle the sulfur in the system is simply recycled as sulfuric acid and returned to the leaching operation. On the other hand (although we have no precise data) it seems clear that much more sulfuric acid is used for leaching than is recovered at the EW stage. The residue presumably reacts with other minerals in the ore or concentrate. Since most sulfates are somewhat soluble, they presumably find their way into surface waters or ground water. The literature does not discuss this point. Reactions with carbonates, in particular, may result in indirect emissions of carbon dioxide.

2.6.4. *Smelting wastes*

The next category of wastes is generated at the smelter/convertor. Smelter recovery of copper from concentrate averages 98 percent. About 2 percent of the copper (and some sulfur) remains in the slag. Apart from fuel, smelter inputs include silica (800 kg/t of copper) and limestone (250 kg/t). These materials are subsequently lost in the slag, which amounts to about 1550 kg/t (or 1.55 t/t), assuming concentrate with a grade of 25 percent. For higher grade concentrates, the quantities of slag are reduced. Slag is marketable as a road surface or concrete additive, at least in Europe.

Apart from silicates, and copper, the slag typically contains 100 percent of the iron, 70 percent of the zinc, 30 percent of selenium and tellurium, 25 percent of tin and nickel, 20 percent of cadmium, cobalt and antimony, and 10 percent of the silver, gold, platinum, and lead that was originally present in the concentrate. The gaseous waste stream from the smelter/convertor typically contains 100 percent of the germanium, 95 percent of the bismuth, 90 percent of the mercury, 85 percent of the lead, 80 percent of the cadmium, 65 percent of the tin, 60 percent of the antimony, 30 percent of the zinc, and 10 percent of the selenium and tellurium that were originally present (Biswas and Davenport 1976). These figures are old, but probably do not change much. The smelter vapor effluents must be captured and treated in any case. The metal fractions in the gas stream can be recovered easily for use by condensation or other means (depending on the actual composition of the concentrate, and the price of the metal in question), or it can be trapped and immobilized as treatment sludge.

Discarded slag heaps from smelters that have long ceased operation can be sources of elevated copper levels in nearby water bodies for many years. As a case in point, New York City and its environs (the Hudson-Raritan Bay) was the center of the world copper smelting and refining industry in the early part of the 20th century.[11] As recently as 1960 there were still four primary

[11] This may be a surprise to many people who think of New York as a trade entrepot, financial and publishing center, rather than as an industrial city. However, the telephone and electrical industries (starting with the Edison companies, which became General Electric Co.) were essentially created in the New York city area and the Hudson Valley. It was logical for wire and cable manufacturers to locate there, and for copper refiners (using imported concentrate from Mexico or Chile) ditto. Of course much of the refined copper (and wire) was exported.

copper refineries in the harbor area, plus several secondary recovery operations; the last refinery (at a place called Laurel Hill in Queens) closed in 1986. Yet copper concentrations in the waters of the New York harbor remain substantially higher than in other comparable locations, and very much higher than in seawater.

The copper concentration in the open ocean is about 1 microgram (μg) per liter, or 1 part per billion (ppb). Near shore the copper concentration varies from 2 ppb to 4 ppb (Spencer and Brewer 1969; Fitzgerald 1970). In industrialized estuaries much higher values have been recorded. For instance 27 ppb has been measured in heavily polluted Boston harbor (Gilbert et al. 1972). However copper concentrations as high as 65 ppb were measured in the western Raritan Bay, the highest ever recorded up to that time (Waldhauer et al. 1975).[12] This is undoubtedly due, in large part, to the existence of old slag heaps that continue to leach (and will do so for many decades to come.) On the other hand, the leaching rate from old dumps does decline over time. Swedish data indicate that the leaching rate from old mine/smelter dumps declined by half between 1990 and 1998 (Landner and Lindeström 1999, p. 89).

The unaccounted for fractions of trace metals are left in the molten copper. The blister copper still contains 90 percent of the silver, gold and platinum, 80 percent of the cobalt, 75 percent of the nickel, 60 percent of the selenium and tellurium, but only small fractions of tin (10 percent) and bismuth and lead (5 percent). Blister copper is, of course, refined further by an electrolytic process. On average, the anode furnace generates about 18.8 kg of slag per metric ton of copper. This material is discarded. In the next (electrolytic) step, the remaining (metallic) impurities are concentrated in a slime known as 'anode mud', which averages 6 kg per metric ton of cathode copper. The anode mud is typically 25 percent Cu; the remainder consists of other impurity metals, including gold, silver, platinum, selenium, tellurium and some others. The anode mud is quite a rich source of by-products, and the recovery of these metals accounts for a significant fraction of the profits of the industry.

Arsenic (as sulfide) constitutes about 0.3 percent of typical concentrate and 0.44 percent of smelter slag (Goessling 2001). Arsenic losses from copper smelting (in the United States) have been calculated quite carefully for the year 1989. In that year, smelter feeds contained 6100 metric tons of arsenic, of which 5300 t was disposed of in slag, while 800 t was emitted to the atmosphere (Loebenstein 1994). As it happens none of the arsenic in US ores was recovered for use in the US. All US consumption of arsenic in 1989 and since, roughly 23,000 tpy, has been imported – mostly from China.

As mentioned previously, sulfur recovery from concentrates is fairly high

[12] Significantly higher concentrations (up to 200 ppb) have been recorded near some Swedish cities, as noted later.

(around 90 percent) in the smelters and refineries of the industrial countries, albeit much lower elsewhere. As noted earlier, sulfur recovery from reverberatory furnaces is uneconomic because of the low concentration of sulfur dioxide in the combustion products. Fortunately these old furnaces are rapidly being replaced. The unrecovered sulfur is lost in a variety of ways, including some sulfur dioxide emissions, but also to some extent in solid waste (slag) or liquid wastes, where it is probably harmless. As of the late 1990s, airborne emissions of sulfur dioxide from copper operations in Europe amounted to about 20 kg/t or roughly 2 percent of the sulfur originally in the concentrate (International Copper Association).

A summary of emissions from smelting and refining 1 metric ton of cathode copper from a mixture of concentrate and scrap, derived from German data for 1992, was given in *Table 2.5* (Bruch et al. 1995 Table 8). Globally it was estimated that non-ferrous copper-nickel smelting operations in 1983 emitted 22.525 (14.45–30.6) kMT of copper into the atmosphere (as particulates), along with 1.063 kMT of antimony; 10.625 (8.5–12.75) kMT of arsenic; 2.550 (1.7–3.4) kMT of cadmium; 0.021 kMT of indium; 16.575 (11.05–22.1) kMT of lead; 2.55 kMT of manganese; 0.122 (0.037–0.207) kMT of mercury; 7.65 kMT of nickel, 0.854 (0.427–1.280) kMT of selenium; 1.063 kMT of tin and 6.375 (4.25–8.5) kMT of zinc (Nriagu and Pacyna 1988).

A recent update of the 1988 study summarized above estimated global copper emissions from primary copper production as follows: copper 17.708 kMT; antimony 0.319 kMT; arsenic 3.183 kMT; cadmium 1.319 kMT; mercury 0.069 kMT; indium 0.027 kMT; lead 6.27 kMT; manganese 0.059 kMT; nickel 8.854 kMT; selenium 0.319 kMT; tin 0.319 kMT and zinc 3.644 kMT (Pacyna and Pacyna 2001 Table 15). The fact that nickel and indium emissions appear to have increased is probably more a reflection on the quality of the data than on the reality, although changes in the composition of ore being exploited could be partly responsible.

For the most part, the 1983 figures have been reduced significantly by 1995 in the industrial countries, by as much as a factor of three in some cases (e.g. arsenic). In Sweden, for instance, copper emissions to the atmosphere have fallen from 280 metric tons in 1977/1978 to 70 tpy in 1985, 27 tpy in 1990 (of which 18 tpy were from primary metal works) and 14 tpy in 1994–1996 (of which 5 tpy were from metallurgical sources (Landner and Lindeström 1999, pp. 90–91). However, the Swedes have been much stricter enforcers of environmental regulations than most other countries. Globally, the emissions from primary smelters are undoubtedly still rather large.

Atmospheric emissions of copper and other metals, as particulates, do not remain in the atmosphere, of course. They are deposited mainly on land, and mainly within a few kilometers or tens of kilometers of the source. There is a long-term buildup in soils which has caused some worry, although it is increasingly evident that copper concentrations in soil do not correlate with mobilization. Moreover, copper is essential for plant growth, whence the concentration in plants is normally much higher than the soil concentration. (This implies that copper is actually scarce as far as plants are concerned.) Thus the soil buildup so far is probably not a serious concern, at least in agricultural soils which are limed and kept at a fairly low pH. The major problem would arise from mobilization resulting from increased acidification or – in the case of agriculture – from a cessation of liming.

2.6.5. *Wastes from finishing operations*

Wastes and emissions from copper finishing and forming operations are largely attributable to energy consumption for melting and electrical power. Details for German operations have been compiled and tabulated by the Technische Hochschule, Aachen (Bruch et al. 1995). Another (and growing) source of waste copper emissions arises from use in the electronics industry. For instance, copper is now replacing aluminum as a conductive plating material in the semiconductor industry, which means that copper chemicals are increasingly contained in wastewater from the industry. Currently the quantities are very small however.

2.6.6. *Recycling (secondary recovery) wastes*

Secondary scrap melting and refining is one of the minor sources of copper-related emissions. Recycling copper-containing wastes is covered in detail in *Chapter 5*.

2.6.7. *Toxic releases*

Both metallic copper (as particulates) and some copper compounds are listed as toxic by the USEPA. It is of some interest to note that in 1992 copper releases reported to the EPA's Toxic Release Inventory (TRI) amounted to 305 kMT (reported as 675,138,852 lbs). The releases increased by 3.6 percent to 317 kMT in 1994 (INFORM 1995 Tables 15, 16). This covers all media (air, water and earth) and probably includes mine wastes, although mining companies were originally exempt from reporting. Copper compounds released during the same period increased from 228 kMT in 1992 to 259 kMT in 1994 (ibid.). Since this greatly exceeds the copper content of all copper chemicals produced (mostly chromated copper arsenate or CCA, used as wood preservatives) it is hard to interpret these numbers.

2.6.8. *Global estimates of airborne emissions*

Two estimates of global anthropogenic copper fluxes to air, water and soils in 1983 and 1995 are shown in *Table 2.6*. Data are taken from the two surveys previously cited (Nriagu and Pacyna 1988; Pacyna and Pacyna 2001). In both cases national surveys from a few countries were used, with the gaps being filled by means of emissions coefficients derived largely from national sources

Table 2.6: Global copper emissions estimates 1983 and mid-1990s (metric tons)

Source	1983 to air	1983 to water	1983 to soil	mid90s to air
Coal combustion: electric utilities	950–3100		93–335	
Coal combustion: domestic + industry	1390–4950	3.6–1.3	(ash)	7081
Oil combustion: electric utilities	548–2320	(only		
Oil combustion: domestic + industry	179–1070	utilities)		
Wood combustion	600–1200			
Metals mining	160–800	0.1–9	262–787	
Copper and nickel production	14450–30600			18071
Lead production	234–312	2.4–17	395–790	
Zinc and cadmium production	230–690			
Iron and steel production	142–2840			142
Secondary non-ferrous production	55–165			
Cement and other manufacturing		11–56	0.95–7.6	
Municipal waste incineration	980–1960			
Sewage sludge incineration	30–180			
Phosphate fertilizers & peat for agriculture	137–685		0.05–0.58	
Municipal waste and other urban			13–40	621
Sewage sludge & organic wastes		2.9–22	4.9–21	
Atmospheric deposition		6–15	14–36	
Domestic wastewater		8.7–48		
Agriculture and food wastes			3–38	
Animal wastes, manure			14–80	
Logging and other wood wastes			3.3–52	
Misc. wastage of commercial products			395–790	
Total emissions	19860–50870	35–190	1198–2944	25915
Median emissions	35370			

Sources: 1983 data adapted from Nriagu and Pacyna (1988).
1990s from Pacyna and Pacyna (2001).

supplemented by expert judgments. This table is nevertheless helpful insofar as it indicates the range of sources and their relative magnitudes, as well as recent changes. For instance, although much more than half of global copper (and zinc) emissions to the atmosphere are attributable to the mining and metallurgical activities *per se*, there are still rather significant (albeit quite uncertain) emissions of copper from other sources, especially coal combustion (Pacyna and Pacyna 2001 Table 15). These sources cannot be ignored in a global overview.

It would appear that global copper emissions from all sources have fallen from 35.4 kMT in 1983 (median estimate) to 25.9 kMT (point estimate) in the mid 1990s. Given the uncertainty of some of the data sources, the lack of an overall uncertainty range in the later Pacyna study is unfortunate. Error ranges are given for major source categories, however. The mid-90s estimates for copper are well within the error range of the 1983 estimates, although the conclusion that emissions have declined is probably justified. In addition, the more recent Pacyna estimates provide a breakdown by regions that was absent from the earlier work. In the case of copper, 50 percent of the total emissions are attributable to Asiatic sources, with another 21 percent from South America. North America contributed 11 percent, Europe (including European Russia) contributed 8.7 percent and Africa 7.7 percent (Pacyna and Pacyna 2001 Table 16).

The 1983 study illustrates some of the problems with using data from a variety of sources. In 1983 the authors estimated that 20–51 kMT of copper were emitted to the atmosphere, but only 1.2–3 kMT was deposited as fallout to surface waters and soils (Nriagu and Pacyna 1988). This is inconsistent, since all of the atmospheric emissions of copper (and most other metals) must be deposited locally in a relatively short time. While some of the atmospheric fallout is over the oceans, the bulk of it is over land, and most of that is certainly accumulating in the soils. In the more recent survey, there is (wisely) no attempt to account for atmospheric fallout in the same way, nor any attempt to account for emissions directly to water or soil (Pacyna and Pacyna 2001). For comparison, *Tables 2.7*, and *2.8* present different sets of estimates for copper fluxes to air and water, respectively (Landner and Lindeström 1999 Tables 3.4, 3.5).

2.7. Optimal extraction/production

Economists generally like to assuming a long-term balance between supply and demand, with growth (in equilibrium) where demand is determined by

Table 2.7: Estimated global copper fluxes to the atmosphere (1000 metric tons per year)

Source	Median fluxes 1983	Estimated fluxes 1996
Volcanoes	9.4	9.4
Airborne soil particles	8.0	8.0
Saltspray from the sea	3.6	3.6
Forest fires	3.8	3.8
Biogenic sources	3.3	3.3
Total, natural sources	**28**	**28**
Smelting, refining	23.2	11.6
Mining, concentration	0.4	0.4
Energy generation	8.0	6.0
Other industrial processes	2.0	2.0
Waste incineration	1.6	1.5
Total, anthropogenic emissions	**35**	**21.5**
Fraction of natural copper	44	57

Source: Landner and Lindeström (1999) Table 3.4.

Table 2.8: Estimated global fluxes of copper to the aquatic environment (chiefly oceans) in the early 1980s (1000 metric tons per year)

Source	Anthropogenic fluxes	Natural fluxes	Total
Domestic sewage	28		
Electrical power stations	13		
Metal mining & production	14		
Other industrial sources	34		
Atmospheric fallout	11	20	
Dumping of sewage sludge	12		
Diffuse sources, dissipative uses	200		
Volcanic, hydrothermal activity		≈ 20	
Seabed erosion		probably no net input	
River transport of erosion products		1200	
Total (order of magnitude)	**310**	**1240**	**1550**
Percent	20%	80%	100%

Source: Landner and Lindeström (1999) Table 3.5.

prices and incomes. Utility is generally equated to consumption, which is maximized over time subject to budget or other constraints and with an assumed discount rate to reflect the general preference for current consumption over future consumption. There are many 'dynamic' economic models based on this scheme.

With few exceptions [e.g. (D'Arge and Kogiku 1973; Ayres 1988)] these models do not take into account energy or material resource depletion or thermodynamic constraints. However, Ruth has published a model of optimal future copper extraction, based on maximizing the difference between cumulative copper mine output, which is assumed to be welfare increasing, less cumulative energy (exergy) cost of mining, which is assumed to be welfare decreasing. The optimization is carried out subject to a constraint that the total cumulative mine output over time cannot exceed the initial resource endowment. (It is also assumed, at least implicitly, that the resource endowment is exactly known at the beginning.)

The problem, thus simplified, can be solved analytically by using a sophisticated branch of mathematics known as 'optimal control' theory.[13] This method involves creating a new function of copper ore mined $Y(t)$, exergy

[13] The first real applications of the Hamiltonian theory were to determine the equations of motion in physics, subject to the conservation laws of energy and momentum. The method was later generalized applied to missile guidance, i.e. to determine the optimal path between two points in a three dimensional gravitational field (Pontryagin et al. 1962). (This may be why some models of stock market behavior have been referred to as 'rocket science') For a detailed description of applications in economics see (Dorfman 1969).

consumption $E(t)$ and crude copper ore output $J(t)$ multiplied by a 'shadow price' $\lambda(t)$. This function (called a Hamiltonian), consists of the expression for welfare to which is added an additional expression which vanishes at the constraint boundary. The Hamiltonian is then maximized with respect to each of these variables, yielding two so-called 'first order conditions' for the existence of a maximum and an adjoint equation for the shadow price.

In view of the crudeness of the model assumptions, especially with regard to the assumption of advance knowledge of unchanging reserves and the poor correlation of past energy efficiency trends with cumulative production, the model cannot be used with any confidence as a forecasting tool. Hence the complete set of equations is not worth reproducing here. However some of the results are interesting. First of all, the simulation model predicts that the shadow price of crude copper ore will rise exponentially to a maximum value when the ore reserves are exhausted. This is intuitively plausible. However according to the model the optimum extraction rate actually declines at first – contrary to any real world evidence – and then rises later, due to technological improvements and finally declines to zero at the terminal time. For an assumed discount rate of 4 percent per annum, an assumed reserve of ore containing 50 million metric tons of copper – roughly 15 percent of estimated global reserves as of 1987 – and initial values of all other variables at 1987 US levels, the model predicts that the US copper reserves will be exhausted in 110 years (Ruth 1995). The author notes that this time frame is consistent with other independent estimates (Yoshiki-Gravelsins et al. 1993). The interested reader is referred to the original article and references therein.

CHAPTER 3

COPPER: DEMAND AND DISPOSITION

3.1. Consumption patterns and trends

As noted already in the historical introduction, the modern industrial era began with the advent of the electric telegraph and the associated use of copper wire. Telegraph lines and undersea cables were the initial users, followed in the 1880s by electric power generating and distribution systems, which grew even more rapidly.

Today, electrical conductivity and thermal conductivity – together with ductility and malleability – are the most critical properties of copper for electrical and electronic products and wire-consuming industries (such as building construction, industrial machinery, and motor vehicles). Thermal conductivity is important mainly for air-conditioning and large scale heat transfer applications. Corrosion resistance is important mainly for roofing, facing and pipes used in buildings. Copper is the best electrical and thermal conductor of all metals (except silver). Nowadays electrical applications account for between 75 percent and 80 percent of US demand for refined copper (up from 50 percent in the 1950s), while brass and certain specialized alloys (notably copper-nickel) account for most of the rest (*Figure 3.1*).

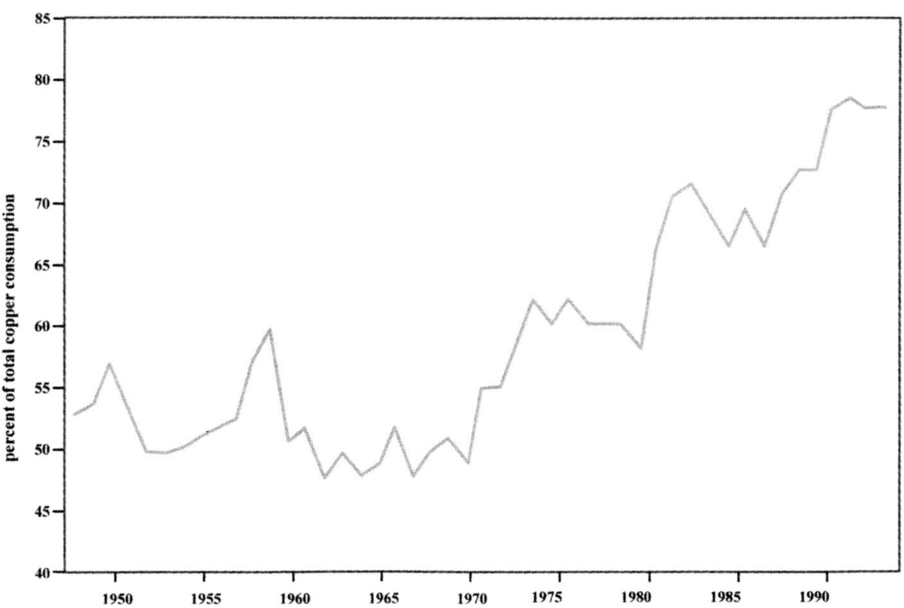

Figure 3.1: Consumption of copper in electrical equipment; US 1947–1993.

The US is the largest consumer of refined copper, accounting for about 20 percent of the global total, followed by China (12 percent) and Japan, slightly less. China recently overtook Japan, and Chinese consumption has been growing in recent years (since 1995) at no less than 15 percent per year. This reflects the fact that China is now investing in a number of major national infrastructure projects, including the Three Gorges Dam and power complex, and a national cable TV network. Other copper-intensive capital projects include power generation and transmission, railways, highways and petrochemical plants, together with an east-west natural gas pipeline. Apart from these, Chinese industrial activity in copper-intensive sectors rose 22 percent in the first quarter of 2001, year-on-year: motor vehicles were up 21 percent, power generating equipment up 15 percent, AC electric motors up 14 percent, air conditioners up 45 percent, integrated circuits up 21 percent and switchboards up 90 percent (*Recycling Today*, summer 2001).

Electrical applications seem to be comparably important in other countries. In Germany wire accounted for 69 percent of refined copper in 1983 (DKI 1984) remained at that level (70 percent) in 1995 (Henstock undated). In Japan, wire and cable accounted for about 80 percent of total refined copper production in 1997 (a normal year) and was nearer 90 percent in 1998 due apparently to sharply decreased demand for other copper tubing and brass products from southeast Asia (Simada et al. 1999).

It must be emphasized that not all electrical uses are in the electrical industry per se. The breakdown by sector is roughly similar in all industrial countries. Taking Japan (1998) as an example, 37 percent of cable and wire produced went to the building and construction sector, 25 percent went to electrical machinery, 13.2 percent to electric utilities, 7.5 percent to automotive, 4 percent to telecommunications, 8.3 percent to 'other domestic demand' – mainly appliances and consumer electronics – and 5.2 percent to exports (Simada et al. 1999, Figure 3).

In Sweden the breakdown between wire and other forms is not available, but 34 percent of all copper goes to buildings, 28 percent to electrical and electronic goods, 14 percent to machinery and accessories, 13 percent to consumer goods and 10 percent to transportation (vehicles), with chemicals taking 1 percent (Landner and Lindeström 1999, Figure 5.5, p. 76).

Brass, an alloy of copper and zinc, has a wide range of non-electrical uses such as industrial valves and fittings, where corrosion resistance and machinability are important (as well as some electrical uses), although it also has many substitutes. Brass is particularly valued for decorative hardware applications (such as locks and latches), partly because of its resistance to corrosion and its beauty. However, the market for brass is largely driven by the supply of secondary copper that cannot be purified sufficiently for use as wire. Bronze is used mainly as castings for ships propellers and commemorative statuary.

The near and medium term future demand for pure refined copper is going to be even more determined by demand for electrical and electronic goods, while these markets also demand higher and higher levels of purity. Even in this market, aluminum is now preferred for high voltage overhead transmission lines because of its lower weight.[1] For underground lines, however, copper is preferred because weight is unimportant but volume matters (RoskillCu 1990) overhead transmission lines are ugly and in time most of them are likely to be replaced, especially in urbanized areas (where most people live). There are also concerns – whose scientific basis is still extremely controversial – that low frequency electromagnetic radiation from overhead power lines may cause health problems for persons living or working nearby. If this is confirmed by future research, it could have an adverse impact on the use of overhead lines (with aluminum conductors) in built-up areas, *vis à vis* underground lines with copper or superconducting cables.

At one point it was thought that aluminum would substitute for copper more widely in household electrical applications. However, since the 1970s this trend has actually reversed. Aluminum has been largely replaced by copper for interior wiring, due to problems of overheating at contact points and a number of fires that were attributed to the use of aluminum wiring. Copper is now displacing aluminum and tungsten for metalization in semiconductors, again because of superior electrical and thermal conductivity (SIA 1997). Copper has always been used for motor and generator windings, where volume is also important because efficiency depends upon maximizing the density of current carrying elements (windings) that intercept the magnetic field lines. Heavy duty electric motors and transformers account for 50 percent of all electrical uses, with power transmission at 20–25 percent and consumer goods and communications 10–15 percent.

In fact, the need for economizing on energy, in general, and electricity in particular may actually result in increased demand for copper in some applications (Lovins 2001). For example, premium efficiency asynchronous motors now use 20 percent more copper than their predecessors, although they may also be smaller (to match reduced loads). Another example is interior building wiring. Doubling the diameter of copper wire quadruples its cross section and cuts resistive losses to a quarter. Such losses now account for 6 percent plus or minus 2 percent of electric power generated. In many cases the extra copper would pay for itself in energy savings) in a year or two. Pipes and heat exchangers are still another case in point. Lovins notes that viscosity friction losses increase as the inverse 4.83 power of the pipe diameter, and each unit of friction corresponds to about ten units of energy consumption at the thermal power plant where the electric power is generated. This translates into a strong economic incentive to use more copper in water pipe.

---

1 The volume conductivity of aluminum is 0.63 that for copper, while its density is only 30 percent of that for copper. For equal conductance, aluminum wire must have 25 percent greater diameter than copper wire, but because of lower density, only 50 percent of the weight. This, in turn, means that the necessary investment in towers is substantially lower.

In the telecommunications market the main challenge to copper wire has been glass fiber, beginning in 1976. In the 1980s glass fiber was restricted to new high capacity trunklines, which at most accounted for only 10 percent of copper consumption in that sector. The advantage of glass fiber is reduced weight and volume, resulting in sharply lower costs per message. As of the late 1980s a single kilogram of glass fiber sufficed to carry about 1500 voice channels over a distance of 5 km, whereas 3 metric tons of copper were needed for the same purpose (RoskillCu 1990, p. 68). Moreover, signal repeaters were needed only at intervals of 50 km, vs. 1.6 km for copper cable.

These advantages have since been extended dramatically. According to one source 18 optical fiber pairs weighing 117 kg/km are equivalent in information carrying capacity to a cable consisting of 2400 copper wire pairs, weighing 7225 kg/km. The energy consumption per voice channel is 0.10 MJ, as compared to 23 MJ per channel for the copper cable. Moreover the energy consumption for production of the glass fibers is 15.88 GJ/km as compared to 827.4 GJ/km for the copper (Pinhao 1996).

The demand for telecommunications channel capacity has exploded since the Internet became a significant market presence in the early 1990s. See *Figure 3.2*. On the other hand, demand for copper wire in the telecom sector declined by 30 percent between 1975 and 1989 in the US (Zeltner et al. 1999). New trunk capacity and even a significant fraction of local loop capacity, is now based on glass fiber (and the manufacturers thereof became glamor stocks in 1998–2000, albeit sadly deflated in the subsequent market crash. Some existing copper cables are also now being replaced by glass. It is likely that within twenty years, and possibly less, virtually all telecom lines will be optical and demand for copper in this sector will become negative. In other words, as happened in the high voltage transmission sector in the 1960s and '70s, the telecom sector will then become a significant source of high quality scrap.

Incidentally, the so-called 'last mile' connections from households to high capacity communications cables, hitherto reserved for copper wire, may not be replaced by optical fibers, but rather by wireless devices. Nevertheless, fiber-optic systems utilize electric power, which cannot be carried through glass fibers. Thus the rapid growth of fiber-optic networks also carries with it an increased peripheral demand for electrical wiring, most likely copper-based. (A recent breakthrough by Corning implies the strong possibility that future glass fibers will be air-filled hollow tubes, rather than solid cores, resulting in dramatic reductions in the need for amplification and associated energy consumption.) Copper foil is also used for sheathing undersea fiberoptic cables. This is a new use.

Non-electrical uses of copper as metal or alloys include heat exchangers (especially air-conditioners), valves, pumps, pipes and tanks. Copper water pipe is still widely used. Advantages as compared to alternative materials (e.g. galvanized steel, stainless steel or plastics such as polyethylene, polybuty-

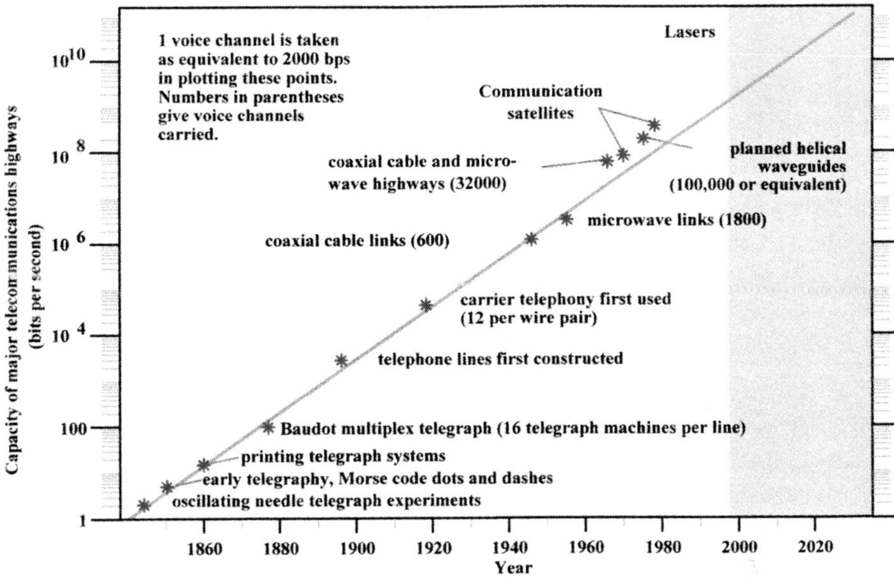

Figure 3.2: Telecommunications capacity.

lene or PVC) include long life, 100 percent recyclability[2], and relative ease of installation and/or repair. (Malleability is a major factor here.) Also, copper water pipe is used, in part, because of its effectiveness in suppressing pathogenic bacterial growth, such as the organism responsible for so-called 'Legionnaire's Disease'. However some corrosion does occur in pipes, especially if sulfates or chlorides are also present. For this reason, municipal wastewater in Swedish towns with unusually large numbers of copper roofs and/or water pipes often contain significantly more copper than natural water (Landner and Lindeström 1999, Chapter 6; Bo Bergbäck, personal communication 2000).

Copper nickel alloys are used extensively for heat exchangers in industrial applications such as desalination plants. Titanium is preferred for heat exchangers through which sea-water flows, but copper-nickel is preferred for other (secondary) exchangers. Copper-nickel alloys also have useful antifouling properties, suggesting the possibility of cladding for some ships hulls. It is worth noting that there is a significant tradeoff between heat exchanger efficiency and area. Bigger heat exchange surfaces, other factors remaining equal, result in a lower temperature gradient and thermal loss. Doubling the exchanger area, hence the copper content, would decrease the temperature

2 Recyclability is regarded as a social good; however PVC has been shown to have a lower installation cost and overall cost for a number of applications where it competes with copper (USBuMines 1991).

differential – thus increase the transfer efficiency – by a factor of four to eight (Lovins 2001).

The use of copper in facing and roofing continues, despite its high cost as compared with alternatives such as aluminum, galvanized iron, and various other materials. The advantages of copper as a roofing or facing material are long life, 100 percent recyclability and low maintenance, as well as attractive appearance.

A surprising new market for copper is in automotive brake linings, where copper or brass powder in a plastic matrix has now largely displaced asbestos, which is widely banned (in the US and Europe) for this application in automobiles (Landner and Lindeström 1999). The high thermal conductivity of copper is the primary reason for this choice of materials. Unfortunately it is a dissipative use.

Chemicals altogether account for less than 1 percent of aggregate demand, and possibly no more than 0.5 percent (RoskillCu 1990). A moderately detailed breakdown of copper chemicals used in Sweden is available for 1988–89, and a more restricted list for 1994–96, from the Swedish chemical inspectorate (Landner and Lindeström 1999, Tables 5.2 and 6.8). Curiously, there is almost no overlap between the two lists, so we have reproduced them separately as *Table 3.1* and *Table 3.2*. The ranges in the 1988–89 data reflect the confidentiality of data on non-toxic chemicals or chemicals used in non-toxic products. Overall the minimum weight of copper chemicals used in Sweden was 1230 kMT (gross), while the maximum was several times higher. The above figures are consistent with either the 0.5 percent (corresponding to 0.7 kMT copper content) or 1 percent (1.4 kMT copper content) estimates for 1989, because of the uncertainties. However the 0.5 percent estimate is close to the lower limit, whence the 1 percent estimate is somewhat more plausible.

Table 3.1: Uses of copper compounds and quantities in some important chemical products imported and/or manufactured in Sweden 1989/88 (metric tons per year)

Compound	Area of application	Quantity	Year
Copper chloride hydroxide	Fertilizer additive	10–97	1989
Copper chloride hydroxide	Fungicide	58	1989
Copper naphthenate	Impregnating agent, wood	80	1988
Copper naphthenate	Paints, lacquers, drying agents	37–425	1989
Copper oxides	Impregnating agent, wood	415	1988
Cupric oxide, CuO	Antifouling paint	8–140	1989
Cuprous oxide, Cu_2O	Antifouling paint	10–140	1989
CuO and Cu_2O	Pigments, catalysts, developer	7–73	1989
CuO and $CuSO_4 * 5H_2O$	Livestock feed additive	13–116	1989
Copper sulfate	Impregnating agent, wood	154	1988
Copper sulfate	Laboratory chemical	40–62	1989
Copper sulfate pentahydrate	Metal coating	120–1270	1989

Source: Landner and Lindeström (1999) *Table 5.2.*

Table 3.2: Copper compounds used in pigments & pesticides in Sweden: 1994–1996 (metric tons per year)

Compound	Area of application	Quantity	Cu content
Cu-tetramine-dihydrogen-carbonate	Impregnating agent, wood	138	35.2
Cu(II)-oxide	Impregnating agent, wood	304	130.6
Cu(II)-hydroxycarbonate	Impregnating agent, wood	75.7	43.5
Cu(II)-sulfate	Impregnating agent, wood	12.3	2.8
Cu-naphthanate	Fungicide, coating, wood	47.9	16.6
Cu(II)-oxychloride	Fungicide, fruit, berries, potatoes	3.4	2.0
Cu metal	Pigment, antifouling paint for boats	5.3	5.3
Cu-phthalocyanine	Complex dyestuff, industrial plant	26	2.9
Cu-phthalocyanine, chlorinated	Complex dyestuff, industrial plant	7.6	0.4
Cu(I)-oxide	Antifouling paint, ships and boats	48	42.6
Cu(II)-thiocyanate	Antifouling paint, ships and boats	17	6.0

Source: Landner and Lindeström (1999) *Table 6.8.*

Copper flows through the economy have been characterized for several countries. For instance a Swedish copper balance for 1990, calculated by the authors, is shown in *Figure 3.3*. A more detailed Japanese copper balance for 1997, is shown in *Figure 3.4* (Kohmei Halada, personal communication, 2002). The Japanese example is interesting because it shows the trade flows at all stages of processing from ore to finished goods and scrap. Regional copper flows have been estimated recently for Europe by a group at Yale University. The results for c. 1993–1995 are shown in *Figure 3.5* (Spatari et al. 2002, Figure 5).

Globally, future growth in different copper markets have been estimated by one industry consultant as follows (Dewison 1999):

- Sheet and strip, used by a variety of original equipment manufacturers (OEMs), including electronics. Moderate to rapid growth (10 percent p.a.), primarily driven by electronics applications.
- Tube (13 percent p.a.), mainly used in air conditioners and water pipes. Air conditioners will penetrate new markets, especially in south Asia. However the market for water pipes is extremely competitive; copper is losing share to PVC.
- Electric power cable, infrastructure (high capacity, high voltage) segment (5 percent p.a.). Competition is mainly from aluminum.
- Electric power cable, local retail distribution (wiring in buildings) segment (27 percent p.a.). Growth is being driven by the increasing number of electrical connections being demanded in new structures.
- Electric power cable, internal distribution within machinery and vehicles (13 percent p.a.). Moderate to rapid growth, with new uses compensating for increased efficiency.
- Telecom cable, infrastructure segment (7 percent p.a. in the near term,

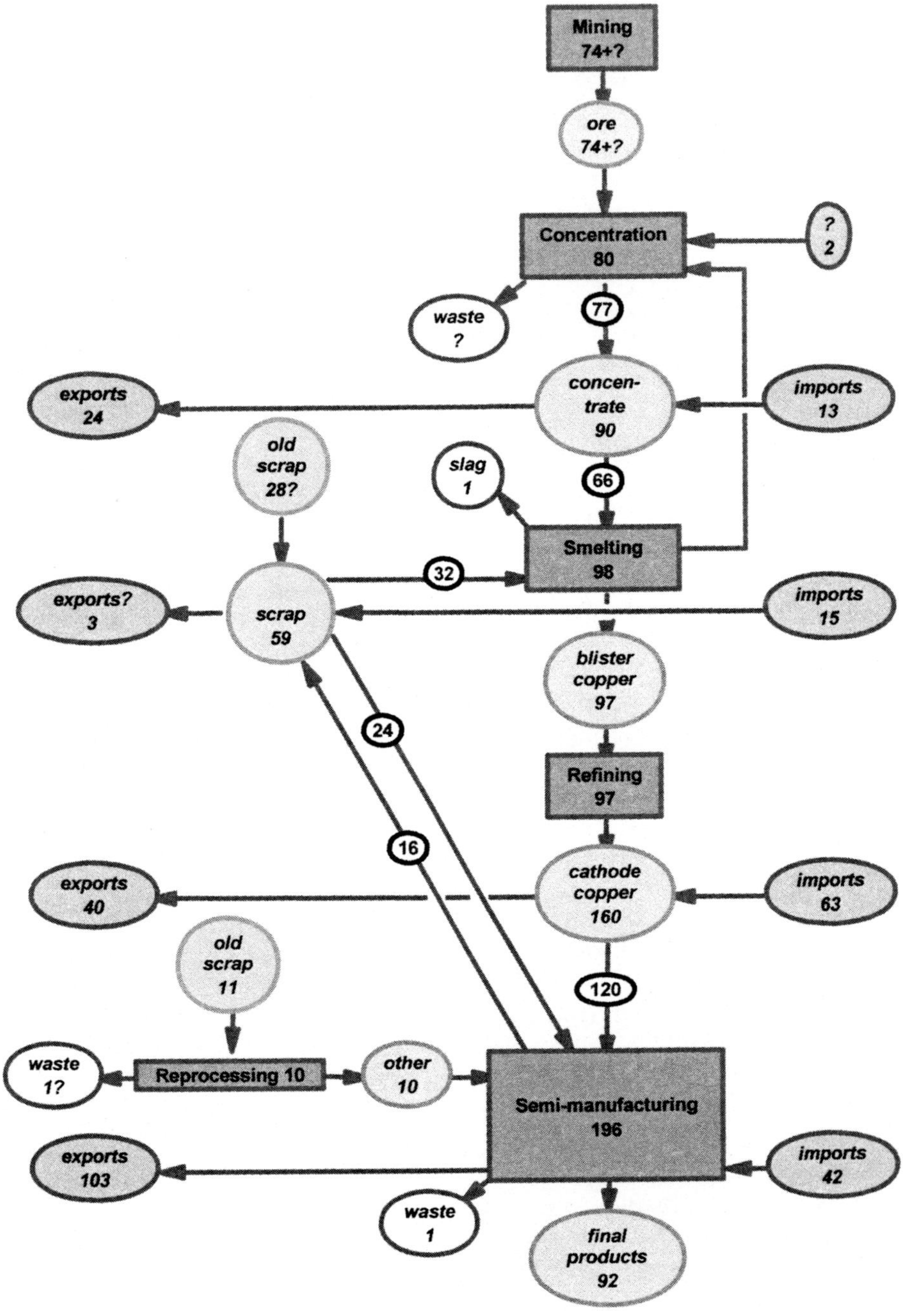

Figure 3.3: A tentative copper balance for Sweden, 1990 (thousand metric tons per year).

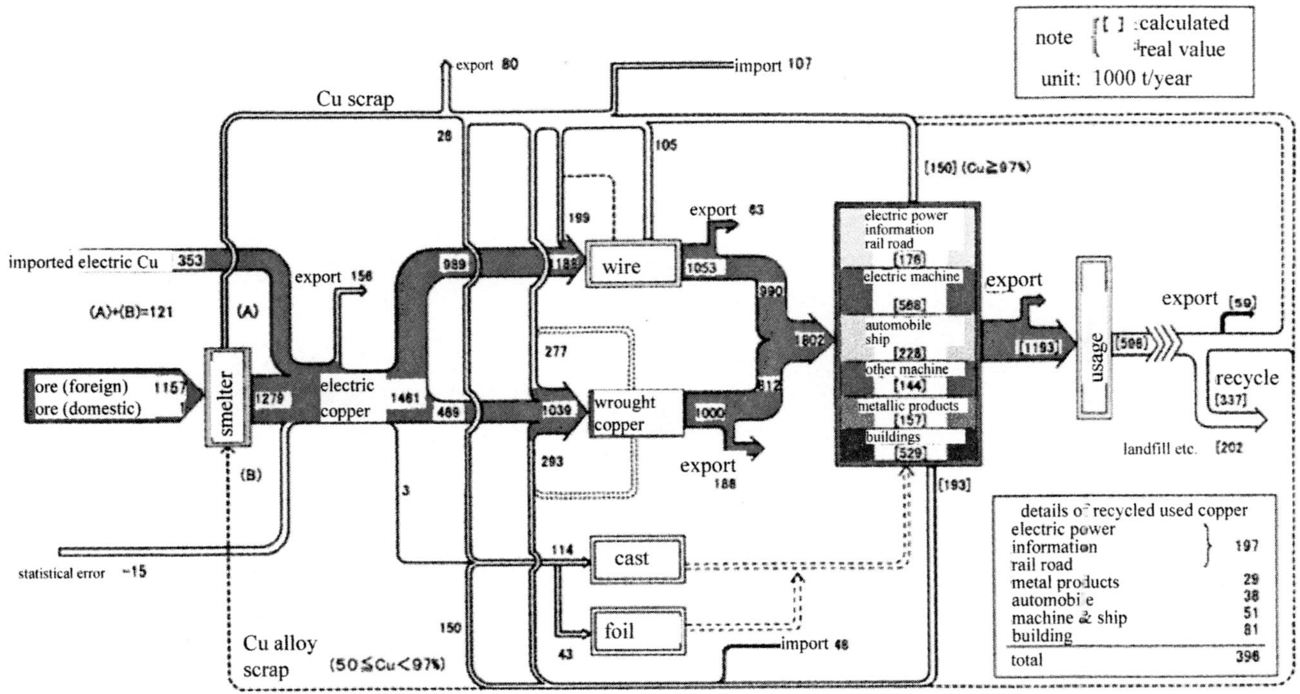

flow data were estimated by Metal Keizai Kenkyusho

Halada's comment

dependency ratio on natural Cu ore = (1157+1+353-156)/(1802-193-150+83+188)=78.3%

contribution of total scrap = 100-78.3 = 21.7% and that of used scrap = 337 / (1802-193-150+83+188)= 18.4%

, while scrap ratio in Cu production = (121+ 199+277+293+114+43+80+59)/(1053+1000+114+43+156)=50%

Figure 3.4: Copper balance for Japan, 1997.

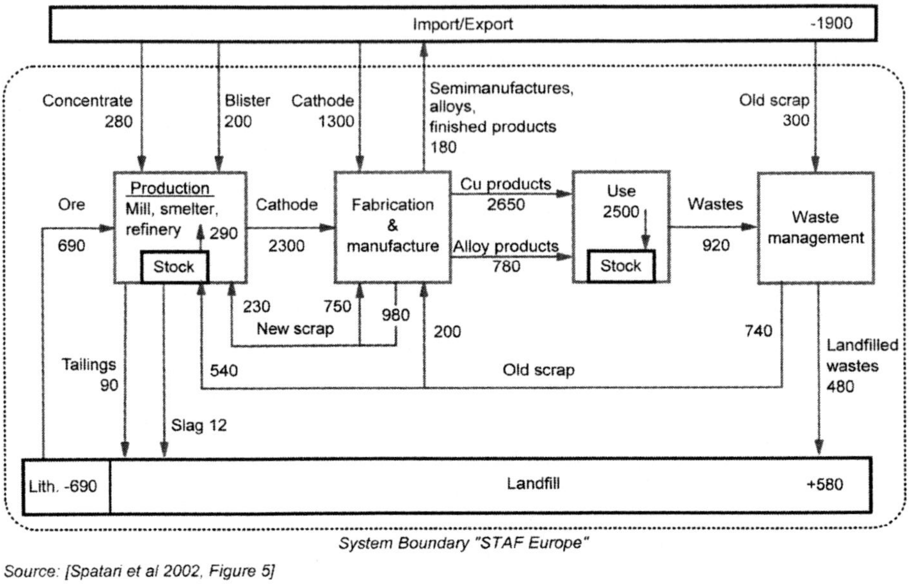

Figure. 3.5: Copper balance for Europe, c 1994 (thousand metric tons per year).

declining). Competition from fiber optics will eventually reduce the market share for copper to zero or negative.

– Winding wire for motors (8 percent p.a.). Rapid growth in both the number of motors and the amount of winding wire per motor of a given rating.

In summary, copper is losing some traditional markets to other materials. One of them is for telecommunications cables (mainly due to increasing use of glass fibers and wireless devices for local distribution – the so-called 'last mile'). It is estimated that the Telecoms replace about 5 percent of their fixed copper wire distribution lines each year, and have done so since the 1980s. (Unfortunately for would-be broad-band internet users, the retail end of the distribution system has been largely ignored up to now.)

Other traditional copper markets, such as copper roofing, copper water pipe, and brass hardware of all kinds are under pressure from cheaper – albeit not necessarily better – materials, including aluminum, galvanized iron and plastics. (These substitutions are not environmentally neutral. There could be significant penalties associated with some of them. However we cannot evaluate all the possibilities.)

Until recently, copper was thought to be virtually irreplaceable for local distribution of electric power, household wiring, motor-generator windings, electronic circuitry and (as alloy) for some kinds of heat exchangers. Further substitution between copper and aluminum, in particular, seems unlikely (Kaufman and Drury 1987). However, the use of copper for carrying electric power, even into buildings, is not necessarily irreplaceable. An article in

Superconductor and Cryotronics Magazine published on-line by WestTech Market Information Services (WestTech undated) turned up the following information. Newly available high temperature superconducting (HTS) cables for voltages ranging from 24 kV to 225 kV, based on the so-called quaternary ceramic super-conductors (discovered in 1986), have now reached the stage of large scale demonstration. Detroit Edison Co. in the US is hosting one such test, using cables developed by American Superconductor Division of Pirelli Cables and Systems Inc. Sponsors include the US Department of Energy, and the Electric Power Research Institute (EPRI). Another US project, supported by industry and government, has been under way for a number of years at Southwire Co. in Georgia, the largest US manufacturer of cable. Tokyo Electric Co (TEPCO) is sponsoring projects at Sumitomo Electric and Furukawa Electric. Sumitomo is now developing a 6kV cable. Large scale testing in Japan appears to be likely in the near future. Pirelli is also working with Electricité de France (EDF) on an experimental cryogenic cable project near Paris, and in several other European projects, including one in Berlin recently acquired from Siemens. Finally, a large scale demonstration project has recently been completed in Denmark, to test HTS technology developed by NKT Cables and Nordic Superconductor Technologies, in cooperation with two Danish electric utilities.

The HTS cables must be refrigerated by liquid nitrogen, but that substance is now cheap. (It is a byproduct of liquid oxygen, widely used by the steel industry and for rocket fuel.) Several firms, including Air Products and Chemicals, Linde A.G., and Air Liquide market stand-alone refrigeration plants to produce the nitrogen refrigerant.

It is not possible to make firm projections regarding the penetration of these new technologies but a study carried out for the Oak Ridge National Laboratories a few years ago concluded that commercial introduction could begin about 4 years after successful demonstration projects at a number of different utilities had been initiated. This might conceivably occur as early as 2004, but early users would have to pay premium prices to gain experience with the technology. At present HTS cables are still significantly more expensive than ordinary copper cables for equivalent performance. For instance, an optimistic target for 2004 is \$55 per kAm (kilo-ampere-meter), but a realistic break-even point might be \$10 per kAm.

Once commercialization begins, the study for Oak Ridge estimated that HTS could capture 56 percent of the underground transmission market in a decade. Still, manufacturing costs are likely to fall quite rapidly, and HTS already has significant advantages that could make it competitive in some niche applications at much higher cost/performance levels. For instance, the HTS cables in the Detroit experiment have three times the current-carrying capacity of the old cables, which significantly reduces costs for replacing existing cables where additional capacity is needed but space is limited.

As regards copper consumption, the HTS cables in the Detroit demonstration project still use small amounts of copper, but only 1.3 percent as much copper at the old cables.

A model for projecting future growth in copper demand is discussed hereafter in *Section 3.4*, and in *Appendix C*.

3.2. Accumulation of copper stocks in the anthroposphere

Apart from copper ores remaining in the ground, there are four important categories of copper stocks. First and largest is the copper embodied in long-lived goods and infrastructure still in use. This category comprises copper wiring in structures, telecoms, electrical utilities and industrial equipment, as well as copper roofs and facings, plumbing and brass and bronze in fittings and industrial equipment.

The second category is the copper embodied in short-lived products still in use. These include motor vehicles, 'white goods' (such as refrigerators, stoves and washing machines), smaller electrical appliances and consumer electronics and business electronics. The third category consists of copper in landfills and identifiable mine waste dumps. The fourth and last category consists of copper that has been dissipated irretrievably into soils, groundwaters or surface waters.

There are no direct census data for any of these categories. Estimates of copper in use (or abandoned in place, such as underground cables) therefore depend very sensitively on the assumed lifetime of copper in these uses. Alternatively, they depend on estimates of the recovery rates for various categories of scrap. Copper left in mine or smelter wastes can be inferred from mine production and recovery rates from various steps in the production chain. Estimates of copper in consumer products that have been discarded and dumped into landfills can be based, in part, on municipal waste sample data. Copper left in automobile scrap can be estimated from data on copper used in vehicles and recovery rates from vehicle scrap. Copper used in automobiles manufactured in the US actually increased from 15.9 kg/car in 1980 to 20 kg/car in 1985 and 20.9 kg/car in 1990. Thereafter it decreased very slightly to 19.1 kg/car in 1994 (Pinhao 1996). This copper is partly in the radiator, partly in wires and partly in connectors and other electronic components, including circuit boards. During dismantling the radiator can be removed relatively easily, but most of the rest of the copper is difficult to remove.

There are several different extant studies that have estimated cumulative copper in use. They differ primarily in terms of assumptions with regard to old scrap recovery efficiency. For purposes of clarity, it is important to distinguish between the recycling rate, defined as the ratio of old scrap consumption to total consumption and the recycling efficiency or separation efficiency, defined as the ratio of old scrap consumption to old scrap generated domestically plus old scrap imported, net of exports. The availability of recoverable secondary metal is defined as the ratio of old scrap generated (domestic plus net imports) to total apparent consumption. The latter measure is the ratio of the first two, viz. *rate* divided by *efficiency*. Evidently availability reflects, in part, the length of life of the applications of the metal. Low availability (other factors remaining equal) implies long lived applications and conversely.

The figures fluctuate considerably from year to year. In the past twenty years (since 1980) the overall scrap percentage of US refined copper consumption has ranged from a high of 28 percent in 1980 and again in 1982 to a low of 11 percent (in 1998), and is generally declining (Jolly 1999, Table 6). But the scrap content of US refined copper supply apparently rebounded to 33 percent in 1999 (Edelstein 1999).

According to Jolly's data the percentage of old scrap in all scrap averaged about 42 percent, with minor fluctuations, from 1980 through 1993, but dropped gradually to 32 percent in 1998 (Jolly 1999) and 30 percent in 1999 (Edelstein 1999). Jolly's data imply that the old scrap recycling rate between 1993 and 1998 fluctuated around 12 percent. This is much less than the 22 percent figure reported by Sibley, Butterman et al. (Sibley et al. 1995). Between 1993 and 1998 the old scrap recycling rate for the US was never higher than 8 percent and was less than 3.5 percent in 1998, the last year for which Jolly had data. On the other hand, it seems to have jumped to 10 percent (30 percent of 33 percent) in 1999 (Edelstein 1999).

The domestic old scrap recycling rate does not include scrap exports, which are large but also fluctuate significantly from year to year. *Figure 3.6* shows copper prices since 1880, both nominal (in current dollars) and discounted for inflation. Discounted prices have always oscillated wildly, but the average from 1880 to 1919 or so was rather constant. However from 1919 until 1970 the price trend was distinctly upward. Since then it has reversed and started down again. It reached an all-time low in 1999, resulting in closures of three of the seven US copper smelters that were still active at the beginning of the year. Overall the price trend has been moderately downward, despite peaks in 1988–89 and again in 1993–5. The trend suggests further price declines in coming decades, though the fluctuations are also likely to continue to be very great.

The domestic old scrap recycling rate is very sensitive to (world) refined copper prices, which had fallen sharply since 1995, but began to recover from its low point at the end of the first quarter of 1999. By year-end the price had increased to 85 cents per pound ($1.90/kg), though the average for all of 1999 was only 76 cents per pound ($1.70/kg), the lowest since 1986 (Edelstein 1999). The main explanation for both the low international price for refined copper and the low US old scrap recycling rate may be that Russia was exporting extraordinarily large amounts of old scrap during that period (it was the largest exporter in 1998, displacing the US). The turnaround in 1999 was partly related to sharply declining Russian and German scrap exports. However there remains a considerable discrepancy between Jolly and

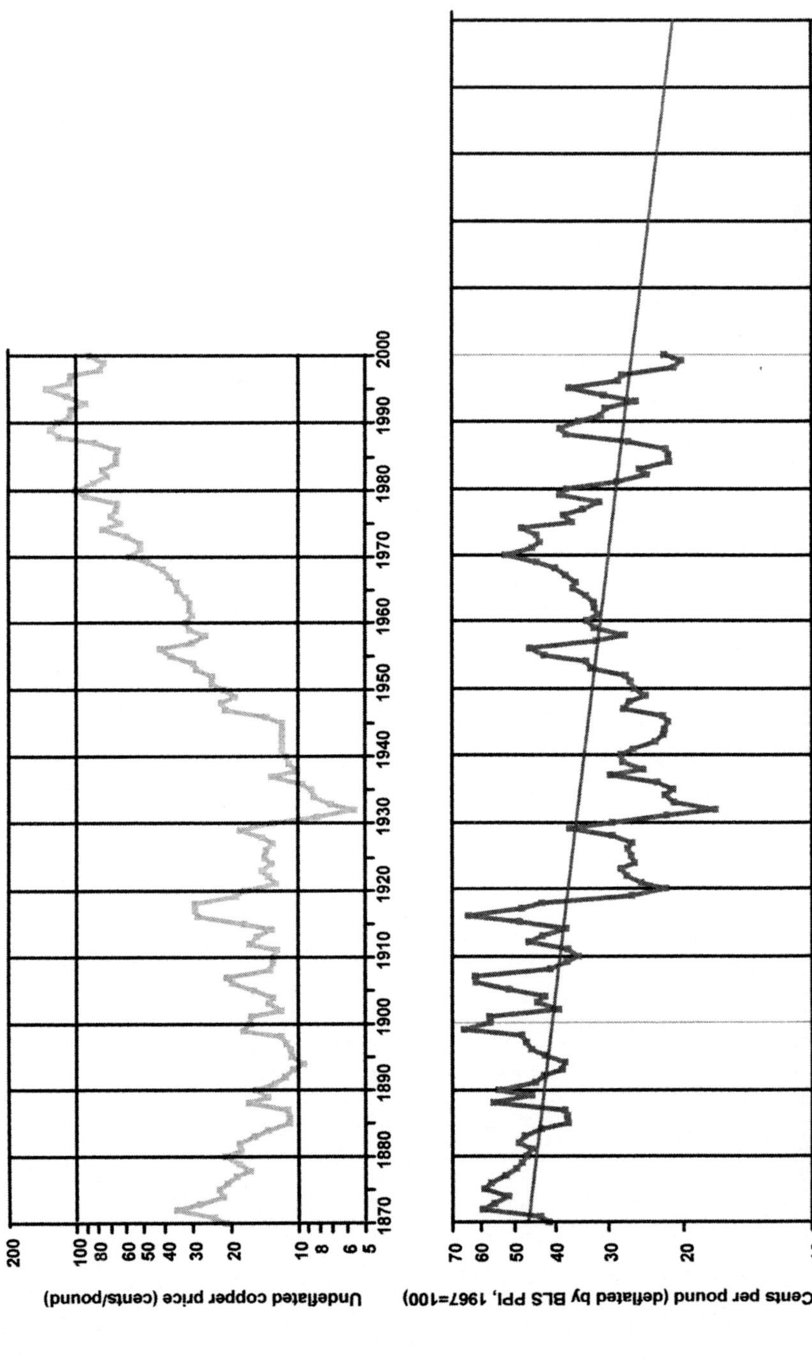

Figure 3.6: US copper prices 1870–2000.

However there remains a considerable discrepancy between Jolly and Edelstein on the apparent old scrap recycling rate for the US.

The recovery (or separation) efficiency is not directly measured – or measurable – for the copper cycle as a whole. There are fairly good estimates for particular industries, especially for products with relatively short useful lives, such as automobiles and electrical goods. In Germany, it is estimated that 40 percent of the copper in an end-of-life vehicle reaches the (old) scrap market, while 35 percent is transferred to the combustible waste fraction where it ends up in incinerator ash, and 25 percent remains with the ferrous scrap (where it is a considerable problem for the steel industry) (Frei et al. 1993). There is some tendency to extrapolate these figure to other end-use categories (DKI 1990; Arpaci and Vendura 1992).

A recent Japanese study reports recovery efficiencies for a variety of short-lived end-use product types, e.g. 17 percent for TVs, audio visual equipment and other major home electric appliances, 35 percent for fax machines and PCs, 40 percent for other office equipment, 85 percent for industrial air conditioners and freezers (after dismantling), 10 percent for automobiles that are shredded without dismantling, 90 percent for automobiles that are dismantled prior to shredding; 80 percent for industrial machinery, ships and railway equipment; 100 percent for heavy electrical machinery and equipment; 50 percent for residential household wiring; 80 percent for wiring in non-residential buildings (Simada et al. 1999, Table 5). However, this study does not venture an overall estimate.

An overall copper recycling (separation) efficiency of > 80 percent has been estimated for Germany, by assuming a mean residence time of 33 years for all products and comparing current old scrap flux with copper consumption 33 years ago (DKI 1990; Arpaci and Vendura 1992). By contrast, recovery efficiency for copper in the US has been estimated at only 30 percent by one Bureau of Mines study (Carrillo et al. 1974), which assumed a lifetime of 35 years for housing, 30 years for electrical plants, 15 years for non-electrical machinery and 10 years for transport equipment and all other uses. Curiously another USBM study, barely ten years later, assumed a 40 year average lifetime in use and arrived at a 60 percent recovery efficiency (USEPA 1983, p. 611). A more recent estimate by Bureau of Mines experts reconfirmed the earlier estimate of 30 percent (Sibley et al. 1995). This is a relatively low figure but, as noted, it depends mainly on the assumed lifetimes of products in use.

Recognizing the ambiguities, a very sophisticated recent model-based analysis offers six different scenarios for the US, starting from 1900 (Zeltner et al. 1999). These scenarios assumed copper recycling (separation) efficiencies ranging from 31 percent (based on Sibley et al. 1995) to 74 percent, with a 'preferred' rate of 42 percent (their scenario 2). Our own global model described in the next section (and in *Appendix C*) is conceptually based on the latter model. Zeltner et al scenario 2 'predicts' that the copper stock in long term use in the US in 1990 would have been 68.4 million metric tons

(MMT), while another 4.6 MMT was embodied in short-lived products. The total comes to 73 MMT, or 270 kg/capita.

A slightly different approach has been used by the US Bureau of Mines and (more recently) the US Geological Service (USGS), namely to assume that 75 percent of the annual consumption of refined copper, less the contribution from recycled old scrap, is accumulated in-use, while the other 25 percent is lost (Jolly 1999, Figure 10). The US data are shown in *Table 3.3* and plotted in *Figure 3.7*. Neither of the above studies allowed for imports or exports, which would have greatly complicated the calculation. It is interesting to note that the calculated copper-in-use for 1990 by Jolly's method is 60.4 MMT or 225 kg/capita. For 1998 she obtains 73 MMT (81 million short tons). As a matter of interest, Jolly uses still another method to estimate the global pool of copper in use, which she calculates by cumulating the annual production less 40 percent, to allow for both unrecoverable losses and new scrap recirculation. Her estimate of the pool of copper-in-use for the world as a whole in 1998 is 210.5 MMT (234 million short tons).

We find the Zeltner et al. scheme slightly more plausible methodologically, but the main point to be emphasized is that there is indeed quite a bit of uncertainty in the estimates. It is also worth noting that independent estimates of total copper in use in Sweden (discussed below) amount to only 140 kg/capita (170 kg/capita in Stockholm), which is far lower than the US estimates, though Sweden has been a major copper producer and user for a longer time (Landner and Lindeström 1999, Chapter 5). This fact raises the suspicion that copper

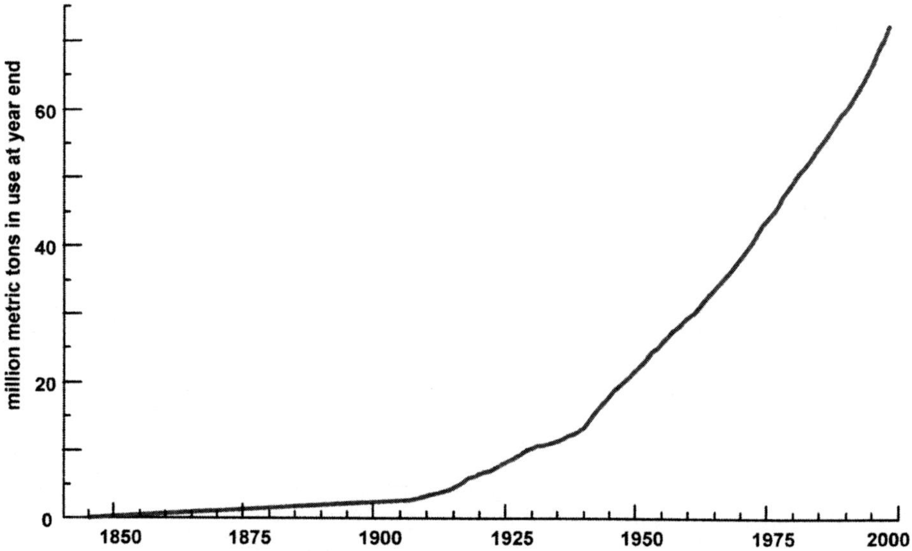

Sources: McMahon 1964, Historical Statistics, and Minerals Yearbooks

Figure 3.7: Estimated accumulation of copper-in-use in USA, 1845–1998.

Table 3.3: Estimated accumulation of copper-in-use in the US, 1845–1998

	Apparent consumption			increase = 75% of total minus old scrap	Copper-in-use Total at end of year		Apparent consumption			increase = 75% of total minus old scrap	Copper-in-use Total at end of year
	New copper	Old scrap	Total				New copper	Old scrap	Total		
1845–1907		new	5518.4								
In use through 1907					2721.6	1953	1301.8	389.5	1691.3	879.0	24406.3
1908	217.7	45.4	263.1	152.0	2873.5	1954	1120.4	369.3	1489.7	748.0	25154.3
1909	312.3	49.9	362.2	221.8	3095.3	1955	1212.0	466.8	1678.8	792.3	25946.6
1910	332.2	58.5	390.7	234.5	3329.8	1956	1240.1	425.0	1665.1	823.8	26770.4
1911	309.3	68.9	378.2	214.7	3544.5	1957	1124.0	403.2	1527.2	742.2	27512.6
1912	352.0	97.1	449.1	239.7	3784.3	1958	1049.6	373.2	1422.8	693.9	28206.5
1913	368.4	83.0	451.4	255.6	4039.8	1959	1073.2	427.3	1500.5	698.1	28904.6
1914	318.2	79.7	398.0	218.7	4258.6	1960	1041.4	389.5	1431.0	683.7	29588.3
1915	515.6	109.9	625.5	359.2	4617.7	1961	1078.0	373.0	1450.9	715.3	30303.5
1916	670.8	158.8	829.5	463.4	5081.1	1962	1246.5	377.1	1623.6	840.6	31144.1
1917	632.7	176.8	809.5	430.3	5511.4	1963	1290.9	382.4	1673.3	872.6	32016.7
1918	753.7	160.3	914.0	525.2	6036.6	1964	1356.2	429.6	1785.8	909.8	32926.5
1919	414.8	138.4	553.2	276.5	6313.1	1965	1384.4	465.8	1850.1	921.8	33848.3
1920	478.0	153.3	631.3	320.2	6633.3	1966	1446.1	485.2	1931.3	963.2	34811.6
1921	277.1	119.7	396.9	177.9	6811.2	1967	1197.5	437.9	1635.3	788.6	35600.2
1922	406.7	184.0	590.7	259.0	7070.3	1968	1429.7	472.4	1902.2	954.2	36554.4
1923	589.9	245.8	835.6	381.0	7451.2	1969	1526.8	521.5	2048.3	1014.7	37569.1
1924	614.5	241.5	856.0	400.5	7851.7	1970	1421.6	457.3	1878.9	951.9	38521.0
1925	635.5	264.0	899.5	410.6	8262.4	1971	1482.6	403.8	1886.4	1011.0	39532.0
1926	712.2	306.0	1018.2	457.7	8720.0	1972	1726.8	415.7	2142.4	1191.2	40723.2
1927	645.4	307.9	953.3	407.1	9127.1	1973	1782.3	441.1	2223.4	1226.4	41949.6
1928	729.6	331.6	1061.2	464.3	9591.4	1974	1706.3	438.6	2144.9	1170.1	43119.7
1929	806.8	421.3	1228.0	499.8	10091.2	1975	1138.5	334.9	1473.4	770.2	43889.9
1930	573.8	310.4	884.2	352.7	10443.9	1976	1543.6	380.2	1923.9	1062.7	44952.5
1931	409.2	237.0	646.2	247.6	10691.5	1977	1659.8	409.9	2069.7	1142.3	46094.9
1932	235.5	164.2	399.7	135.6	10827.1	1978	1867.9	501.6	2369.5	1275.5	47370.4
1933	307.9	236.1	544.0	171.9	10999.0	1979	1829.9	604.3	2434.2	1221.4	48591.8
1934	292.7	282.0	574.7	149.0	11148.0	1980	1565.4	613.5	2178.8	1020.7	49612.4
1935	400.4	328.1	728.5	218.3	11366.3	1981	1679.6	591.8	2271.4	1111.8	50724.2
1936	595.3	347.2	942.5	359.7	11725.9	1982	1244.7	517.7	1762.4	804.1	51528.3
1937	630.4	370.9	1001.4	380.1	12106.0	1983	1563.3	449.5	2012.7	1060.1	52588.3
1938	369.2	242.5	611.7	216.3	12322.3	1984	1655.4	460.7	2116.1	1126.3	53714.7
1939	648.5	260.3	908.8	421.3	12743.6	1985	1641.0	503.4	2144.4	1104.9	54819.6
1940	915.2	302.9	1218.1	610.6	13354.3	1986	1659.0	479.2	2138.2	1124.5	55944.1
1941	1489.2	374.4	1863.6	1023.3	14377.5	1987	1698.6	497.9	2196.5	1149.5	57093.5
1942	1458.8	387.5	1846.2	997.2	15374.7	1988	1695.6	518.2	2213.8	1142.1	58235.7
1943	1362.6	387.8	1750.4	925.0	16299.7	1989	1633.3	547.6	2180.8	1088.1	59323.8
1944	1364.4	414.3	1778.7	919.7	17219.4	1990	1632.2	535.7	2167.9	1090.3	60414.0
1945	1283.7	451.0	1734.6	850.0	18069.5	1991	1572.0	518.0	2090.0	1049.5	61463.5
1946	1261.9	368.7	1630.6	854.2	18923.7	1992	1749.8	555.0	2304.8	1173.6	62637.1
1947	1166.6	456.7	1623.3	760.8	19684.5	1993	1967.0	543.0	2510.0	1339.5	63976.6
1948	1101.3	458.5	1559.9	711.4	20395.9	1994	2190.0	500.0	2690.0	1517.5	65494.1
1949	972.5	347.9	1320.5	642.4	21038.3	1995	2098.0	442.0	2540.0	1463.0	66957.1
1950	1312.7	440.2	1752.9	874.5	21912.7	1996	2402.0	428.0	2830.0	1694.5	68651.6
1951	1183.0	415.6	1598.6	783.3	22696.1	1997	2452.0	498.0	2950.0	1714.5	70366.1
1952	1233.8	376.2	1609.9	831.3	23527.4	1998	2554.0	466.0	3020.0	1799.0	72165.1

Data sources: 1845–1960: McMahon (1964) Table 10, p. 77.
1961–1969: calculated from USCensus (1975) series 236–240.
1970–1998: calculated from USBuMines (annual) 'Copper', Table 1.

remaining in use in the US may actually have been over-estimated by both of the methodologies cited above.

Copper mined in the US between 1900 and 1990 cumulatively amounted to 112.6 MMT. Copper cumulatively 'lost' by all routes therefore amounted to 39.6 MMT if the Zeltner et al estimate is correct. The loss would be 52.2 MMT if the US Bureau of Mines/US Geological Survey (USBM/USGS) approach is correct. In percentage terms, the lower of the two estimates of lost (unavailable) copper is 35 percent and the larger of the two estimates is 46.3 percent.

The long-term picture of copper production, consumption, waste and accumulation has also been studied in depth for Sweden, using a different methodology based on direct estimates of historical production, waste flows and losses (Landner and Lindeström 1999, Chapter 5). Cumulative Swedish copper mine production since the 13th century is known, viz. 2.8 million metric tons (MMT). This figure presumably refers to gross (mine) output.

The copper content of various types of mine-related wastes is also known, thanks to a detailed survey carried out by U. Qvarfort of Upssala University for the Swedish Environmental Protection Agency (SEPA) (Qvarfort 1997 cited in Landner and Lindeström 1999, *Table 5.3*). They are as follows:

- Mining waste rock from 'Kopparsberg' (copper mountain): 4 MMT, with an average copper content of 1.5 percent, or 62 kMT (Cu). This is not recoverable because the city of Falun is built over it.
- 'Sterile' waste rock from underground mines: 200 MMT with a grade of 0.05 percent, or 100 kMT of Cu-content.
- Concentrator tailings: 430 MMT, with an average Cu-content of 0.1 percent, or 430 kMT
- Slags: 22 MMT with an average Cu-content of 1.36 percent, or 300 kMT (Cu).The sum total of all these wastes is 660 MMT, containing 892 kMT (0.89 MMT) of Cu.

Cumulative waste from industry and municipalities (landfills) has also been estimated, though with somewhat less precision. Currently 300 active municipal landfills in Sweden apparently contain between 40 and 130 kMT of copper, accumulated over the past 30 years (1965–1995) (Landner and Lindeström 1999, Chapter 5). A 'mean' estimate of 85 kMT has been chosen. The copper content of industrial landfills and dumps (including incinerator ash) was estimated to be 20 kMT over the last 30 years. The contents of some 6000 older landfills and industrial tips is unknown, but an 'educated guess' might be 50 kMT. Thus the copper that has accumulated in solid wastes in Sweden appears to be of the order of 85 + 20 + 50 = 155 kMT, although this number is uncertain by at least 50 percent.

The copper accumulated in the Swedish anthroposphere over the last 100 years (1895–1995) and embodied in products still in use in Sweden has been estimated by several different methods (Landner and Lindeström 1999, Chapter 5). One approach, starting in 1950, was to calculate mine production plus imports, less exports less domestic use of recycled copper. The difference should correspond to domestic accumulation in the anthroposphere (with no allowance for unrecoverable and/or dissipative losses). The annual data, which fluctuate widely, are shown in *Figure 3.8*. Evidently, there was net accumulation virtually every year from 1960 through 1987, followed by an apparent reversal of this trend since then. The accumulation since 1950 is plotted in *Figure 3.9*, which also indicates a significant reduction in the Swedish anthropospheric stock in the late 1950s and again since the late 1980s. This reversal is fundamentally implausible, absent a known cause. There may be data problems.

Another calculation can be based on the Swedish copper cycle (see *Figure 3.3*). Total copper used for production of semi-manufactures since 1895 (7.0 MMT), less secondary copper from recycled scrap (3.2 MMT), less copper contained in net exports of semi-manufactures (1.54 MMT) less returns from manufacturers of machinery and equipment (new scrap) equals 1.210 MMT, the amount of copper presumably remaining in use or in landfills in Sweden. This amounts to about 140 kg/capita (170 kg/cap in Stockholm). This is significantly less than either of the two calculations previously cited for the US. Given the significantly larger use of copper for water pipes and roofs in Sweden, the supposed difference in per capita use between Sweden and the US is hard to account for at first glance.

Even so, it would appear that between 80 percent and 90 percent of the copper that has been produced and consumed in Sweden since the middle ages is still either in use or in long-lived products that are no longer in use but have not been discarded to known landfills. Thus the recent decline in old scrap recycling rates is a very disturbing development, insofar as long-term sustainability is concerned.

The difference between the quantity of copper mined in Sweden and the quantity accounted for in use, above is 2.8 – 0.89 – 1.2 – 0.16 = 0.55 MMT (with an uncertainty of perhaps 0.1 MMT). It is possible that the amount copper remaining in use is greater than calculated above. This would bring the Swedish estimates closer to the US estimates on a per capita basis. On the other hand, some of this 'missing copper' may have been exported as scrap (probably to refineries in Germany).[3] Assuming copper exported from Sweden has been used by importing countries in similar ways, the bulk of it must have been accumulated somewhere in the anthroposphere.

To be sure, some copper was dissipated into the Swedish environment via corrosion and dissipative chemical and other uses. How much? Assuming 1

---

[3] Swedish mine production remained roughly constant during WWII, but domestic consumption (and recycling) dropped drastically. The implication is that a great deal of copper was exported during 1939–1944. The 'missing' quantity appears to be around 250 kMT, most of which presumably went to Germany. Incidentally, a fairly large amount of copper was probably lost during that war, especially in ships that sank.

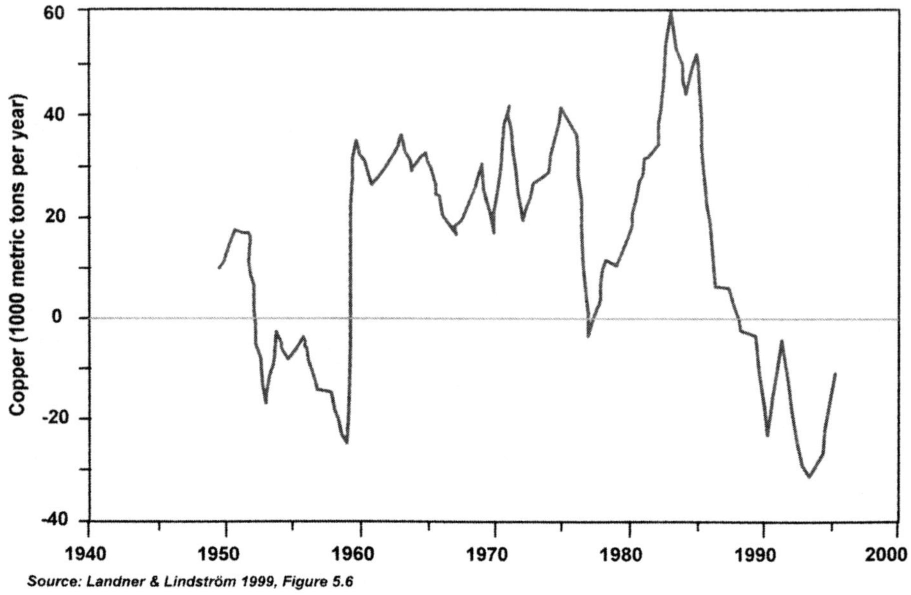

Figure 3.8: Annual change in the copper store in the Swedish technosphere: 1950–1995.

Figure 3.9: Cumulative evolution of the copper reservoir in the Swedish technosphere (1950 base).

percent of copper consumption was in chemical form, chemical dissipation (in Sweden) would have been about 0.01 × (1200 + 160) = 14 kMT, over the past century. This would include agricultural uses, paints, wood preservatives and so on. Current losses due to corrosion of copper roofing and pipes has been estimated at 2 tpy (Landner and Lindeström 1999, pp. 86–87). The corrosion rate may have been higher earlier in this century, but in the long run it must be roughly proportional to the copper in the anthroposphere, which was certainly less in previous centuries. An educated guess for the maximum total loss by this route (authors) might be 500 metric tons, or 0.5KT.

Dissipative losses have also been estimated explicitly for Switzerland and Denmark. The current loss rate in Switzerland by all routes appears to be of the order of 0.1 kg per capita per year (Von Arx 1996). Extrapolated to Sweden (population 8.8 million), this would amount to 880,000 kg/y or 0.88 kMT/y. Extrapolated to the US (pop. 275 million) the current rate of annual dissipative loss would be 27.5 kMT/y. In all cases, dissipative losses appear to be of the order of, or somewhat less than, 1 percent of annual consumption.

The Danish Environmental Protection Agency has estimated total annual copper emissions from all sources including chemicals, fertilizers, and energy production to be between 4.3 kMT/y and 8.2 kMT/y, of which by far the largest component (and uncertainty) is solid waste incineration (between 3.6 kMT/y and 7.1 kMT/y). There are no smelters or refineries in Denmark. All copper flows into and out of the country are embodied in products (including scrap). Based on a population of 5 million, dissipative consumption-related losses per capita appear to be of the same order as Switzerland and Sweden.

Another view of the situation is shown in *Table 3.4* adapted from Azar et al. (1996). The two critical measures are in the last two columns. The ratio of total metals mined since 1900 to natural flows is an indicator of the extent to which human activity is accelerating the mobilization of the metals. For copper, zinc and lead this ratio is quite high (24 for copper) according to the authors. Yet this, too, is uncertain. The ratio of anthropogenic emissions of copper to natural emissions (atmosphere only) is barely greater than unity based on natural emissions data from an earlier published study by Nriagu (Nriagu 1989). We note that the Azar et al. work uses more recent data on natural emissions.

The other ratio in *Table 3.4*, metal mined since 1900 to metal contained in the topsoils of the human inhabited area, is also significant. While most of the metals that have been mined are either still in use or have been disposed of in ways that do not result in soil buildup, it is evident that if even a small fraction of the non-ferrous metals that remain in the anthroposphere finally end up in the soil, the composition of the topsoil will be altered significantly. It must be emphasized, however, that soil buildup of trace metals is not necessarily threatening to humans, since uptake into the food chain depends on

Table 3.4: Indicators for elements extracted from the lithosphere, c. 1990

Element	Concentration in soils (mg/kg)	Weathering[a] and volcanic (kMT)	Mining (kMT)	Fossil[b] fuels (kMT)	$I_{1,1}$[c] (ratio)	$I_{1,2}$[d] (ratio)
Zinc	60	910	7300	260	8.3	6.9
Copper	25	380	9000	55	24	23
Lead	19	290	3300	85	12	19
Arsenic	7.2	110	19	18	0.33	
Germanium	1.2	18	0.27	17	0.96	
Antimony	0.66	9.9	54	10	6	
Selenium	0.39	5.9	2.1	12	2	
Cadmium	0.35	5.3	20	3.4	3.9	3
Silver	0.05	0.75	15	1.7	22	

[a] Weathering mobilization is calculated as average concentration in soils (column 1) times a suspended sediment flux of 1.5×10^{16} grams per year.
[b] Data on the trace element content of crude oil are for the US (Yen 1975).
[c] Indicator $I_{1,1}$ is calculated as anthropogenic flows from the lithosphere to the ecosphere divided by the natural flows. Anthropogenic flows are defined as mining and flows associated with fossil fuels. Natural flows are defined as weathering and volcanic processes.
[d] Indicator $I_{1,2}$ is calculated as accumulated mining since 1900 divided by the amount in the top soil layer in the human area. If flows from fossil fuels had been included, the indicator value would increase substantially for many of the elements.
Source: Adapted from Azar et al. (1996) Table 2, p. 96

bio-availability, which is now known to be nearly independent of soil concentration. See *Appendix B*.

3.3. Dissipative uses and losses of copper

Wastes and emissions associated with mining, smelting and refining operations were discussed in the previous chapter. Most of the losses end up in mine wastes (gangue) or slag. Some of this copper could conceivably be recovered in future as technology improves.

However there are four routes by which copper from the stock-in-use is converted into unrecoverable forms and dissipated into the environment. These are via food, chemicals, corrosion and in copper-containing products that are discarded. The latter route is discussed later in Chapter 5 under the topic of recycling.

Human beings require 1.5 to 3 mg of copper per day with food. This intake is not accumulated in the body and must be discarded in feces. For instance, it has been estimated that 21.3 mg of copper per household is discharged into the Palo Alto (California) sewage treatment plant. Of this amount, 18B35 percent originates in food. However, this copper is mostly taken up by plants from the topsoil, whence it can actually be regarded as a mechanism for copper concentration (and potentially, for future recovery).

The second source of non-recoverable losses is via chemical uses of copper. Essentially all chemical uses are dissipative and eventually become emissions. The major categories are wood preservatives, fungicides, pigments and dyestuffs and antifouling paint for ships. A compilation of quantities used in Sweden during the period 1994–1996 was shown in *Tables 3.1* and *3.2*. CCA wood preservatives have a fairly long life in use, probably greater than 20 years. At present the ultimate fate of this copper is unclear, but most of it is likely to move into topsoil and/or groundwater. Chromic acid and arsenic are more dangerous to humans than the copper, and future regulation of CCA is not unlikely).

Copper chemicals are all produced from copper sulfate, which is produced commercially (in turn) from secondary sources, notably shredded cable wire, waste wire from cable production or spent electroplating solution. Chemicals account for less than 1 percent of total global consumption, probably less than 50,000 tpy in 1990 (RoskillCu 1990). Uses include pesticides, fungicides (e.g. Bordeaux mixture for protecting grape vines), herbicides, colorants (for ceramic glazes and glass), catalysts, pharmaceuticals and electroplating of circuit boards and metalization of computer chips. The biggest single use of copper sulfate is as a component of the wood preservative CCA (Chromium-Copper-Arsenic) which is widely used in the US and, to a lesser extent, in Europe (but not in Japan). Anti-fouling paints for ships, to replace organotins (which have been found to be toxic to many marine organisms), is another significant and growing use. Actually copper chemicals are used largely as algicides, fungicides, wood preservatives and disinfectants because of their biocidal properties.

Corrosion of copper roofs and copper pipes (especially water pipes) is the third major source of anthropogenic copper dissipation in the environment, at least in some locations. The corrosion rate tends to decline over time, but not to zero. On the other hand the runoff/corrosion ratio apparently increases over time, so that the two are essentially equal after many years. Runoff rates for aged panels seem to average around 1 gm/per square meter of exposed surface per year. The average for central Stockholm is about 1.35–1.5 gm per square meter per year (Landner and Lindeström 1999, p. 96).

The total surface of copper roofs in Stockholm is approximately 622 000 m^2 (Odnavall Wallinder, Royal Institute of Technology Stockholm, personal communication, 2001). Studies have been performed to estimate the runoff rate of copper from roof surfaces in Stockholm giving values at 0.7–1.4 g/m^2/year (Leygraf and Wallinder 1997). Naturally aged copper (> 40 years) exhibits a somewhat higher yearly runoff rate than new copper. For example for roofs from the 18th century the value with worst case conditions is 1.6–2.3 g/m^2/year (ibid). The amount of copper released from copper roofs in Stockholm is about 0.44–0.87 ton/year (calculated from the first-mentioned runoff rates).

To determine how much copper is dissipated in corrosion products is more difficult than to estimate the runoff rate. This is due to the fact that the corrosion rate is time dependent. For example, it is known that the corrosion rate for one specific copper roof in Stockholm – the Vasa Ship Museum – has decreased by a factor of three during the latter part of the 20th century (Sundberg 1998). Studies have also shown that a relatively large fraction of the corrosion product is retained on the surface during initial exposure, but this fraction decreases over time. The copper patina eventually reaches a constant thickness (Leygraf and Wallinder 1997).

The discharge of copper in Sweden due to corrosion of roofs and the tap water system has been estimated, in an earlier study, as 30–40 tpy and 50–60 tpy respectively (Nilarp 1994) based on information on amounts of metals that have been analyzed at municipal purifying plants (Svenska 1990). The institute of corrosion has tried to evaluate the discharge of copper due to corrosion and the result shows discharge of about 5 metric tons of copper from roofs and between 5 and 9 metric tons of copper per year from tap water system in Stockholm. These calculations confirm that the leakage from roofs and tap water system in Sweden is on the order of 100 tpy (ibid).The corrosion rate that has been calculated in these studies includes also that part of the products that, over time, turns into green patina but remains on the surface of the metal and therefore the above estimation may be a bit too high.

The copper concentration in sewage water entering Stockholm's largest sewage treatment facility (Hendriksdal) averaged 53 parts per billion (micrograms per liter) in 1990 and 78 ppb in 1996 (Landner and Lindeström 1999, p. 101). It is estimated that copper plumbing in buildings contribute two thirds of this. By comparison, fresh water entering the city of Stockholm contains 3 ppb of copper from natural sources. The Stockholm figures are not especially high. In Uppsala and Malmo, concentrations average around 200 ppb, with occasional measurements ten times as high (2.24 ppm) (ibid). These very high concentrations are almost certainly due to leftover slag heaps from copper smelters or refineries (as in the case of New York Harbor, mentioned earlier.)

A waterworks in Colorado, USA, carried out an interesting and probably significant experiment. At the outset, measured copper concentrations in tap water ranged from nearly zero to over 4 ppm, with the 90th percentile at 2.26 ppm (meaning that 90 percent of the households had copper concentrations at that level or lower) (Landner and Lindeström 1999, p. 100). Measures were then taken to reduce the corrosion rate of the copper pipes in the system. The main steps in corrosion treatment were to increase the pH (alkalinity) of the water from 7.1 to 7.4 and reduce the average carbonate concentration from 268 ppm to 98 ppm. In response, the 90th percentile of copper concentration readings in the water district fell to 0.31 ppm (310 ppb) (ibid).

Similar measures undertaken by the waterworks at Uppsala and two other Swedish towns have reduced the copper concentrations in tapwater from 0.7 ppm to 0.1 ppm (100 ppb). The optimum conditions for minimizing copper corrosion in pipes have been summarized as follows (Landner and Lindeström 1999, p. 100):

- Alkalinity (as bicarbonate ion) between 70 and 300 ppm
- acidity (pH) between 7.5 and 9.0
- alkalinity to sulfate ratio > 1
- sulfate concentration < 100 ppm
- chloride concentration < 100 ppm

These conditions are not difficult to achieve in a modern water treatment facility.

3.4. The future of demand for copper

Long term trends are often the best basis for understanding the future. With this in mind, global copper production has been plotted since 1880 against the US manufacturing index, on a logarithmic scale (*Figure 3.10*). The correlation is obviously very close, suggesting that future demand may also tend to follow overall industrial production. However, during most of the past century (up to 1980 or so) electric power production and consumption was a major driver of economic growth. Electric power usage is clearly complementary to the total stock of copper in use and it is plausible to imagine that annual production, too, follows the demand for electric power. However, there are reasons to think that the global economic system is now undergoing a sort of phase-change, as information technology begins to replace electric power per se as the driver of economic growth.

There are other patterns and trends that can be helpful for purposes of long-term forecasting. The one we depend upon mostly hereafter is the historical intensity-of-use (IU) vs. income pattern observed in the OECD countries and Asia (*Figure 3.11*) and extrapolated (subject to some modifications to be explained) to other world regions. Regional groupings and future population and income trajectories have been chosen to coincide with Intergovernmental Panel on Climate Change (IPCC) scenarios to facilitate comparisons.

The basic forecasting model we adopt hereafter is based to largely on the work of Zeltner et al. (Zeltner et al. 1999)for the US, but extended globally. A methodologically similar model was developed by Dutch researchers at the public health research center (RIVM) in the Netherlands, except that they lumped all non-ferrous metals (including aluminum) together as a single material MEDalloy, for which they conducted a number of scenario exercises (Van Vuuren et al. 2000). The results of the Dutch study are qualitatively similar to ours, except that they explicitly introduce scarcity and rising prices into their model whereas we do not.

We define four world regions: OECD90, REF, ASIA and ALM. (Roughly speaking, REF refers to countries of eastern Europe and the former USSR now undergoing economic reform; ASIA refers to non OECD Asian countries and ALM refers to Africa and Latin America. For a definition of the regions,

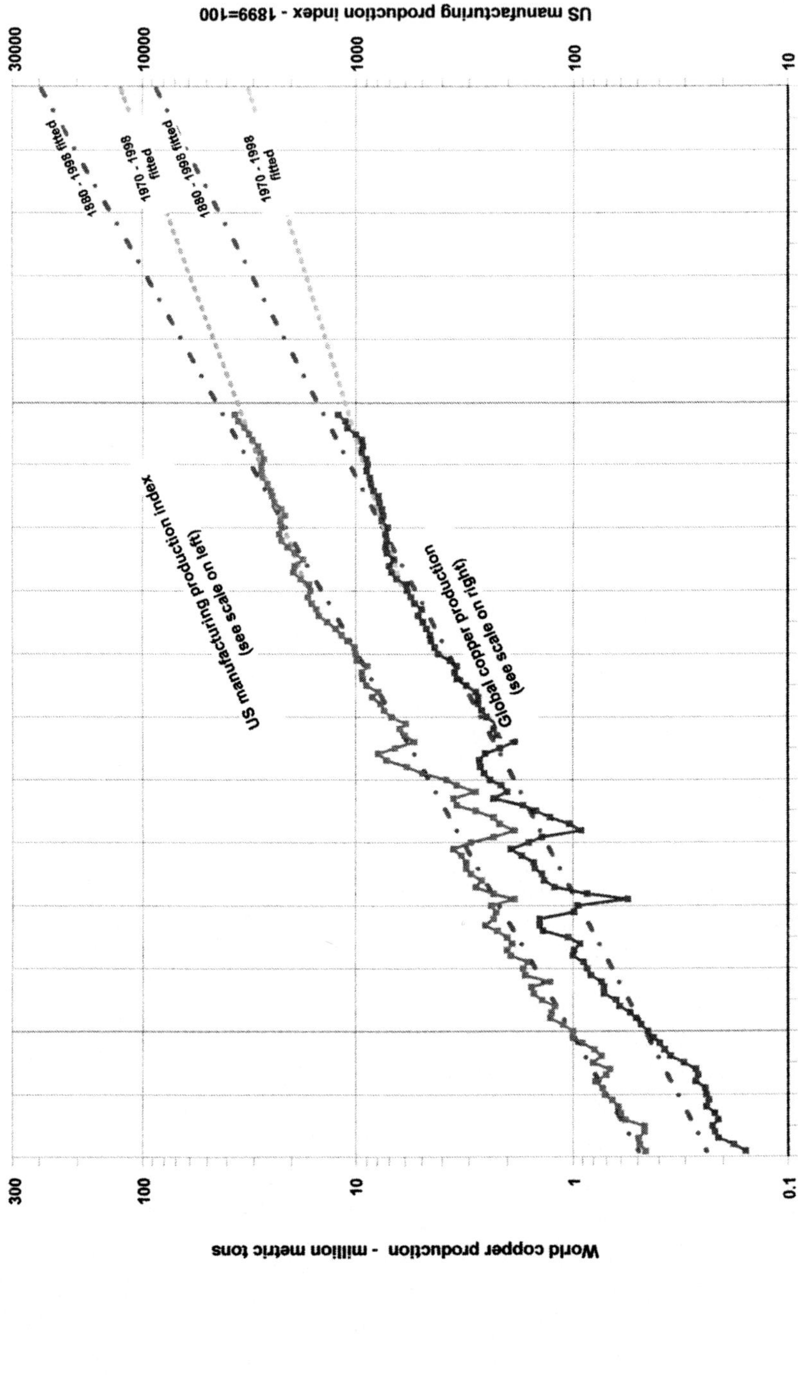

Figure 3.10: World copper production and the US manufacturing production index, 1880–1998 with profections to 2050.

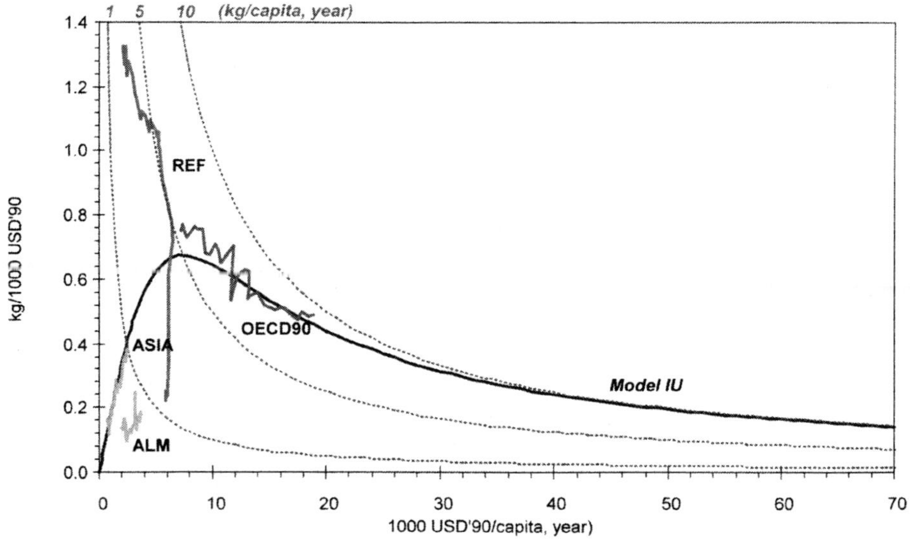

Figure 3.11: Historical and modeled intensity of use (consumption of refined copper) as a function of GDP/capita in 1960–1997.

see *Appendix C, Table C2*) These are the same regional definitions that are used in IPCC (IPCC 2000). Each region is modeled separately.

The copper cycle used for the model is shown schematically in *Figure 3.12*. The model consists of six processes or quasi-processes and eight stocks The six processes are identified by the number of the corresponding node in the schematic diagram:

P3) Smelting and refining
P4) Production of semi-manufactures
P5) Production of finished goods
P9) Long-term usage
P10) Short-term usage.

Similarly the eight stocks are identified by node number, viz.

S1) Primary resources (ore not yet mined)
S6) New scrap from manufacturing
S7) Long-lived goods (e.g. wiring in buildings, pipes, roofing)
S8) Short-lived goods (electrical and electronic goods, cars, etc.)
S11) Gangue (concentration waste)
S12) Slag (smelting/refining waste)
S13) Waste from short-term usage
S14) Wastes from long-term usage

There are five types of losses included in the model, namely concentration losses $l_c(t)$, smelting and refining losses l_s, losses associated with long-

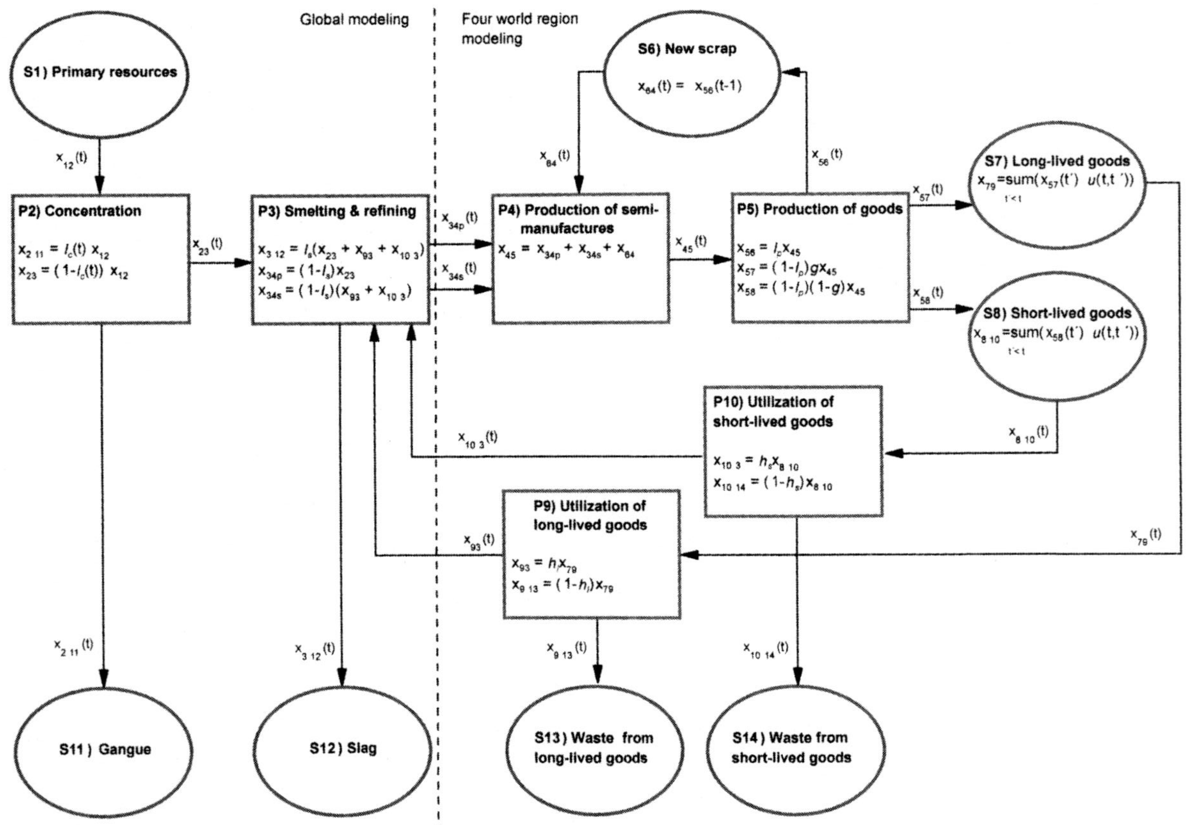

Figure 3.12: Model of the global copper system.

term ($1-h_l$) and short-term use ($1-h_s$) products. There are also losses in the production of goods l_p (new scrap). However, it is assumed that all new scrap is collected and returned to the production of semi-manufactures, with a delay of one year (in reality it is likely to be less.) No losses are assumed in the production of semi-manufactures. The only loss process that is time-dependent is in concentration.

The share of all produced copper-containing goods destined for long-term usage is g. The residence times of copper products in the stocks of long-lived and short-lived goods (S7 and S8) are assumed to be normally distributed. The 'outflow' $x_{57}(t')$ from the long-lived stock at time t of products produced at time t' is

$$x_{57}(t)u(t, t') = \frac{1}{\sigma_l\sqrt{2\pi}} e^{-\frac{1}{2\sigma_l^2}(t - t' - \tau_l)^2} \qquad (t > t') \tag{1}$$

where τ_l is the mean residence time and σ_l is the standard deviation. The total outflow from the long-lived stock $x_{79}(t)$ at time t is thus

$$x_{79}(t) = \sum_{t' < t} x_{57}((t)u(t, t')) = \sum_{t' < t} x_{57}(t') \frac{1}{\sigma_l\sqrt{2\pi}} e^{1-\frac{1}{2\sigma_l^2}(t - t' - \tau_l)^2} \tag{2}$$

There are corresponding formulae for short-lived products.

The driving variable, i.e. the input to the model, is the annual consumption of refined copper (primary and secondary) in the four regions, respectively, viz. ($x_{34p}(t) + x_{34s}(t)$). However, primary resource stocks, concentration, and smelting and refining are not modeled on a regional basis, but on a global basis. This seems reasonable, given that the copper mining companies (which also operate the concentrators, smelters and refineries, are global concerns.

For the years 1900–1997 historical statistics of the consumption of refined copper, by region, are used (*Table C3, Appendix C*). For the years after 1998 scenarios of the consumption of refined copper are created, based on population and assumptions about the economy. The scenarios are also based on the basis of the intensity-of-use (*IU*) hypothesis, mentioned previously, viz.

$$IU(y(t)) = \left[\frac{a}{y(t) + \frac{b}{y(t)^c}}\right] d^{(t - t0)} \tag{3}$$

where $y(t)$ is GDP/capita (thousand US\$/capita) and a (kg/capita), b (thousand US\$/capita) and the exponent c are parameters, d is a factor that scales down the intensity of use with time and t_0 is the first year. If c and d are equal to unity, then for low values of $y(t)$, IU grows linearly with the slope a/b and for high values of $y(t)$, IU declines as $a/y(t)$. The parameter a thus represent the asymptotic per capita level of consumption of refined copper.

The consumption of refined copper $z(t)$ is then

$$z(t) = IU(y(t)) \times w(t) \tag{4}$$

where $w(t)$ is Gross Regional Product (GRP) or Gross World Product (GWP) in billions of 1990 US dollars/year at purchasing parity price (PPP) exchange rates.

The model is run in discrete time steps of one year from 1900 until 1997, and then from 2010 to 2100. The gap between 1997 and 2010 is interpolated linearly to avoid discontinuities. In the beginning of year 1900 all stocks, except for primary resources, are assumed to have been zero, for convenience. The gap during the years 1997-1010 is interpolated linearly to avoid discontinuities that would otherwise occur. These (avoided) discontinuities result from the fact that the 'real' IUs for OECD, REF, ALM and ASIA in 1997 do not exactly coincide with the model assumptions.

No regional trade of scrap, semi-manufactures or finished goods is allowed in the model. That is, it is assumed that the entire consumption of refined copper and recovery of new scrap in a region ends up in products utilized in the same region. This assumption is quite unrealistic, of course, since there is a large and growing trade in scrap (China being the major importer) but there is reason to think it has less effect on global totals. A model properly reflecting inter-regional trade would be far more complicated and – in consequence – less transparent and probably less plausible.

The intensity of use of refined copper in OECD90, REF, ASIA and ALM in 1960–1997 was shown graphically in *Figure 3.11*. It resembles a so-called Kuznets curve, with rising consumption at low levels of per capita GDP and declining consumption after some peak. The 'standard' IU is calculated from OECD statistics on consumption of refined copper (*Table C3, Appendix C*), and statistics on population and economy in 1960–1990 (*Table C6, Appendix C*). The future extension of the curve is based on forecasts of population and GDP after 1997 from IPCC Scenario B2 (see below).

During the fitting period, in the poorest region ASIA, the intensity of use has risen from 0.1 kg/thousand US$ to 0.4 kg/thousand US$. In the next poorest region, ALM, the intensity of use has been lower. It has risen from 0.1 to 0.25 kg/thousand US$. The intensity of use in REF has actually declined drastically, due to the collapse of the USSR, from above 1.4 kg/thousand US$ to just above 0.2 kg/thousand US$. In OECD90, the intensity of use has declined from around 0.75 kg/thousand US$ to almost 0.45 kg/thousand US$.

We set the parameters as follows: a = 10 kg/capita, b = 75 thousand US$/capita and c =1. The curve fits ASIA quite well, but it does not fit REF or ALM It falls slightly below OECD90 in recent years (and even more so for earlier years). It is reasonable to assume a lower peak of the IU-curve for ASIA and ALM, than has historically been the case for OECD90, due to a combination of technological evolution and resource scarcity.

We have developed 8 scenarios, labeled SC-1 through SC-8. These are based on two alternative treatments each of three main assumptions: economic growth, evolution of consumption v. income behavior (the IU curve) and old scrap recycling (separation) efficiency. The different scrap recycling efficiency assumptions are tantamount to different assumptions about product lifetimes, as discussed in the previous section.

As regards population and future economic growth, we have chosen to borrow from work by the Intergovernmental Panel on Climate Change (IPCC). The recent IPCC report (IPCC 2000) used four different scenarios, known as A1, A2, B1, B2. For a more complete description see *Appendix C*. Scenario A1 is characterized by a very high growth in GDP/capita for the next century and a very slow growth followed by a decline, in population. Scenario B1 has the same population evolution as Scenario A1, but a slower growth in GDP/capita. Scenario A2 is characterized by a high growth in population and a relatively slow growth in GDP/capita. Scenario B2 comes in between, with a medium growth in both population and GDP/capita. Therefore, Scenario B2 was chosen as our base case scenario. We have also used IPCC scenario B1 as an alternative economic growth case. Scenarios A1 and A2 are not used and are not discussed in *Appendix C*.

In one set of 4 scenarios (the ones labeled 1,2,5,6) we assume that the IU curve is fixed once for all, as above. In the other set of scenarios (3.4.7.8) the entire curve is scaled down by a parameter d ($d = 0.9975$) so as to decrease uniformly at the rate of 0.25 percent per year. Scenario pairs (1,3), (2,4), (5,7) and (6,8) differ only in terms of this scaling factor. Over the next century, the differences are significant, roughly 22 percent globally.

Summary, Scenarios 1,3,5,7 are all based on the population and economic growth assumptions of IPCC scenario B2. The other set of four (2,4,6,8) are based on IPCC scenario B1. As noted already, scenarios 1,2,5,6 are based on a fixed unchanging IU curve with parameters noted above, while the group of scenarios numbered 3,4,7,8 are based on an IU curve that is scaled down over time at the rate of 0.25 percent per year from 1997 on.

Finally, we have made two different sets of assumptions with regard to product lifetimes. It will be recalled from the previous section that the efficiency of old scrap recycling (or separation) is not directly measurable. It can be inferred only by fitting a model to past history, and the results are strongly dependent on assumed product lifetimes in use. The detailed parametric choices are shown in *Table C1, Appendix C*. However the differences between our two cases can most easily be seen graphically (*Figure 3.13*). The first four scenarios (Sc1–Sc4) all have current recycling efficiencies of 50 percent increasing very slowly to about 55 percent at the end of the next century. By contrast, the second group of four (Sc5–Sc8) start with efficiencies of about 70 percent rising toward 80 percent by the end of the century.

As noted in the previous section, the recycling rate is not the same as the recycling efficiency. The rate as usually defined is simply the fraction of refined copper derived from scrap, which we would expect to rise more or less indefinitely; it does so in all scenarios, as shown in *Figure 3.14*. In the

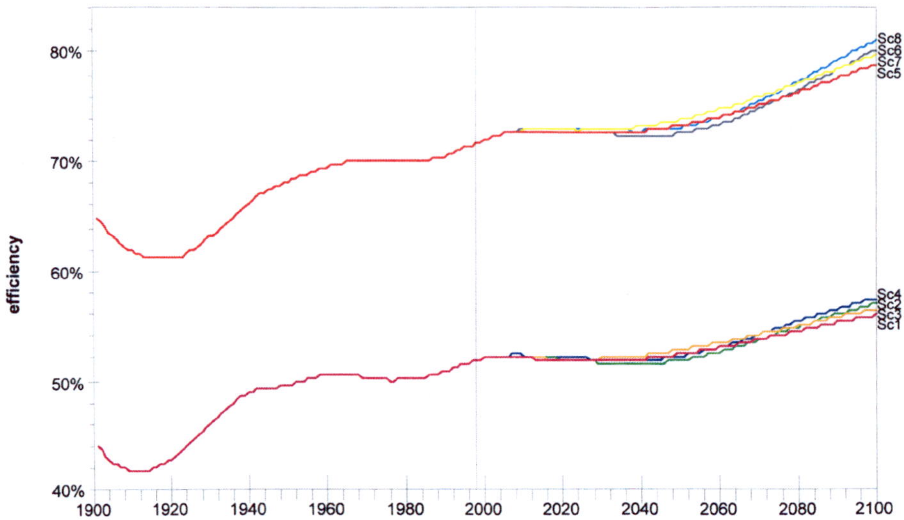

Figure 3.13: Global copper recycling (spearation) efficiency: 1900–1998. Comparison of copper system scenarios Sc1–Sc8.

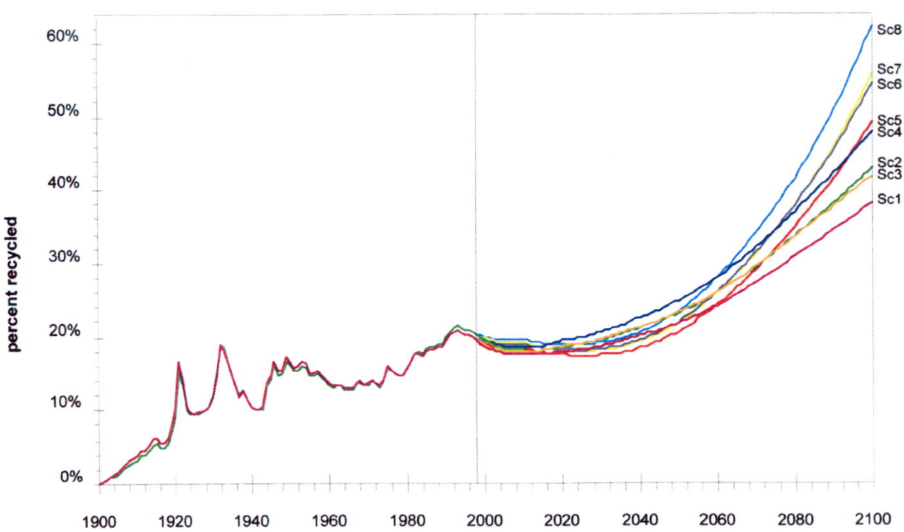

Figure 3.14: Global copper recycling rate: 1900–1998. Comparison of copper system scenarios Sc1–Sc8.

model a slightly different definition is used which gives equivalent results as long as copper consumption does not change sharply from year to year.

Model calibration procedures are described in detail in *Appendix C*, which also gives detailed tabular results of model runs and graphics for each individual scenario. The main results are shown in graphic form hereafter. The

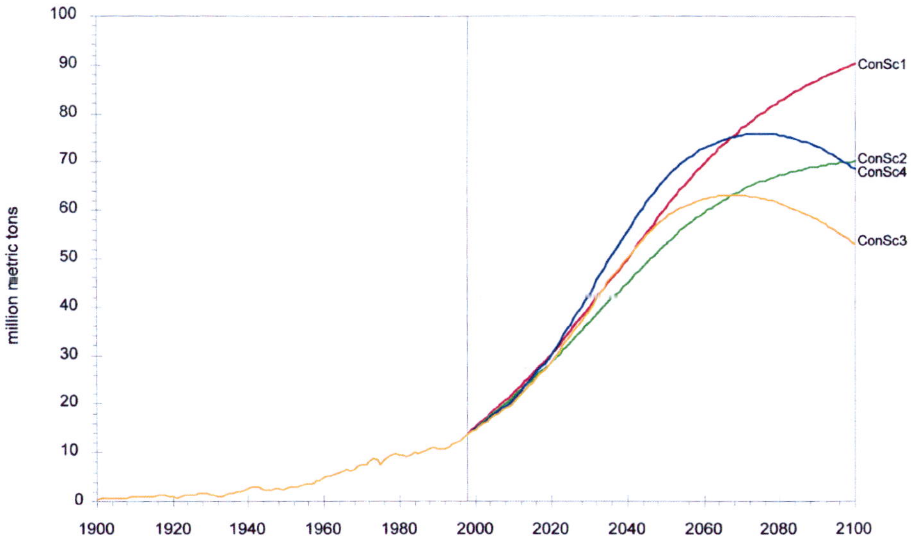

Figure 3.15: Global comsumption of refined copper, scenarios ConSc1 through ConSc4.

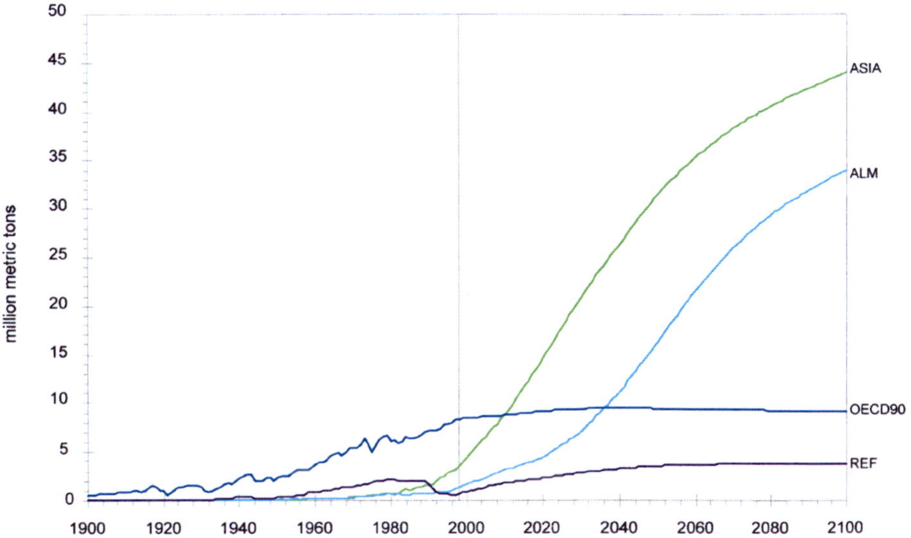

Figure 3.16: Regional consumption of refined copper; scenarios Sc1 and Sc5.

next five graphs (*Figures 3.15* through *3.19*) show global consumption of refined copper from 1900 to 2100 for scenarios ConSc1–ConSc4 (low recycling efficiency) and by region for four pairs of scenarios. Copper consumption is determined only by the IU curve, of course.

The last set of five graphs (*Figures 3.20* through *3.24*) show global mine

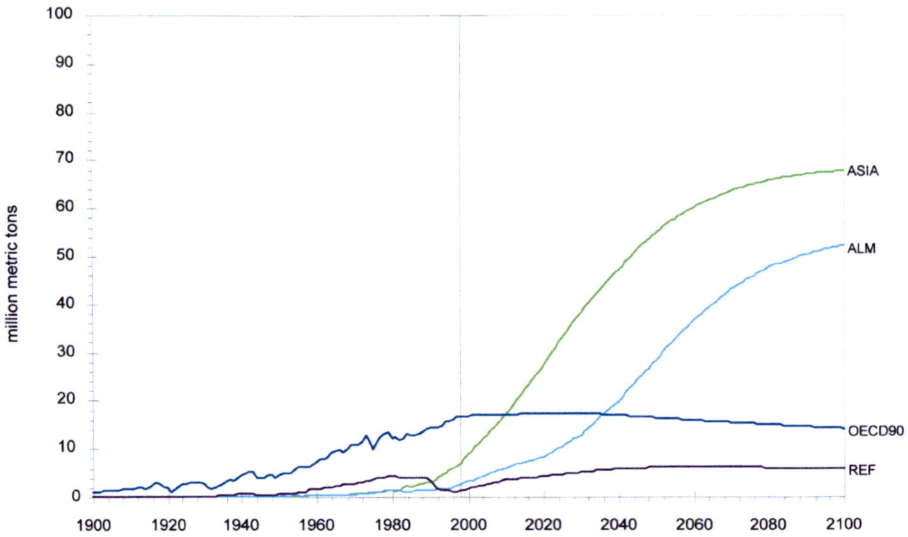

Figure 3.17: Regional consumption of refined copper; scenarios Sc2 and Sc6.

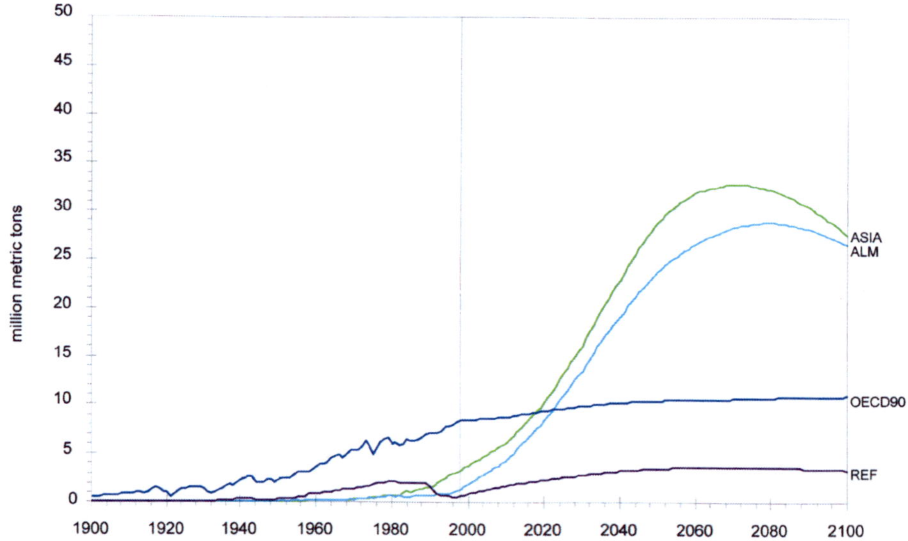

Figure 3.18: Regional consumption of refined copper; scenarios Sc3 and Sc7.

production, global cumulative mine production, global stock of waste copper and global stocks of long-lived and short-lived products in use from 1900 to 2100. Total and per capita results for 1900–1997 are based on historical consumption data. Figures for 2010, 2025, 2050 and 2100, from the model, are also given in *Table 3.5*. More detailed results are also in the *Appendix C*.

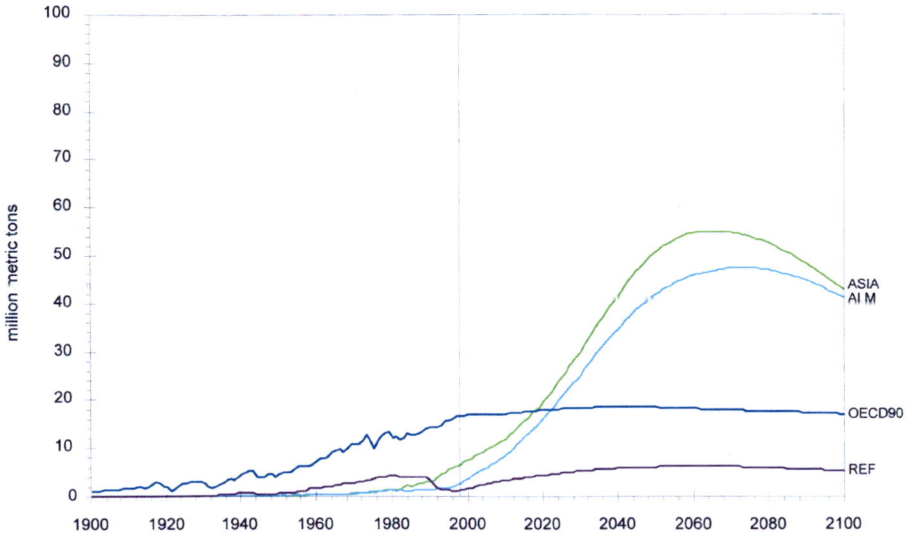

Figure 3.19: Regional consumption of refined coper; scenarios Sc4 and Sc8.

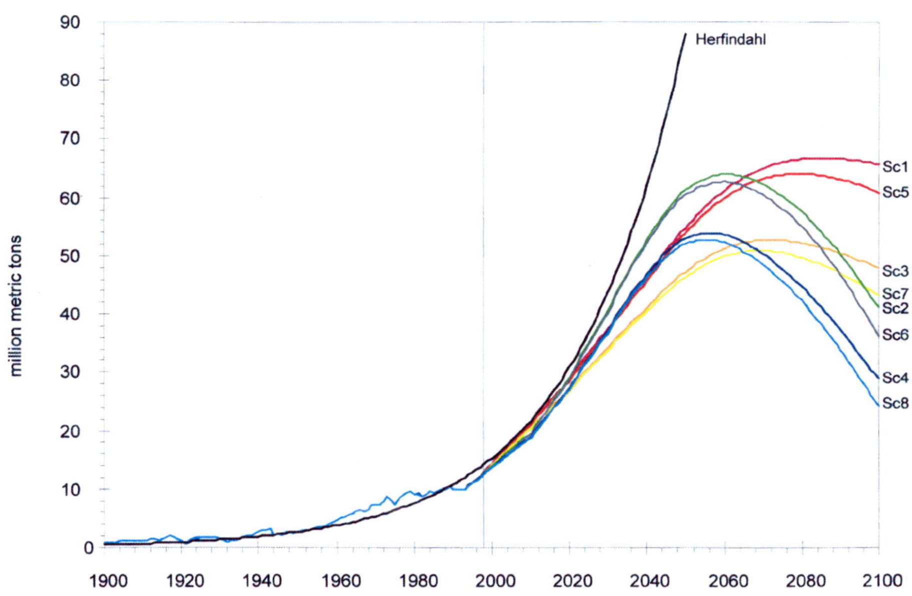

Figure 3.20: Global mine production of copper: 1900–1998. Comparison of copper system scenarios Sc1–Sc8.

In each of *Figures 3.20* and *3.21* an additional curve is shown, labeled 'Herfindahl' This is extrapolated from the historical production curve displayed in *Figure 3.10* (Herfindahl 1967). It is noteworthy that all scenario curves lie below the Herfindahl curve. This is to be expected, since even the

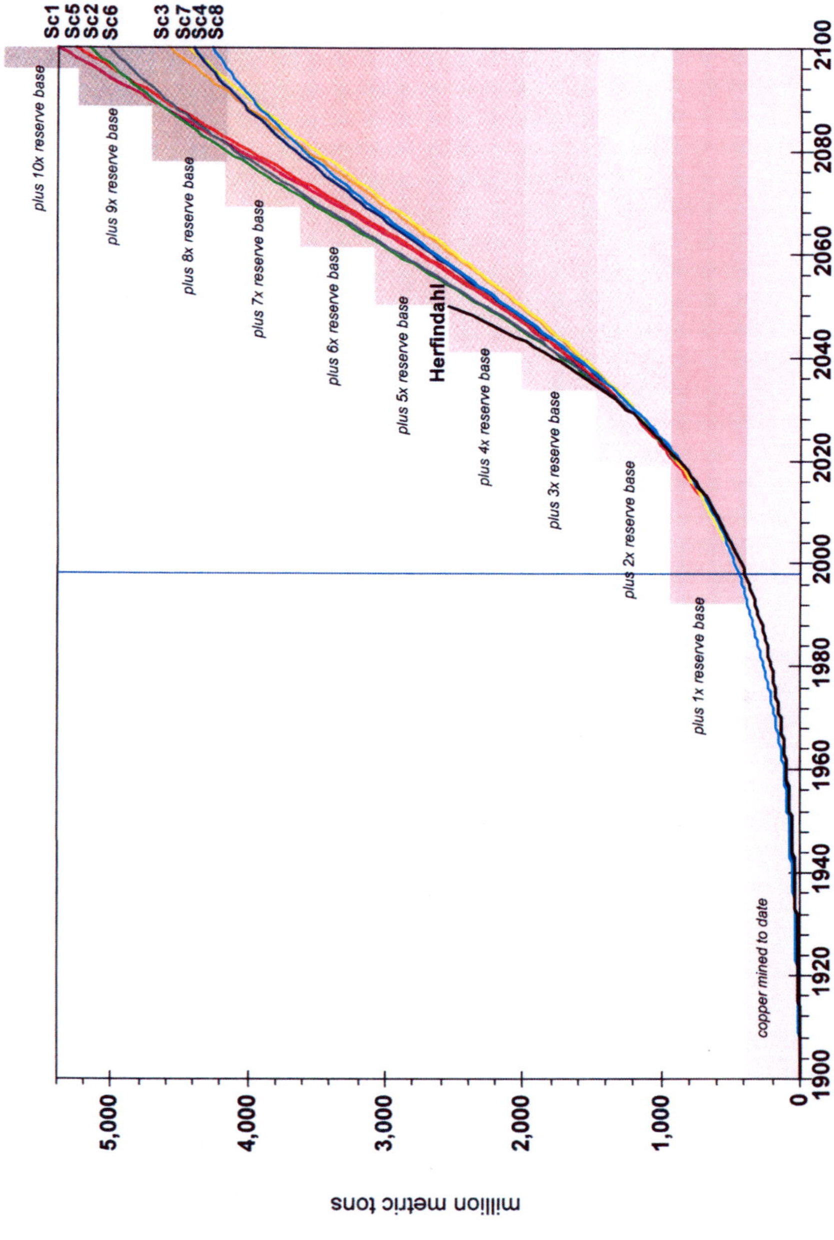

Figure 3.21: Global cumulative mine production of copper: 1900–1998. Comparison of copper system scenarios Sc1–Sc8.

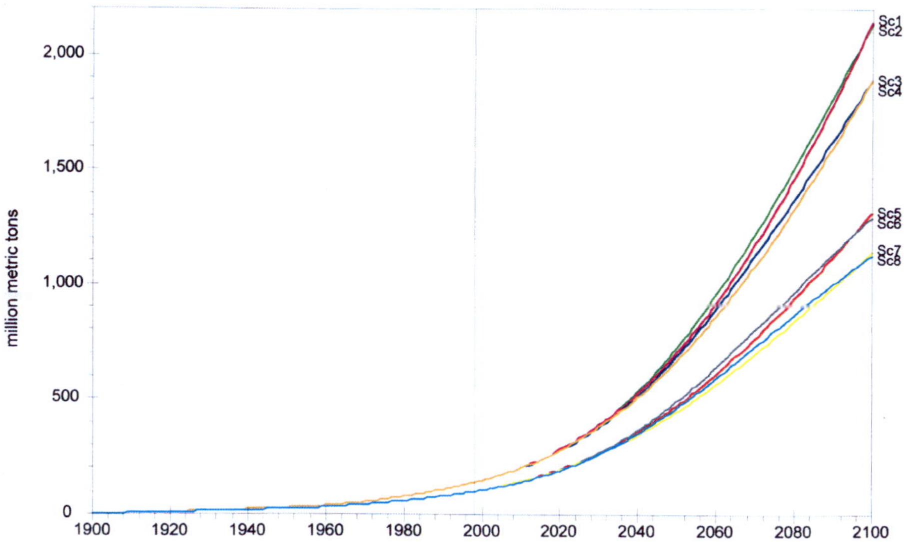

Figure 3.22: Global stock of waste copper: 1900–1998. Comparison of copper system scenarios Sc1–Sc8.

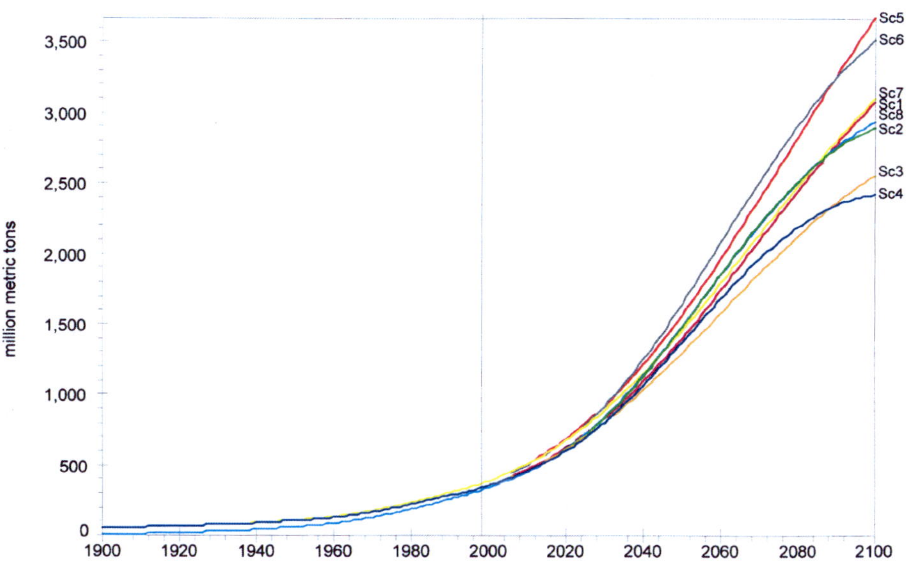

Figure 3.23: Global stock of long-lived copper products: 1900–1998. Comparison of copper system scenarios Sc1–Sc8.

historical global production trend curve exhibits some (albeit small) tendency to saturate. Even so, in view of the geological facts outlined in *Chapter 1* of this book, there is a very serious question as to whether recoverable global copper resources can sustain future consumption indefinitely – say to the end

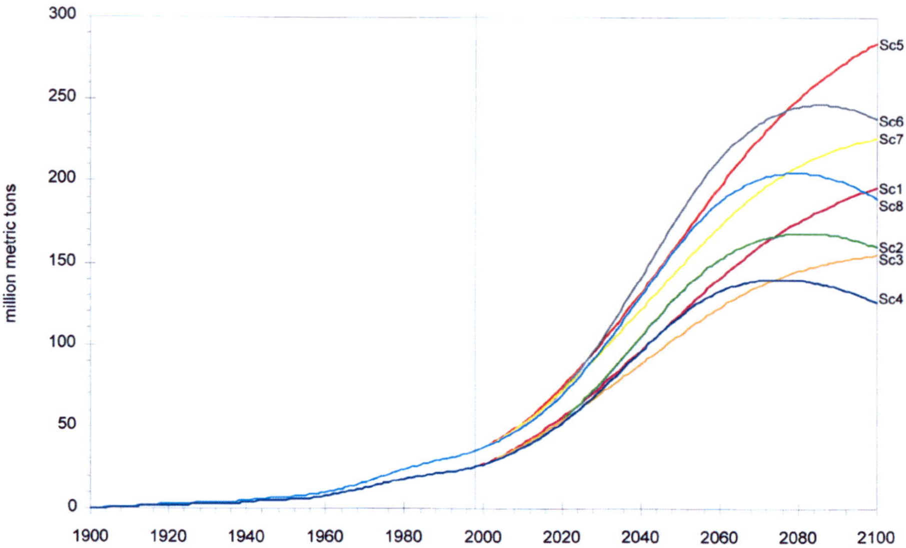

Figure 3.24: Global stock of short-lived copper products: 1900–1998. Comparison of copper system scenarios Sc1–Sc8.

of the present century (2100) – at anything like the rates projected by the model.

All model scenarios suggest that mining will peak sometime in the 21st century. The earliest peaks occur between 2050 and 2060 (scenarios 4,8) at levels greater than 50 MMT/y while the latest one peaks after 2080 at a level above 60 MMT/y (scenario 1). Given that total global reserves are 310 MMT, or just 5 to 6 years production at that rate, while the reserve base is order-of-magnitude 10 years production (*Table 2.1*), our educated guess is that even the lower peaks will not be achievable in practice. In short, we think that global copper production will actually peak considerably earlier in the 21st century. We also suspect that the maximum mine output will probably be not more than twice or possibly three times the current rate of mine production. This behavior is more or less according to the pattern first described by M. King Hubbert for petroleum (Hubbert 1956; Hubbert 1962; Hubbert 1973) and already confirmed for the US.

While the Hubbert-type depletion model must be taken somewhat seriously, if only in view of its success in predicting the 1969–70 peak output of domestic petroleum in the US, it is by no means a complete model, nor has it proved especially successful as applied to minerals and metals. In particular, an attempt to do so was published in 1976, entitled 'The Metals and Minerals Fuels Crisis' (Arndt and Roper 1976). The methodology was, in brief, to fit historical production data to a logistic-type curve. (Several different functional forms were considered.)

The form of curve that best fit historical data was then used to predict the trends of future (or past) peak production for the US and for the world.

Table 3.5: Copper consumption scenario results in 2010, 2025, 2050 and 2100; consumption of refined copper

	World				OECD90				REF			
	2010	2025	2050	2100	2010	2025	2050	2100	2010	2025	2050	2100
(1000 metric tons)												
ConSc1	22 232	35 061	60 230	90 394	8 706	9 228	9 326	9 110	1 767	2 508	3 452	3 679
ConSc2	20 794	36 510	66 520	68 491	8 759	9 657	10 456	10 960	1 694	2 528	3 405	3 326
ConSc3	21 574	32 770	52 879	70 026	8 448	8 625	8 188	7 057	1 715	2 344	3 030	2 850
ConSc4	20 178	34 124	58 401	53 058	8 500	9 026	9 180	8 490	1 644	2 363	2 990	2 577
(kg/capita)												
ConSc1	3.2	4.4	6.4	8.7	9.1	9.3	9.6	9.8	4.2	6.0	8.5	9.7
ConSc2	3.0	4.6	7.6	9.7	9.1	9.4	9.7	9.9	4.0	5.8	8.1	9.8
ConSc3	3.1	4.1	5.6	6.7	8.9	8.7	8.4	7.6	4.1	5.6	7.5	7.5
ConSc4	2.9	4.3	6.7	7.5	8.8	8.8	8.5	7.6	3.9	5.4	7.1	7.6
					ASIA				ALM			
					2010	2025	2050	2100	2010	2025	2050	2100
					(1000 metric tons)							
ConSc1					8 687	17 698	31 331	43 784	3 072	5 626	16 122	33 821
ConSc2					6 066	13 244	28 869	27 632	4 274	11 081	23 789	26 573
ConSc3					8 430	16 542	27 507	33 918	2 981	5 258	14 155	26 200
ConSc4					5 886	12 378	25 346	21 406	4 148	10 357	20 886	20 585
					(1000 metric tons)							
ConSc1					2.4	4.3	6.7	8.8	1.6	2.3	4.9	8.2
ConSc2					1.7	3.3	6.8	9.6	2.3	4.6	8.0	9.7
ConSc3					2.3	4.0	5.9	6.8	1.6	2.1	4.3	6.3
ConSc4					1.6	3.1	6.0	7.4	2.2	4.3	7.0	7.5

In the case of copper, Arndt & Roper noted that high grade ores were already depleted early in the 20th century and that copper mining since then has been limited to low grade porphyry ores. Using this double-peak scheme, Arndt & Roper predicted that the US production peak would occur at or near 2017. Curiously, using the same methodology, they also predicted that the global peak would occur in 1988 (ibid). Obviously this was too pessimistic (as, to their credit, they suspected might be the case (ibid, p.30)).

The implications of an early peak in mine output are qualitatively clear, albeit difficult to model. Prices will end their long downward drift, probably quite soon, and scrap prices will start to rise, probably sharply. This will encourage more efficient recovery of old scrap. Whether the old scrap recycling efficiency is 30 percent or 70 percent today (as noted, authorities differ widely on this) it can rise. Old mine wastes can be 'mined' again, and so on.

Also, the nature of demand can change. As noted several times already, copper is almost irreplaceable in some applications, especially for electrical wiring and motors. But copper is definitely replaceable (and already being replaced) in some others. Copper roofs, copper pipes and brass hardware will be increasingly rare and costly in the future. A few decades hence, copper may even be recovered from secondary steel operations.

A very significant implication of the model projections is that the accumulation of all copper byproduct metals in the earth's anthroposphere will also increase dramatically. The most important of these byproduct metals, from an environmental point of view, is arsenic (see Section 4.8.2). Arsenic is an extremely toxic metal, used almost exclusively for its toxicity. It has been used as an insecticide (e.g. lead arsenate), as a herbicide (cacodylic acid) and most recently as a wood preservative (chromated copper arsenic, or CCA). The latter use immobilizes the arsenic for some time in the wood fibers, but not forever. Eventually the arsenic in CCA-treated wood will leach out, or it will be released by burning or by some other mechanism. Ultimately this arsenic will be dispersed in the natural environment; some will end up in silt or topsoils, where it is likely to be immobilized – mainly by clay – as long as the soil pH is low enough. But acid deposition (due to sulfur dioxide emissions and nitrogen oxide emissions, for the most part) can re-mobilize toxic metals, including arsenic and cadmium (a byproduct of zinc).

A further point of considerable importance is that the projection model described above not only ignores any possibility of copper resource exhaustion, it also neglects the potential environmental implications of sharply increased copper mining. These environmental implications are not directly of concern in this report, but as communities and governments become wealthier they will be less and less tolerant of large-scale open-pit mines and associated operations. This will result in tighter regulations and increased costs that will be reflected, of course, in higher prices and lower demand than our model has suggested.

Evidently the absence of prices and price effects from the model is a major weakness. On the other hand it must be acknowledged that it is extraordinarily

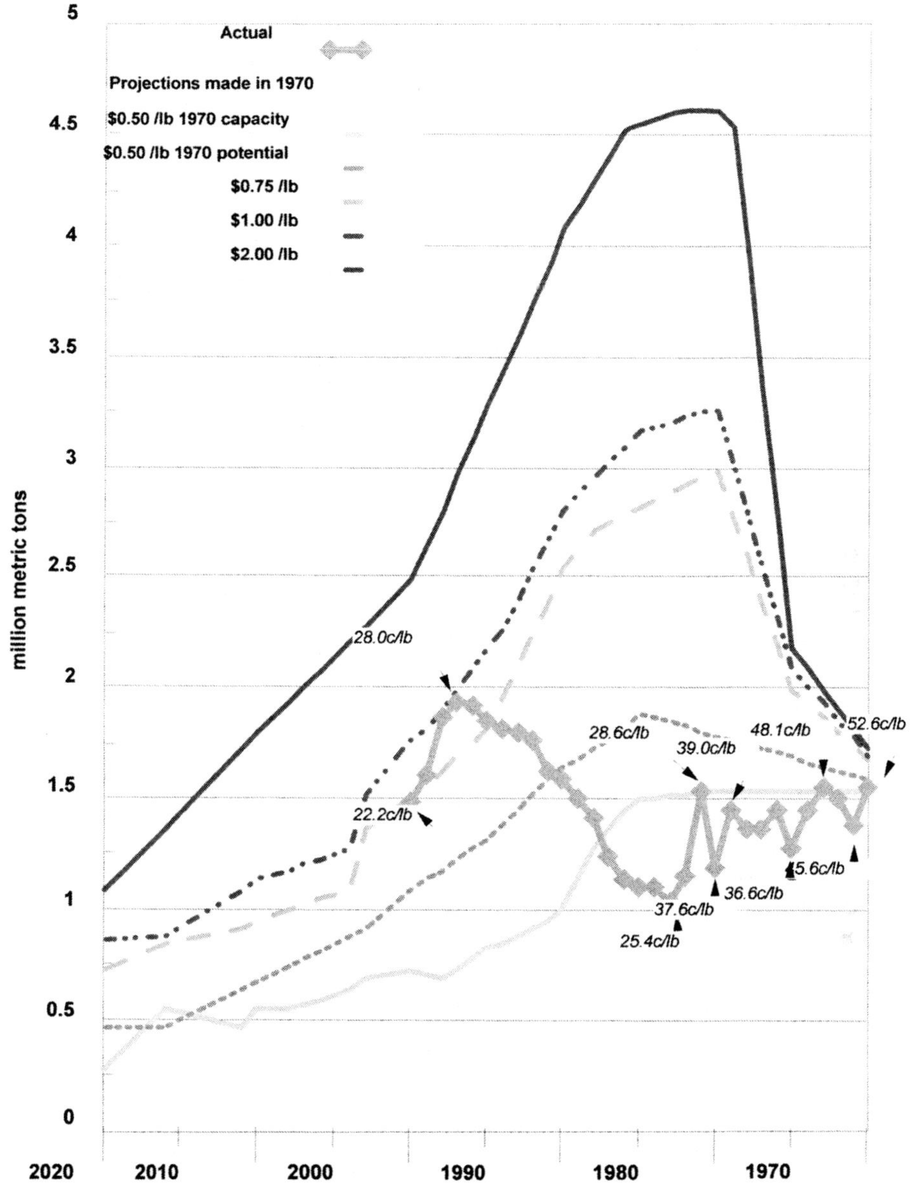

Sources: Actual [USGS Minerals Yearbooks]; Projections [COMRATE 1975]
Dollar values in constant 1967 dollars

Figure 3.25: Actual and projected mine production of copper for the US 1970–2020.

difficult to forecast prices. As an illustration of the hazards of forecasting, we reproduce in Figure *3.25* a model-based forecasting effort for the period 1970–2020 carried out in the early 1970s by a prestigious team of experts assembled by the US National Academy of Sciences (COMRATE 1975). The study considered a range of prices (in 1967 \$) from 0.50 \$/lb to 2.00 \$/lb. Actual copper prices after 1970 fell almost monotonically from 0.52 \$/lb in 1970 until 0.22 \$/lb in 1999. However, whereas US mine production also declined along with prices from 1970 until the early 1980s, production began rising in 1983 and continued to increase for 14 years, until 1997 even though prices remained nearly constant. The price and production decline resumed again in 1997.

CHAPTER 4

LEAD, ZINC AND OTHER BYPRODUCT METALS

4.1. Context

Just as metals are rarely found in pure form, the same is true of minerals. Moreover, certain combinations are apparently preferred by nature. That is to say, certain minerals tend to be found together or in each other's ores. This is especially true of the sulfide ores of heavy metals. Lead and zinc (and a number of other metals, discussed in this chapter) are almost invariably found in copper ores (in trace amounts, to be sure) and conversely. Most lead mines also produce zinc, and conversely. Molybdenum, nickel and cobalt are also commonly found with copper, though copper is generally a byproduct of these metals and not the primary product.

Moreover the technology of ore concentration for all sulfide ores (by flotation) is very similar, and there are other similarities in the later stages of reduction and refining. To a significant extent, the reduction and refining processes involved will tend to converge in the future, as each byproduct will have its uses (and dangers) and the entire system will necessarily evolve into a system of extraction and separation of a suite of metals, from a family of ores.

4.2. Physical properties and chemistry of lead and zinc

Lead is one of the rarer metals (16 ppm in the earth's crust). It is 36th in order of abundance. However US lead ores being mined as recently as 1988 averaged about 5 percent in grade (Woodbury et al. 1993, Table 1). It is a soft, heavy metal with a low melting point (327.5 °C) and a low boiling point, making it one of the easiest metals to cast and also one of the most volatile metals. It is the most corrosion resistant common metal. It has unusual electrical and chemical properties, accounting for some of its uses. (Lead is a relatively poor conductor of both electricity and heat, about 8 percent of the conductivity of Cu.)

Zinc is five times more common than lead (80 ppm) making it the 23rd most abundant metal in the earth's crust. It is found in trace quantities in all rocks. Zinc also has a low melting point (419.5 °C) and, like lead, an unusually low boiling point. It is the sixth best electrical and thermal conductor among metals, at 29 percent of the conductivity of copper. Chemically zinc is similar to cadmium, which accounts for its presence in zinc ores and as a contaminant of most zinc products. Cadmium is, however, far more toxic than zinc.

Lead and zinc are radically different insofar as their biological effects are concerned. Lead has no known role in biological organisms and is toxic at quite low levels. It causes physiological effects in humans, including nausea, anemia and (at high doses) coma and death. It causes anemia by replacing iron in the biosynthesis of hemoglobin. It can cause calcium deficiency by accumulating in the bones. Neurological effects include reduced nerve function, loss of coordination and encephalitis.

Zinc, by contrast (like copper) is essential for life processes. It is a constituent of several enzymes involved in carbohydrate metabolism and cell growth. It is widely used as a fertilizer additive and animal feed supplement. However, it is also toxic to some plant and animal species, mainly in aquatic systems. (Lead in biological systems can interfere with zinc-based enzymes.)

As regards zinc (as well as copper and some of the other byproduct metals that are also essential to plant growth) there seems to be no consistent relationship between eco-toxicity and concentration in the environment. It is also important to recognize that the natural concentrations within cells may be much higher – as much as 1,000 times higher – than concentrations found in soils. Thus uptake is unlikely to be driven by spontaneous diffusion through cell walls.

The exact mechanism for uptake is not completely understood, but the process is known to be adenosine-triphosphate (ATP) driven. This means that only electrically charged free ions can be transported through cell walls. For this reason bio-availability is closely related to ionization rates in soil pores, not to total concentration. Metal ions in solution are first bound to organic molecules such as fulvic or humic acids. This complexation process may occur in minutes or hours, depending on pH and other factors. Acidification is particularly important; liming increases soil pH (reduces acidity) and reduces bio-availability. Gradual additions of metals to soils can usually be accommodated (up to some limit) without increasing bio-availability. On the other hand, sudden oxygenation of anaerobic sediments or sludges (for example after a flood) can reduce pH and oxidize the sulfides, mobilizing toxic metals. This phenomenon has been called the 'chemical time bomb' (Stigliani et al. 1991).

The difference between an essential trace element and a non-essential one is reflected in the biodynamic accumulation of metals in various soil layers, resulting from uptake via roots, concentration in plants and deposition on the surface. Zinc and copper, being essential to growth, are actually depleted in the lower horizons and concentrated in the upper ones, as time passes. The degree of enrichment is a measure of the essentiality of the metal, though a very slight enrichment seems to occur even for the most toxic metals such as cadmium, lead and mercury. In Dutch soils the biodynamic enrichment process of the top layer of soil has increased zinc concentrations by 40 percent and copper concentrations by 10 percent over the natural background concentration (van Tilborg 1998, Table III). (The surface enrichment for cadmium is

6 percent, probably due to its chemical similarity to zinc; for lead the enrichment is only 0.1 percent.)

This soil enrichment phenomenon has been misinterpreted by some researchers as evidence of accumulation from atmospheric deposition. To clarify the situation comparisons of 'clean' soils and measured increase in soil metal content due to atmospheric deposition have been undertaken by the Dutch environmental health research organization, RIVM. The results are surprising and at odds with some figures that have been published in the media. It turns out that, at current deposition rates, it would take around 1,000 years to double the metal concentration in the plough layer of Dutch soils (van Tilborg 1998, Table V). During the past century the biodynamic accumulation effect has been an order of magnitude greater than the deposition effect. And, as noted above, as long as the soil concentrations do not exceed the storage capacity of the soil, the ionization rate in soil pores – and hence, the ecotoxicity – will be unaffected.

4.3. Lead process technology

Lead is found in nature almost exclusively as the sulfide galena (PbS) which is its primary sulfide. Oxidized ores can also occur, such as anglesite ($PbSO_4$) and cerussite ($PbCO_3$) which are the weathered products of galena. Lead ore often contains iron, copper and zinc sulfides or sulfates, as well as a small percentage of precious metals (especially silver) which can be recovered during the lead smelting. The average lead content of the material currently mined is approximately 6 percent (USBuMines annual 1991). Antimony and bismuth, as well as copper, zinc, silver and gold are byproducts of lead mining.

The process for the production of primary lead from ore consists of five basic stages, namely: ore mining and beneficiation, sintering, smelting, drossing, and final refining and casting. Each subsidiary process can be split again into several different operations. A general mass flow diagram for lead production is shown in *Figure 4.1*. Quantitative data were obtained from several sources (Thomas 1977, vol. 2, pp. 299–303; PEDCoPb 1980; Morris et al. 1983; Szargut et al. 1988; USBuMines annual 1991) and integrated with gross mass balance estimates.

4.3.1. *Ore mining and beneficiation*

The ore as mined has a lead content too low for direct use in the lead smelter. Lead ore is crushed, ground, and then concentrated by means of chemical and/or mechanical processes. In the chemical process ore dust (in water as a transport medium) is treated with chemical additives that create a froth in which the mineral particles are floated from the gangue. The process is similar to that for copper. In the mechanical process, used for Missouri (galena) ores, separation is achieved thanks to the difference in specific gravity of the lead

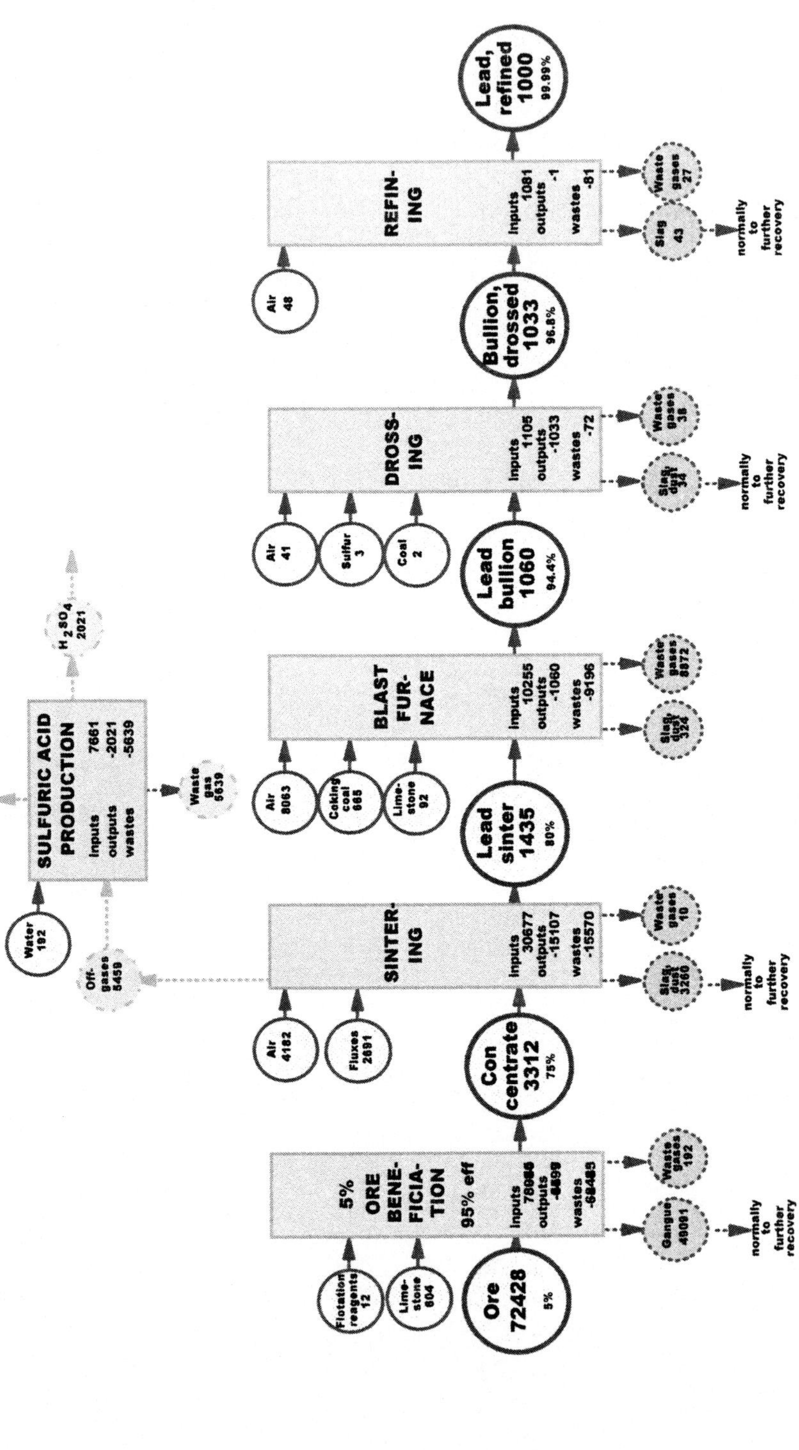

Figure 4.1: Mass flows (kg) in the production of 1 MT lead (simplified processes, typical material mixes).

ore and the gangue particles. After concentration, the lead content is increased from 6 percent to about 75 percent by weight. Western ores are concentrated by the chemical froth flotation process, achieving 45–60 percent lead content.

4.3.2. *Sintering*

Sintering consists of roasting the concentrate to obtain a material suitable for subsequent reduction in the blast furnace. Metal sulfides and sulfates are transformed to oxides, suitable to be reduced by coke in the blast furnace. The sinter machine is charged with ore concentrate, limestone and silica. Approximately 45 percent of the produced sinter is recovered and added to the charge in order to control the sulfur content of the charge. The reaction is exothermic. About 85 percent of the sulfur is removed from the concentrate during sintering, as sulfur dioxide. The latter is conveyed to the acid plant where it is used to produce sulfuric acid.

4.3.3. *Smelting*

There are three main cases. Untreated scrap lead from battery recycling is usually melted and partially purified in a reverberatory furnace. (Sulfur dioxide loss is not a serious problem for scrap.) The molten lead can be sent directly to a refinery or cast into ingots. The insoluble impurities form a slag which floats over the liquid metal so that it can be easily removed. Slag is cooled and crushed, then partially recycled to the sinter machine. The remaining part is discharged.

Slag from a reverberatory furnace (as above), or ore concentrates low in zinc are smelted in a blast furnace. In the blast furnace lead oxides are carbothermically reduced to lead bullion. The process is very similar to iron smelting. According to one study, about 4,800 kg of sinter are needed to produce 1 metric ton of lead (Morris et al. 1983). The reducing agent is coke (about 650 kg/t of lead) and propane. Approximately 2,900 kg of slag leaves the blast furnace for each ton of lead and about 50 percent of it is recycled internally. Other output streams are dust and off-gases. Despite of their heat content, the latter are not usually re-used in the process. Blast furnace utilities are steam and electricity (approximately 1.5 GJ /t of lead). Sulfur dioxide is mostly recovered from blast furnaces and sent to a contact acid plant, as in the case of copper. Reverberatory furnaces (as in the case of copper) are not good for recovering sulfur dioxide.

Lead concentrates high in zinc content are normally sent to a so-called 'Imperial smelter', which removes the zinc in vapor form for subsequent condensation (retorting, see below). About two thirds of worldwide lead output is obtained from mixed lead-zinc ores.

Both reverberatory and blast furnaces generate significant lead emissions, due to the volatility (low vaporization temperature) of the metal. The vapor condenses into particulates. The particulates are captured by means of baghouses (in older plants), wet scrubbers or electrostatic precipitators. Nevertheless, losses remain significant and emissions controls constitute a fairly large cost element for modern smelters.

4.3.4. *Drossing and final refining*

Lead bullion from the blast furnace or imperial smelter needs to be further refined to remove sulfur and other metallic impurities for most industrial applications, including battery manufacture. Therefore, smelted lead bullion passes through a multi-step process, whereby metal impurities are removed in the form of slag. For that purpose silica and caustic soda (17 kg and 78 kg per metric ton of lead, respectively) are added to the liquid metal, together with small quantities of zinc, in order to accelerate slag formation. The slag is partially recycled to the blast furnace. A matte rich in copper, a silver-gold alloy (known as doré metal) and small quantities of antimony, bismuth and zinc oxide are also byproducts of lead refining.

Approximately 284 kg of copper matte are recovered per metric ton of lead in the US. This is transferred to copper convertors. No quantitative data were available to evaluate the amounts of other byproducts produced from lead smelters, though silver is a significant byproduct.

4.3.5. *Exergy and exergy flows*

As in the case of copper, lead and zinc ores embody significant amounts of exergy, which contributes to the exergy required for the smelting and refining stages. Thus, to produce a metric ton of lead with an embodied exergy of 1.124 GJ, ore containing 11.867 GJ/t is consumed. The difference contributes to the exergy consumed in the process. Note that waste solids embody about 4,16 GJ/t, waste gases contain 5.745 GJ/t while waste heat accounts for 39.663 GJ/t according to our calculations. Fuel and utility inputs add up to 61.1 GJ/t, which is considerably higher than figures published elsewhere (typically about 30GJ/t. (Forrest and Szekely 1991). We suspect that the most modern lead smelters by-pass the energy-intensive sinter stage. Also, it appears that the SX-EW process is growing in importance for the lead industry.

Figures 4.1 and 4.2 are simplified diagrams of mass flows, and exergy flows of the main material inputs and outputs per metric ton of product for each stage of the primary lead industry. The overall mass and exergy flows of inputs, outputs, byproducts and wastes for the primary lead sector are illustrated in *Figure 4.3*. See *Appendix A* for details on the assumptions used regarding the composition of ores, concentrates and compounds.

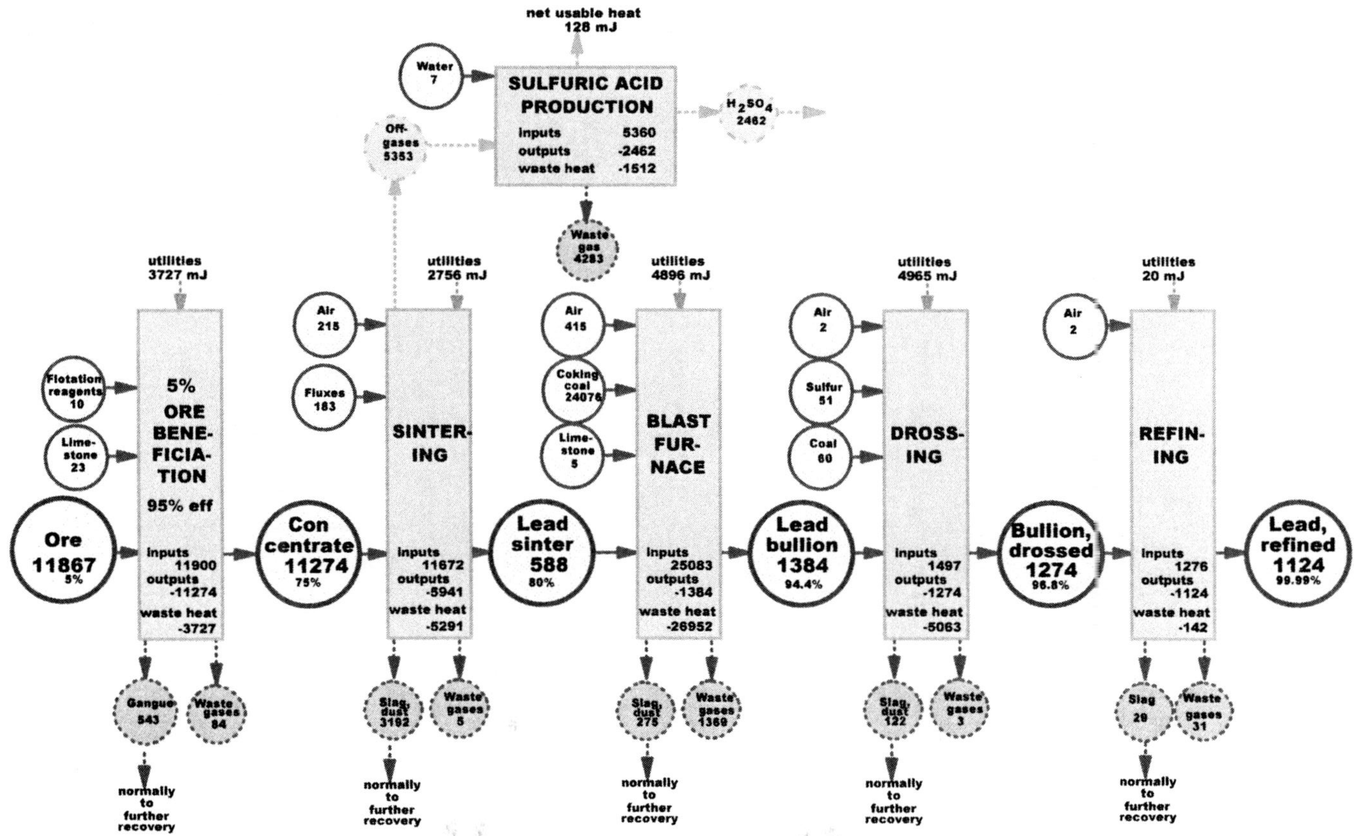

Figure 4.2: Exergy flows (mJ) in the production of 1 MT lead (simplified processes, typical mate-ial mixes).

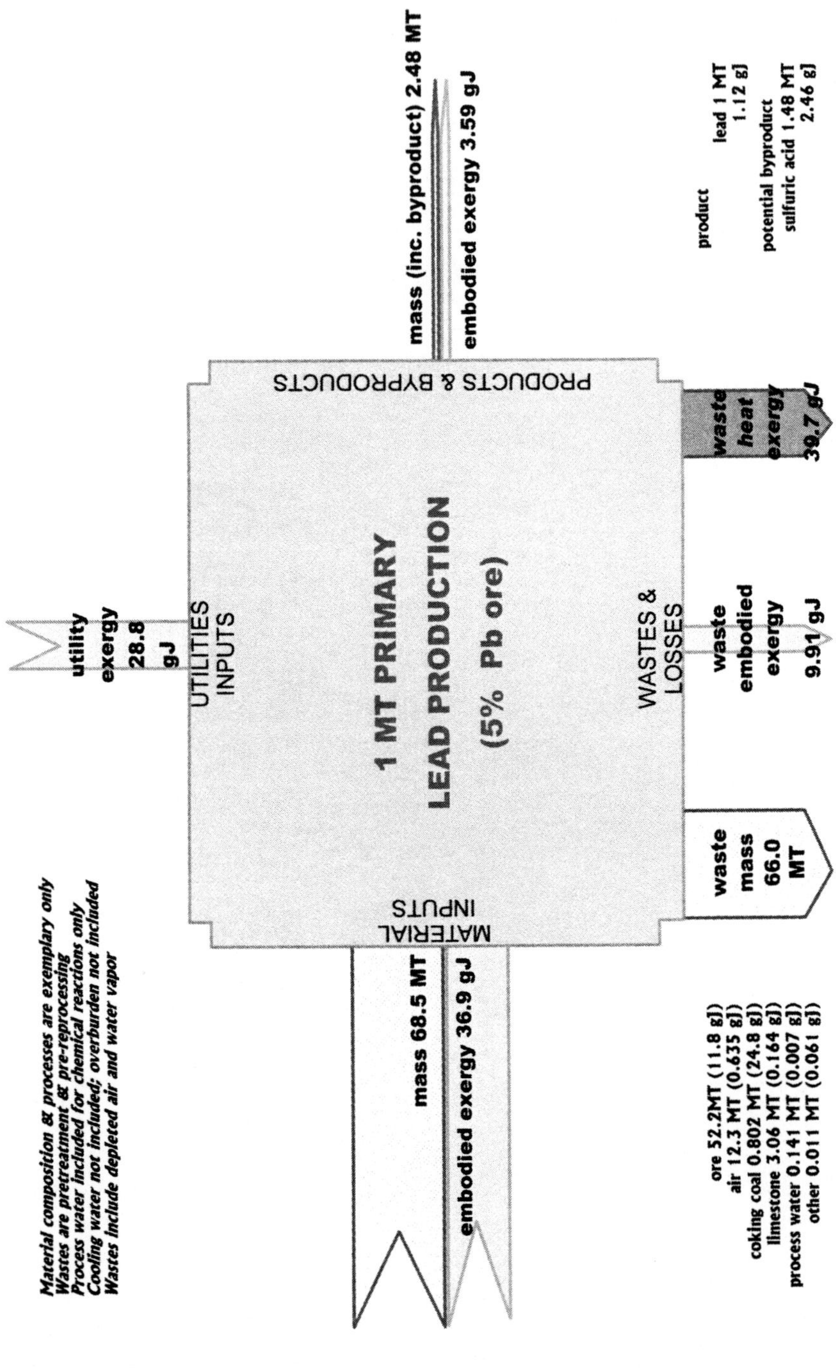

Figure 4.3: Primary lead unit mass and exergy flows.

4.4. Lead sources and uses

Lead has been mined as long as copper, i.e. throughout human history. An early peak level of use (about 100,000 tpy) was reached during Roman times, about the beginning of the Christian era, when lead was mined largely in Spain. The Spanish mines were exhausted within 300 years, however, and use declined more than 10-fold. The discovery of silver in Germany (associated with lead) created a new source of lead, and consumption has been rising ever since, at least until very recently (*Figure 1.5*). Evidence confirming this early history is found in Greenland ice cores (Hong et al. 1994). See *Figure 1.6*.

Global mine production of lead actually peaked (at 3,650 kMT) in 1977 and has been declining steadily since the early 1980s. Actually, mine output has been increasing since 1994. Total global refinery output seems to have stabilized. It may have peaked (at 5,673 kMT) in 1989 (Roskill 1997). The biggest current producers (1998) are China (710 kMT), Australia (583 kMT), US (458 kMT), Peru 254 kMT) and Canada (189 kMT). Chinese output has grown very rapidly. By contrast, production in the former Soviet Union has fallen from a peak of 600 kMT in 1977 to just 47 kMT in 1998 (ILZSG 1999).

Refinery output has continued to increase, reaching 6 MMT in 1998. The growth has been entirely from secondary (recycled) lead, which accounted for about 60 percent of the total in the western world by 1998, as compared to 43 percent in 1980 and 33 percent in 1965 (ILZSG 1999) The US is by far the biggest producer and consumer of refined lead, followed by China.

Modern uses of lead are extremely diverse. Metallic lead has been used for water pipes, as sound-proofing, protection against radioactivity and X-rays and as a protective shield for undersea cables, and for bullets and shotgun pellets (still a major use, 2.4 percent). It is also an important component of solder. Metallic uses of lead – mainly for pipe – accounted for 18 percent of total consumption in 1960, but only 6 percent in 1997. Cable sheathing accounted for 3 percent of demand in 1997. Alloys (including solder) accounted for 10.5 percent in 1960, down to 2.4 percent in 1997 (ILZSG 1999) A possible future use for lead (with bismuth) is as a coolant for molten metal-cooled (Generation IV) nuclear power reactors.[1]

Chemical uses for pigments, stabilizers (for PVC), catalysts and pesticides added up to 9.9 percent of consumption in 1960, and 10.9 percent in 1997 White lead was once used extensively as an exterior house paint, until it was displaced by zinc white (lithopone) and subsequently by titanium dioxide. Red lead, mainly used to protect bare iron or steel surfaces (such as bridges), is

---

[1] This is the application for which liquid metal sodium is currently preferred, but sodium has the problem that it reacts violently with water (or steam), whence for safety reasons a sodium-cooled reactor would have to incorporate an extra heat exchanger, thus reducing its maximum potential thermal efficiency. No reactor has been built using molten lead as a coolant, however.

still used for that purpose, though other protective agents are gradually making inroads. Lead arsenate was also a major insecticide, especially for orchard trees. However it has been banned for many years in Europe and the US, and is no longer used for that purpose. The major new market for lead compounds (oxides) is in glass (see below).

Lead compounds (tetraethyl or tetramethyl lead, known as TEL and TML) are uniquely efficient as an anti-knock additive for gasoline. This inexpensive additive adds significantly to the so-called 'octane' level of gasoline and enables gasoline engines to operate at compression ratios significantly (up to 20 percent) higher than engines burning unleaded fuel. This, in turn, adds about 10 percent to the fuel efficiency of the spark-ignition engine, as compared to engines operating at lower compression ratios. Octane levels can be raised to some extent by other means, notably by adding aromatics (benzene, xylene, toluene) or alcohols. However these additives are not only more expensive, they are less effective. As a result, 'unleaded' fuel nowadays is typically around 87 octane whereas leaded fuel of typically 95 octane and 100 octane fuel was not uncommon (in the US) in the 1960s. The banning of lead from motor fuel for new cars starting in 1970 in the US and around 1990 in Europe has added significantly to fuel costs while simultaneously increasing fuel consumption slightly over what it would otherwise have been. This use of lead accounted for 9 percent of global lead consumption in 1960, but only 0.9 percent in 1997 (ILZSG 1999).

The other major use of lead is in storage batteries, for non-interruptible power systems and traction systems. The best known application is for starting-lighting-ignition (SLI) batteries. Such batteries have recently been considered seriously as candidates for electric vehicles (thanks to California legislation in the late 1980s requiring that a certain percentage of cars sold after 1998 qualify as 'zero emission' (Illman 1994). However the law has since been modified, largely due to recognition that an electric car based on lead-acid batteries would be severely performance constrained, as well as very expensive. Lead is not the ideal storage battery material, however, and more efficient storage batteries are certainly possible in principle, although the development and engineering problems have been formidable (Ayres and Mckenna 1972). Despite decades of research, the conventional lead-acid battery is still by far the most cost-effective alternative for its main (SLI) purpose. It accounts for 72.5 percent of current demand, up from 27.7 percent in 1960. Thanks to accelerated R&D triggered by the California legislation mentioned above, it is now fairly likely that this application of lead will also be supplanted in another decade or two by more efficient and compact nickel-zinc batteries, already commercially available (Coates et al. 1999).

On the other hand, lead oxide is an important constituent of certain types of glass ('lead crystal') which was formerly a luxury product accounting for an insignificant fraction of the lead. Since the early 1980s the market for lead glass has grown rapidly because of its use for TV screens and computer monitors. Monitor glass is 30–35 percent lead by weight, and each monitor contains on average 6 lbs (2.7 kg) of lead, the primary purpose of which is to protect users from radiation.[2] Glass had became the second most important use of lead in the US by 1995.

Another significant dissipative use of lead is in solder. Apparently there is somewhere between 1 and 3 kilos of lead, mainly in the form of solder, scattered among various circuit boards in every PC. Solder production in 1997 was about 60 kMT (ILZSG 1999). Very little of this material is currently recycled. Electronic materials – including lead – constitute an increasing proportion of materials sent to landfills and leachates may contaminate groundwater. This possibility has resulted in bans on landfilling of electronic wastes in a number of states, which in turn has motivated some efforts to recycle electronic materials. In the European Union, the Waste in Electric and Electronic Equipment Directive proposes to ban the use of lead in solder from 2008. The Japanese Environmental Agency has proposed that electronic materials be disposed of in sealed landfills to prevent leaching, while the Japanese Ministry of International Trade and Industry (MITI) has called for a significant reduction of lead-based solder in automobiles as early as 2003. These initiatives have prompted a considerable research effort to find suitable lead-free solders. Among the combinations being studied are Sn-3.5Ag, Sn-0.7Cu, and Sn-3.8Ag-0.7Cu. Motorola Corp. introduced a line of lead-free products in 2001 (Frear et al. 2001).

The US Bureau of Mines conducted a study of cumulative lead flows in the US from 1940 through 1988 (Woodbury et al. 1993, draft). Lead extracted from domestic ores was 20.2 million metric tons (MMT) of which 1 MMT was lost in mill tailings, 0.6 MMT in smelter slag and emissions, and 0.4 MMT to 'new scrap' (primary drosses). Imported concentrates added 4.2 MMT, while recycled scrap (old and new) added 22.1 MMT, after losses of 0.8 MMT. So 44.5 MMT went into the manufacturing system from domestic refineries, plus 9.8 MMT of imports less 0.6 MMT of stock increase and 0.5 MMT of process wastes. Finally, 53.2 MMT of lead was sold as, or incorporated in final end use products. As of 1990 the authors estimated that 13.25 MMT had been dissipated in use (mainly tetraethyl lead in motor fuel), 6.95 MMT was dissipated as managed process waste and 23 MMT was recycled (of which 0.4 MMT was exported). Overall, 24.4 MMT of lead entered the US economy, of which 14.4 MMT was dissipated into the environment and 10 MMT (41 percent) was still in use by 1988, mainly as automotive storage batteries.

Non-increasing, even declining, demand for lead, together with increasing availability of secondary material have caused overcapacity and downward pressure on prices. this, in turn, adversely affects the competitiveness of all

---

[2] See 'Please dispose of properly' by David Pescovitz in *Scientific American*, February 2000; also reply to a letter to the editor in the June 2000 issue of the same magazine by Robert Knowles (of Technology Recycling Inc.), who was quoted by Pescovitz,

producers in countries with tight environmental standards (Ballance and Forstner 1990).

4.5. Zinc processing

The overall process for the production of primary zinc metal (slab zinc) is very similar to that for copper and lead, albeit with minor differences. It consists of five basic stages, namely: ore mining and beneficiation, sintering, smelting, and final refining. However, zinc ores are more variable than lead ores (but perhaps not more complex than copper ores), and treatment is more complex. A general mass flow diagram for zinc production is shown in *Figure 4.4*. Quantitative data were obtained from several sources (PEDCoZn 1980; Szargut et al. 1988; USGS annual) and integrated with the help of mass balance estimates.

4.5.1. *Ore mining and beneficiation*

The most common zinc mineral is the simple sulfide, sphalerite (ZnS) and the zinc-iron sulfide (Zn, Fe)S known as marmatite, but zinc is also extracted commercially from a number of other minerals, including oxides (e.g. zincite, franklinite), carbonates (smithsonite, hydrozincite), sulfates (goslarite), and silicates (willemite, hemimorphite). See *Table 1.1* for chemical formulae.

Zinc ores being mined range in grade from less than 1 percent to more than 15 percent zinc, with the majority of mines being in the range 3–9 percent (IZA/ILZA 2000). Most mines (80 percent) are underground, although open pit and combination mines are somewhat more productive on average, accounting for 15 percent and 21 percent of output, respectively (ibid.). There were 266 mines in the world in 1997 (IZA/ILZA 2000).

Some zinc sulfide ores are associated with iron pyrites, copper minerals (chalcopyrite), and lead minerals (galena). Cadmium (as cadmium sulfide, or greenockite) is the third most abundant impurity of zinc, after sulfur and iron, having an average abundance of 1 part per thousand of zinc and a maximum abundance of 1–2 percent of zinc.

Zinc oxide ores are mostly not mined for their zinc content, but are byproducts of mines for other metals, and are typically sent to ferro-alloy (e.g. spiegeleisen) operations. Zinc sulfide ores beneficiated by froth flotation. Concentrates range from 52–60 percent zinc, 30–33 percent sulfur and 4–11 percent iron.

4.5.2. *Roasting and sintering*

Zinc concentrates may be roasted, with sodium or zinc chloride, to drive off the sulfur (as sulfur dioxide, sent to a contact sulfuric acid plant) and convert the zinc to zinc oxide for subsequent reduction. The product of roasting is

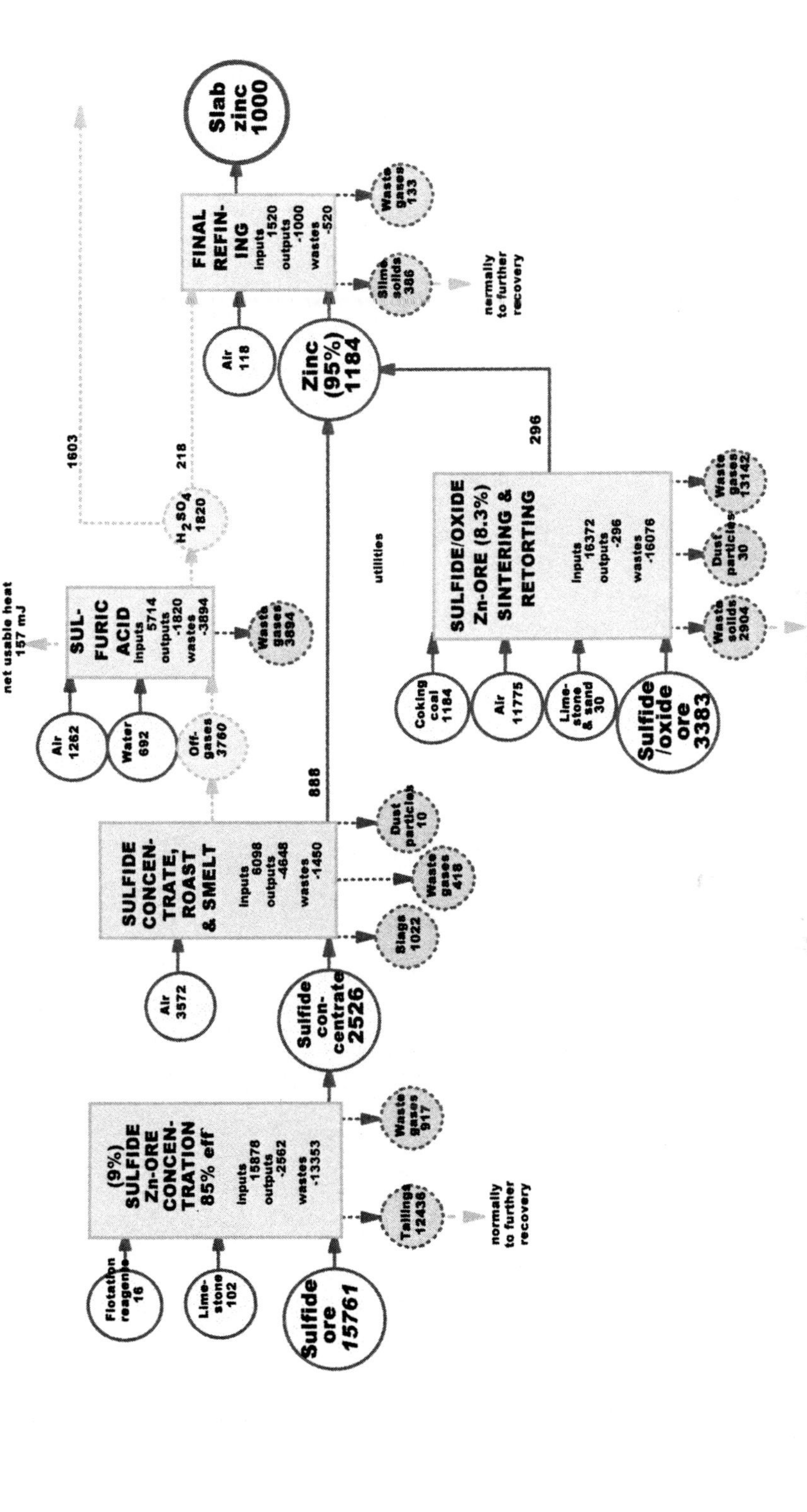

Figure 4.4: Mass flows (kg) in the production of 1 MT zinc (simplified processes, typical material mixes).

known as calcine, with a typical residual sulfur content of 2 percent. The calcine is then either sintered (with sand and coke) or leached by sulfuric acid.

4.5.3. *Smelting and refining*

There are 79 smelters in the world today using (mainly) zinc concentrates (IZA/ILZA 2000). The major process by far is electrolysis (80.6 percent of primary refinery capacity). The calcine is treated by leaching (with sulfuric acid) and the zinc is subsequently recovered by electrolysis as cathode zinc. Flue dusts from roasting and sintering operations, along with leach residues leached again for cadmium recovery and purification. The leaching (hydrometallurgical) process yields a solution of copper, zinc and other impurities such as cadmium. Copper is precipitated from the solution by the addition of zinc dust. Cadmium is then precipitated by a second and possibly a third addition of zinc dust, leaving a pure zinc sulfate solution that can be reduced by electro-winning (EW) to pure zinc. This process is closely analogous to the SX-EW process for copper.

The pyro-metallurgical reduction process for lead-zinc ores is Imperial Smelting, discussed previously. This process accounts for 13.2 percent of current global primary zinc capacity. Zinc-only concentrates are usually smelted in *retorts* (6.2 percent of capacity). Sintered calcine is reduced by reaction of the zinc oxides with carbon monoxide from coke, usually in a vertical retort, which is similar in function to a blast furnace. However the zinc is vaporized and subsequently condensed to molten metal which is subsequently cast into slabs. Impurities are separated at the retort, or by subsequent fractional distillations for byproduct (e.g. cadmium) recovery. Retorts may also be horizontal (formerly the dominant process).

4.5.4. *Exergy and exergy flows*

Figures 4.4 and *4.5* are simplified diagrams of mass flows, and exergy flows of the main material inputs and outputs per metric ton of product for each stage of the primary zinc industry, based on older process descriptions. A composite diagram combining both mass and exergy flows is *Figure 4.6*. In the case of zinc, sulfide ores embody 18.449 GJ/t of slab zinc, which itself has an embodied exergy of 5.182 GJ/t. The remainder is available to contribute to the exergy needed to reduce the ore to metal Based on the older process description (c. 1970) fuel and utility inputs amounted to 42.847 GJ and 63.26 GJ per metric ton, respectively, generating solid wastes (1.182 GJ), gaseous wastes (17.27 GJ) and waste heat 99.16 GJ respectively. See *Appendix A* for details on the assumptions used regarding the composition of ores, concentrates and compounds.

In modern smelters with higher sulfur recovery the need for coal is significantly reduced. A recent Ecoprofile sponsored by the International Zinc Association, Brussels, has concluded that the total energy (roughly equal to

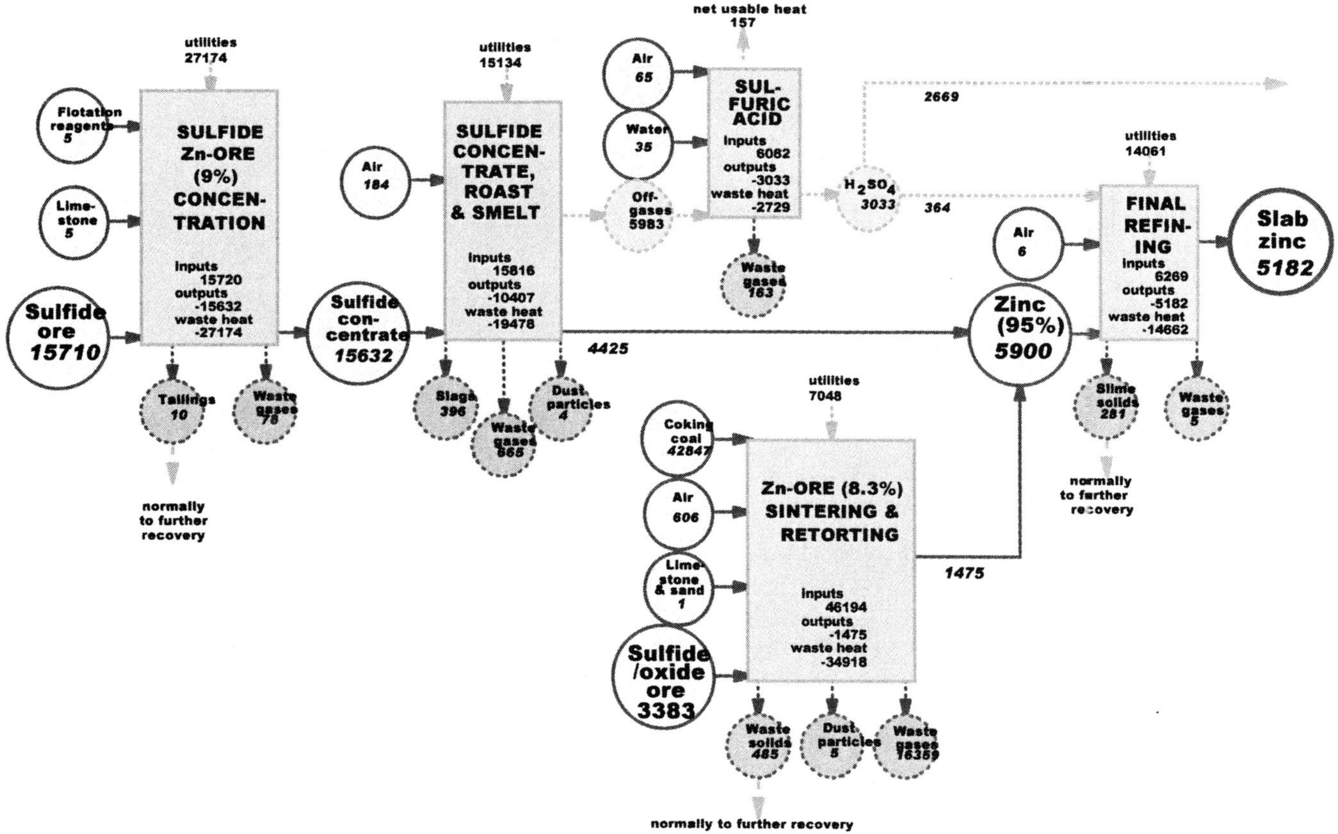

Figure 4.5: Exergy flows (mJ) in the production of 1 MT zinc (simplified processes, typical mater al mixes).

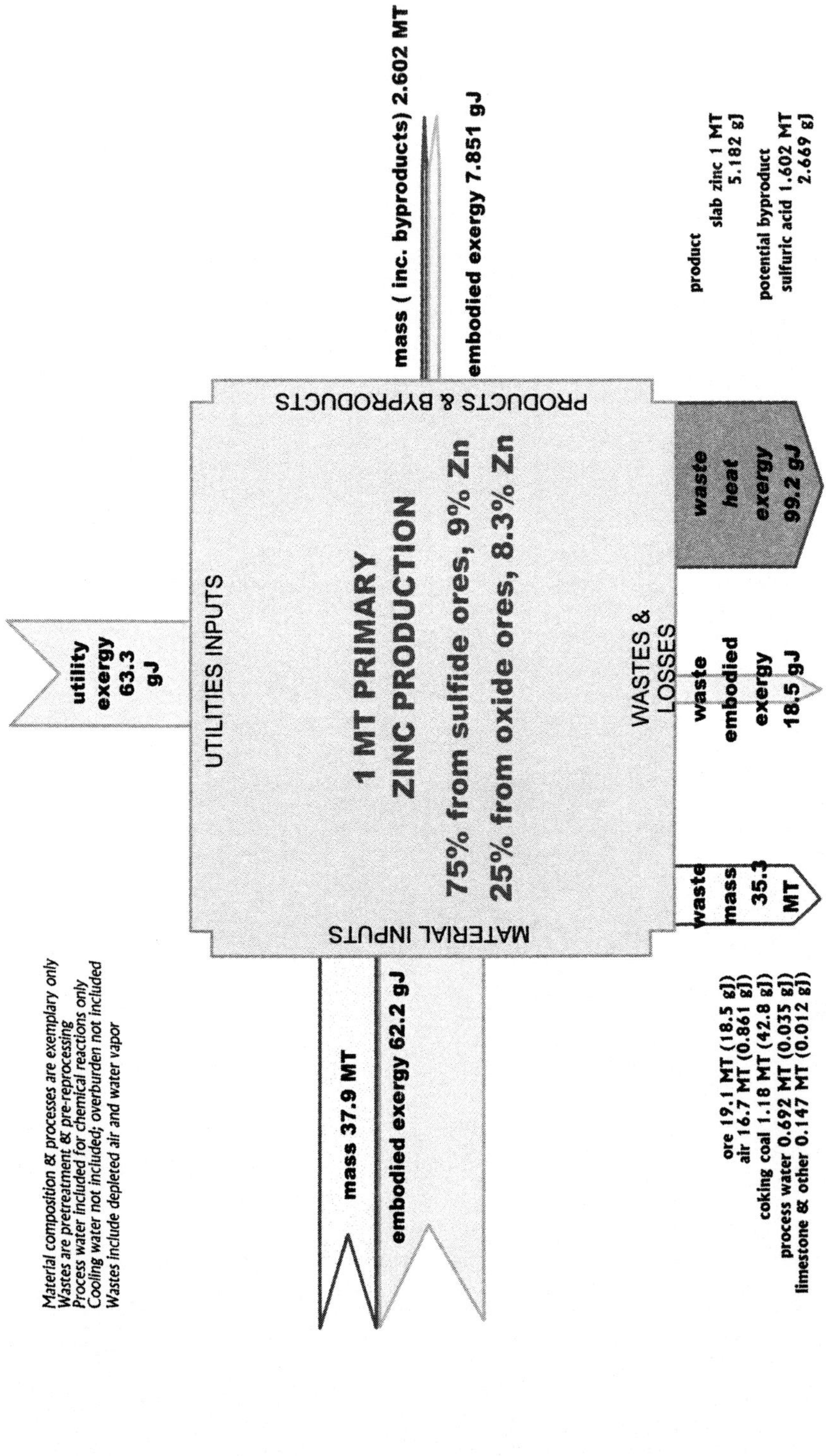

Figure 4.6: Overall unit zinc mass and exergy flows.

exergy) in fuels and electricity required to produce a metric ton of concentrate is 7.6 GJ, while 50 GJ is required to produce a metric ton of slab zinc or zinc alloy (Boustead and Dove 1999, Tables 6 and 14). These figures are significantly lower (and more recent) than our own calculations. The SX-EW process is also gaining in importance. Another source suggests that the fuel and utility exergy requirements for zinc production is nearer 37 GJ/t (Forrest and Szekely 1991).

4.5.5. *Recycling old zinc scrap*

According to the IZA there are 50 secondary zinc operations in the world. Recycling processes are quite varied. New scrap is usually just remelted. In the case of mixed non-ferrous shredded metal scrap the zinc is separated by hand or magnetically before being re-refined by retorting. A major and growing source of secondary zinc is flue dust from secondary electric arc furnaces for iron production, using galvanized iron scrap. For further discussion see *Section 4.5*.

4.6. Zinc sources and uses (WWW.zincworld.org) (Boustead and Dove 1999; ILZSG 2000b; ILZSG 2001; IZA 2001).

World mine production of zinc has risen dramatically since World War II, from a low point of 1.536 MMT in 1946 (RoskillCu 1990) to a peak (at 8.643 MMT) in 2000. There were intermediate peaks in mine production in 1977, 1985, and 1992 (ILZSG 2001). Primary zinc metal production and refinery consumption follow roughly on the same track, with some year-to-year adjustments for stock changes. For example, production in 2001 was 8.928 MMT and consumption was 8.799 MMT.

Regional differences are interesting. The Americas (North and South) accounted for 44 percent of mine production, 22 percent of smelter-refinery production and 26 percent of consumption (largely in the US). By comparison, Asia-Pacific accounted for 40 percent of mine production, 44 percent of production and 40 percent of consumption. Europe accounted for only 12 percent of production, but 32 percent of production and consumption.

Refined zinc metal consumption in recent years has been slightly higher than total metal production, perhaps because a significant amount of secondary zinc is used directly in alloys (e.g. brass) regardless of source material, whether concentrates, residues, slags or scrap. (Brass consumption has grown significantly since 1975.) On the other hand the situation was reversed from 1988 through 1993 (production exceeded consumption), and prior to that for many years the two curves were almost identical.

Galvanizing (and hot dipping) steel now accounts for about 65–70 percent of zinc consumption, and is increasing at an annual rate of about 3.5 percent p.a. There were more than 340 plants producing zinc coated steel products in 1999, mostly by the hot dip process (IZA/ILZA 2000). Demand for galvanized steel has more than doubled since 1975.

Brass accounted for 18 percent of zinc consumption in 1998, while other zinc alloys accounted for 13 percent. Zinc chemicals, mainly the oxides, accounted for around 8 percent of consumption in 1998. Semi-manufactures (e.g. roofing and rainwater gutters) and zinc die castings for many purposes accounted for much of the remainder. On a sectoral basis, 48 percent of zinc goes to the construction industry, 23 percent to transport, 10 percent to machinery/equipment, 10 percent to consumer durables and 9 percent to infrastructure.

Automobiles, on world average, used 10.2 kg of zinc, of which 4.9 kg was die castings, 3.2 coating for steel used in the body and frame and 1 kg was zinc oxide in tires. Cars produced in the US consumed about 13.2 kg of zinc, on average.

The statistics presented by ILZSG imply that very little of the current global refinery production is recycled from old scrap since mine production and smelter production are nearly equal. However according to the USGS, the global recycling rate for 1999, overall, was 30 percent, meaning that 30 percent of refinery output (2.9 MMT for Europe in 1996) was from scrap. Most of this must have been new scrap, which is recycled in the same year. Also, a significant fraction of zinc output is in chemicals (mostly produced from scrap).

The recycling picture for old scrap is quite confused. On the one hand, based on physical properties, most metallic zinc is theoretically recoverable, even after very long periods in use. According to the ILZSG (2000) the recovery efficiency for zinc – meaning the fraction of old scrap available for recycling that is actually recycled – is 80 percent. This is consistent with a very long lifetime of zinc products in use, as much as 100 years. It is also implausible.

For what it is worth, the picture presented by the ILZSG for 1996 was as follows: Primary production was 7.4 MMT of which 6.6 MMT was from mines and 0.8 MMT from scrap feed. Secondary production was 2.1 MMT plus 0.8 MMT scrap feed into primary smelters, resulting on the 2.9 MMT 'recycled' zinc noted above. Of the recycled zinc, 1.4 MMT was derived from end-of-use products (old scrap) and 1.5 MMT was new scrap recycled from manufacturers and the remainder. Gross consumption, before new scrap was 9.6 MMT, leaving 8.1 MMT incorporated in new products for consumption (including 0.1 MMT from stockpiles.).

However, it must be pointed out that the US Bureau of Mines estimates for these recycling relationships are very different. According to Sibley et al. the recycling efficiency for old scrap zinc for 1993 (in the US) was only 12 percent and the recycling efficiency in that year was 21 percent (Sibley et al. 1995). We suggest that the discrepancy between ILZSG and USBM/USGS figures is much too large to be ignored. If the ILZSG has firm data to back up its optimistic assumptions, these data should be published. We have found no consistent long term trends on zinc recycling from any source.

Zinc is used for many purposes, both as metal and as chemicals. Metallic uses include brass (an alloy with copper), zinc castings for a variety of hardware items, sheet zinc for various purposes including coins, primary cells, and zinc coating (galvanizing) for iron and steel to provide corrosion resistance. Chemical uses are also quite diverse. Zinc chemicals of importance include the oxide, the sulfate and the chloride. Other zinc chemicals are mostly made from the oxide. The oxide is used in large quantities by the tire industry as a pigment. The sulfate was formerly used as a white pigment for exterior paint (largely replaced by titanium dioxide nowadays), among other purposes. The chloride has medicinal uses, as well as being a fertilizer and animal feed additive.

Many current uses of zinc – especially as chemicals – are apparently dissipative in practice, even though in principle most of the metallurgical uses of zinc are recoverable in principle. However, as will be noted subsequently, there is a long way to go before this recycling potential is implemented on a wide scale. Few if any uses of zinc are non-substitutable, although its use as a corrosion-resistant coating material for iron and steel comes close, as does its use in primary batteries (dry cells).

There is a 'long shot' potential market for zinc in rechargeable nickel-zinc batteries for hybrid or electric vehicles. This newly commercialized battery type (which competes with so-called nickel-metal-hydride or NiMH batteries) is one of the possible successors of the lead acid battery that currently dominates the market. The reason is that zinc provides the highest specific energy, the lowest cost and the lowest environmental impact of any potential anode material (Coates et al. 1999). The nickel-zinc battery has an energy density at least double that of the traditional lead acid battery (i.e. 50–60 watt-hr/kg) along with performance almost equivalent to the nickel-cadmium battery (which it could conceivably also replace). Thus, while most industry experts are currently skeptical, one cannot rule out the possibility that nickel and zinc may effectively substitute for some major markets for lead and cadmium within the next twenty to thirty years.

Since cadmium is a byproduct of zinc mining, and will therefore be produced as long as zinc is mined, this suggests an emerging dilemma. Up to now, nickel cadmium batteries have constituted a use for cadmium which, in principle, is amenable to recovery and recycling. Other past and possible uses of cadmium are less environmentally friendly.

The US Bureau of Mines has carried out a more elaborate study of zinc uses and losses during the period from 1850 through 1990 (Jolly 1992) During that period, during which the US was the world's largest producer and consumer, cumulative inputs from domestic ores were 56 million metric tons (MMT), plus 13 MMT of imported concentrates. Domestic refinery output – including secondary recovery of 4 MMT – was 59 MMT and zinc imported as refinery products was 13 MMT. Inputs to manufacturing were 73 MMT (of which 1 MMT was imports of semi-manufactures) and 3 MMT was lost. Finally, end-use products placed in use was 73 MMT, of which 71 MMT was from domestic manufacturers and 2 MMT was imported. Of this, 4 MMT was recycled old scrap, 45 MMT was apparently dissipated to the environment and 23 MMT was still in use. Major flows and losses by source were estimated as follows:

– mining and milling	56 MMT of which 8 MMT lost in tailings
– smelting and refining	61 MMT of which 6 MMT lost in slag and emissions
– manufacturing	73 MMT of which 3 MMT lost in process waste
– end-uses	73 MMT of which 46 MMT dissipated and 4 MMT recycled

The cumulative total of zinc inputs to the US economy over that period, according to the author, was about 86 MMT while the inventory of zinc in products still in use in 1990 was estimated to be about 23 MMT, or slightly over 26 percent.The author of the study emphasized that the end-use losses may have been over-estimated to some extent, because of lack of data on some re-use and recycling routes.

Corresponding figures for the world as a whole were also estimated as follows: mining and refining flux 350 MMT, with losses of 80 MMT. Total refined: 285 MMT (including 15 MMT recycled scrap). Manufacturing losses were 10 MMT. End use was 275 MMT, of which 110 MMT still in use, 15 MMT recycled as old scrap and 150 MMT dissipated (Jolly 1992). The fraction of zinc still in use – mainly in brass and galvanized steel – appears to be around 31 percent.

4.7. Lead and zinc wastes and emissions

Tailings from lead and zinc mining and milling operations are one major source of wastes, as in the case of copper. Quantitative data are not readily available outside the US. One small indication might be that tailings from US lead-zinc mines during the decade 1970–1981 totaled about 180,000 metric tons. (The tailings included chemicals used for concentration by flotation. These chemicals differ significantly from one mine to another, based on the ore. See *Table 2.2.* for a list of the chemicals used for copper flotation.)

Lead and zinc are both relatively volatile metals, as compared to copper. As a consequence, there are significant atmospheric emissions of these metals associated with all kinds of combustion processes. It is estimated that 4 percent of lead in primary production is currently lost to the environment, along with 2 percent of secondary production and 1 percent of lead used in manufacturing (Lave et al. 1995). Over the 50 year period prior to 1992, the US Bureau of Mines estimated that 6.5 percent of primary production, 3.4 percent of secondary production and 1.1 percent of lead used in manufacturing were released to the environment, mostly into the atmosphere (Woodbury et al. 1993).

Non-ferrous metallurgy was the second biggest global source of atmospheric emissions of lead in the mid 1990s (14.8 kMT/y) and by far the biggest source of emissions of zinc (40.9 kMT/y) according to a recent study (Pacyna and Pacyna 2001, Table 15). The other major source of atmospheric emissions, for both metals, was fossil fuel (coal) combustion: 11.7 kMT/y for lead and 9.4 kMT/y for zinc (ibid.). Asia was the dominant source in both cases (ibid.).

By comparison, metallic lead releases in the US reported to EPA's Toxic Release Inventory (TRI) actually increased 9.1 percent from 118 kMT in 1992 to 129 kMT in 1994 (INFORM 1995, Tables 5.22 and 5.23).[3] Zinc (fume and dust) releases also increased, from 47 kMT (1992) to 51.5 kMT in 1994 (ibid.). Zinc compounds released increased by nearly 38 percent, from 217 kMT to 293 kMT during the same period. The TRI figures cited above refer to total mass, and to all media, not lead or zinc content to the atmosphere alone. However, even adjusting crudely for this, TRI estimates appear to be significantly larger than the Pacyna estimates for the world as a whole, although TRI usually underestimates.

Alkyl lead anti-knock additives are virtually instant emissions. At least 90 percent of the lead in the fuel is in the exhaust gas (the remainder being trapped in valves, cylinder rings and other internal engine parts. At peak, c. 1970–1973 alkyl lead emissions in the US were close to 240,000 tpy (Woodbury et al. 1993, Figure 4). By 1988 this had fallen to less than 20,000 tpy. According to another authoritative source, total lead emissions in Europe (13 countries, including European USSR) rose from 54 kMT/y in 1955 to a maximum in 1975 at a little over 138 kMT/y. Since then, emissions have declined fairly sharply, due mainly to the phase-out of lead anti-knock additives to motor fuels, to around 100 kMT in 1980, and about 60 kMT in 1990 (Olendrzynski et al. 1995). By 1995 lead emissions from fuel additives had fallen to 19.5 kMT in Europe, and about 13.1 (10.4–15.8) kMT in North America. The biggest source by far was Asia 36.7 (33–44.3) kMT, with Africa and South America accounting for most of the rest in that order. The global total for 1995 was about 89 (76.6–100.8) kMT (Pacyna and Pacyna 2001, Table 4).

A global summary of lead emissions for 1983 and 1995 is shown in *Table 4.1* and for zinc in *Table 4.2*. A more detailed breakdown for Europe in 1992–1993, in metric tons per year (tpy) by source, is shown in *Table 4.3* (Pacyna 1998). In the case of lead, 70 percent of the total emissions in both years is attributable to motor fuel additives. The picture is similar for zinc, except that the peak production (for Europe) was nearer 1965 at about 110,000

---

[3] For mysterious reasons – perhaps to magnify the numbers – TRI data are reported in pounds. We have converted pounds into metric tons at the rate of 2,204 lb/metric ton.

Table 4.1: Global lead emissions estimates 1983 and mid-1990s (metric tons)

Source	1983 to air	1983 to water	1983 to soil	Mid-90s to air
Coal combustion: electric utilities	775–4650	0.24–1.2	45–242	11690
Coal combustion: domestic + industry	990–9900	0.24–1.2		11690
Oil combustion: electric utilities	232–1740	0.24–1.2		11690
Oil combustion: domestic + industry	716–2150	0.24–1.2		11690
Wood combustion				
Metals mining	1700–3400	0.25–2.5	130–390	14815
Copper and nickel production	11050–22100	1.0–6.0	195–390	14815
Lead production	11700–31200	1.0–6.0	195–390	14815
Zinc and cadmium production	5520–11500	1.0–6.0	195–390	14815
Iron and steel production	1065–14200	1.4–2.8		2926
Secondary non-ferrous production	90–1440			
Cement and other manufacturing	18–14240	2.9–26	4.1–11	268
Fuel additives (TEL, TML)	248000			88.739
Municipal waste incineration	1400–2800			821
Sewage sludge incineration	240–300			821
Phosphate fertilizers & peat for agriculture	55–274		0.42–2.3	
Municipal waste and other urban	3900–5100		18.5–66	
Sewage sludge & organic wastes		2.9–16	2.8–9.7	
Atmospheric deposition		87–113	202–263	
Domestic wastewater		1.5–12		
Agriculture and food wastes			1.5–27	
Animal wastes, manure			3.2–20	
Logging and other wood wastes			6.6–8.2	
Misc. wastage of commercial products			195–390	
Total emissions	288700–376000	97–180	808–1893	119259
Median emissions	332350	138	1350	

Sources: 1983 data adapted from Nriagu and Pacyna (1988).
1990s from Pacyna and Pacyna (2001).

tpy. By 1975 the emissions rate was down to about 80,000 tpy. Again, progress was more rapid after that, especially between 1975 and 1980 when the 50,000 tpy level was reached.

A considerably lower set of numbers for lead emissions in 1990 (40,400 tpy) was estimated by the Dutch research group TNO for a group of 38 countries, including western Europe plus eastern Europe and the former USSR west of the Urals (Berdowski 1998), using a different methodology. The average lead emission per capita for this group of countries was 50 gm per year (gpy). The highest by far was Luxemburg (200 gpy), followed by Spain (120), Ukraine (73), Belgium (68), Russia (65), Sweden (60), Yugoslavia (53), France (48), UK (45), Poland (40), Germany (35), Italy (33), and the Netherlands (20). The correlation with automobile density and metallurgical activities is obvious.

Table 4.2: Global zinc emissions estimates 1983 and mid-1990s (metric tons)

Source	1983 to air	1983 to water	1983 to soil	Mid-90s to air
Coal combustion: electric utilities	1085–7750	6–30	112–484	9417
Coal combustion: domestic + industry	1485–11880	6–30	112–484	9417
Oil combustion: electric utilities	174–1280	6–30		9417
Oil combustion: domestic + industry	358–2506	6–30		9417
Wood combustion	1200–6000			
Metals mining	310–620	0.02–6.0		40872
Copper and nickel production	4250–8500	2–20	310–620	40872
Lead production	195–468	2–20	310–620	40872
Zinc and cadmium production	46000–82800	2–20	310–620	40872
Iron and steel production	7100–31950	5.6–24		2118
Secondary non-ferrous production	270–1440			
Cement and other manufacturing	1780–17800	25–145		2670
Municipal waste incineration	2800–8400			1933
Sewage sludge incineration	150–450			1933
Phosphate fertilizers & peat for agriculture	1370–6850		0.26–1.1	
Municipal waste and other urban	1724–4783		25–121.5	
Sewage sludge & organic wastes	2.6–31		18–57	
Atmospheric deposition		21–58	49–135	
Domestic wastewater		15–75		
Agriculture and food wastes			12–150	
Animal wastes, manure			150–320	
Logging and other wood wastes			13–65	
Misc. wastage of commercial products			310–620	
Total emissions	70250–193500	77–375	1193–3204	57010
Median emissions	131800	226	2200	

Sources: 1983 data adapted from Nriagu and Pacyna (1988).
1990s from Pacyna and Pacyna (2001).

Table 4.3: Atmospheric emissions of lead and zinc in Europe, 1992–1993 (metric tons per year)

Source	Lead	Zinc
Stationary fuel combustion	3555	6400
Non-ferrous metallurgy	8251	13775
Iron and steel production	3123	8477
Motor fuel additives	41922	–
Waste disposal (incineration)	1020	3843
Total	58130	32947

Source: Pacyna (1998).

Even more detailed data have been developed for 1996, covering the same territory and distinguishing between emissions and deposition (Gusev 1998). In the case of lead, 76 percent of depositions were on land within the territory, 18 percent on water and 6 percent outside the territory (mainly south and east). See *Table 4.4*. Lead deposition density (kg/km^2-yr) in 1996 was

Table 4.4: European lead deposition in 1996 (metric tons)

	Emissions	Depositions	
		Total	Own
Russia	7266	6623	5183
Spain	4674	2623	2369
France	4414	3132	2302
Ukraine	3878	2610	2008
United Kingdom	2703	1076	850
Germany	2347	1856	1092
Italy	1642	1087	835
Poland	1372	1372	777
Belarus	736	724	321
Belgium	716	311	149
Hungary	639	459	304
Portugal	631	314	235
Yugoslavia	597	419	305
Romania	585	673	327
Sweden	537	463	183
Switzerland	520	361	239
Greece	505	267	164
Croatia	466	357	237
Czech Republic	337	328	122
Bulgaria	317	312	151
The Netherlands	266	213	50
Lithuania	246	201	68
Norway	226	265	80
Latvia	218	180	58
Austria	215	265	103
Finland	215	283	94
Macedonia FYR	210	157	95
Denmark	179	121	27
Estonia	171	113	38
Moldavia	168	105	41
Slovakia	166	216	62
Ireland	134	100	37
Slovenia	123	98	50
Luxembourg	74	25	7
Albania	33	73	15
Bosnia Herzegovina	9	127	6
Iceland	6	16	2
Cyprus	1	1	0

Source: Gusev (1998, p. 115).

highest in Luxemburg (11.1), followed by Belgium (10), Switzerland (8.5), Macedonia (6.2), Croatia, the Netherlands, France and Slovenia were between 5.5 and 6.0; Spain and Germany (5); Hungary, Slovakia, Poland,Ukraine, UK, Czech Republic, and Yugoslavia were all bunched between 4 and 4.5; Belarus, Italy, Portugal, Moldavia, Austria and Lithuania were bunched between 3 and 4; all the rest were below 3. Russia, with its large territory, received about 1.8 kg/km^2-yr but the three Nordic countries were even lower (< 1 kg/km^2-yr).

Significant progress has been made in reducing emissions from metallurgical operations in the OECD countries, especially Japan, US and Europe. The Boliden company of Sweden offers a particularly impressive example. Whereas the firm's output of copper, lead and zinc increased from 150,000 tpy in 1970 to about 250,000 tpy in 1995, its emissions of the three metals from mining and milling operations dropped from 160 tpy to about 5 tpy during the same period (UNEP/ICME 1996, p. 3). Detailed estimates of solid, liquid and airborne wastes for zinc mining and production for Europe have been presented by Boustead and Dove (1999).

Progress in less industrialized countries has been slower. For instance, in 1986, lead emissions from copper smelting operations in Peru and Chile were 2.65 kMT; from Zaire and Zambia another 1.82 kMT and from lead smelting in Australia 1.75 kMT, for a southern hemisphere contribution of 6.22 kMT, comparable to all of Europe (Boutron 1998, p. 161).

Total lead emissions to air from all sources for Europe c. 1995 were only 28.1 kMT and for North America they were was just 17.0 kMT (Pacyna and Pacyna 2001 Table 16). As a matter of interest, US releases of lead chemicals (total weight) reported to the EPA's Toxic Release Inventory (TRI) was 500 kMT in 1992; it fell to just over 400 kMT two years later in 1994 – a 19.6 percent drop (INFORM 1995). It must be noted that lead chemicals have ingredients other than lead, and releases for TRI purposes include all media (air, water and earth. Even so, the TRI numbers dwarf the Pacyna numbers. We cannot explain the discrepancy. Another more realistic estimate by the Council for Environmental Quality (CEQ) for US atmospheric lead emissions for 1995 was 4500 MT (4.5 kMT) (cited by Newell 1998 Table 4.18).

As of the mid 1990s motor fuel additives still accounted for 74 percent of global atmospheric lead emissions of 119.3 kMT (median), much of it in Asia (Pacyna and Pacyna 2001 Table 15). The global total was down, however, from 332.3 kMT (median estimate) in 1983, mostly due to sharp reductions in the use of fuel additives in Europe and North America (Nriagu and Pacyna 1988).

The Pacyna estimates for North America have been cited several times above. It is useful to point out that alternative estimates have been made for the US and the world. These are compared in *Table 4.5* (Pacyna and Pacyna 2001 Table 10); USEPA (NAPET) cited by (Newell 1998). The NAPET data for air emissions differ radically from the Pacyna data in almost every case except for mercury and nickel. (Nickel emissions from metallurgical opera-

Table 4.5: Different estimates of metallic atmospheric emissions (metric tons per year)

Pollutant metal	USA c. 1996 Newell (1998)	North America mid 1990s Pacyna (2001)	Global mid 1990s Newell (1998)	Global 1983 Pacyna (2001)
Arsenic	12000	652	5011	18820
Cadmium	720.91	482	2983	7570
Chromium	352	3284	14730	30480
Copper	2564	2841	25915	35370
Mercury	220	215	2235	3560
Indium	–	6	45	25
Manganese	1500	2670	11047	38270
Molybdenum	–	579	2642	3270
Nickel	4945	11236	95287	55650
Lead	4530	17015	119259	332350
Antimony	–	375	1561	3510
Selenium	–	1086	4601	3510
Tin	–	531	3951	3790
Thallium	0.13	564	1824	5140
Vanadium	–	26660	240255	8600
Zinc (inc ore)	3680	5859	57010	131880

tions are concentrated in Canada and thus included in North America). The arsenic emissions cited by Newell and attributed to NAPET appear far too large in relation to total production of arsenic and may be a misprint. As for the TRI releases, they agree roughly with Pacyna for airborne copper and zinc, but are radically smaller – probably undercounted – for thallium and vanadium.

In general the Canadian estimates for non-ferrous smelters (Skeaff and Dubreuil 1997), cited by (Pacyna and Pacyna 2001) are considerably higher than the Pacyna estimates; in some cases the differences are dramatic. See *Table 4.6*. For instance, the Canadian study estimates copper emissions from Canadian copper-nickel smelters as 618 metric tons, whereas the Pacyna study estimated 149.7 metric tons. In the case of nickel, the two studies estimated 498 metric tons and 73.2 metric tons, respectively. For lead emissions the corresponding estimates were 995 metric tons and 4563.8 metric tons, respectively. Again, we cannot account for the discrepancies. However, it seems clear that none of the available emissions estimates can be accepted as definitive

A recent MIT PhD thesis has summarized a number of attempts to convert emissions data for lead into monetary damage figures (for the US) (Newell 1998, p. 117). The uncertainties are extremely great, not least because of the emissions data problems noted above. As mentioned in chapter 1, it is now standard practice in life cycle analysis (LCA) circles to treat lead as a numeraire (perhaps because it has been studied so intensively for many years). Other metals are then compares to lead, in terms of toxicological data and calculated exposures, based on the USES model (Guinée et al. 1996).

Table 4.6: Comparison of 1993 Canadian non-ferrous smelter emissions estimates from two sources (metric tons)

Trace metal	Pb industry		Cu+Ni industry		Zn industry		Totals	
	S & D	P & P	S & D	P & P	S & D	P & P	S & D	P & P
As	7.6	0.5	71.8	48.2	17.3	18.7	96.7	67.4
Cd	4.1	1.8	44.6	24.1	32.7	35.3	81.4	61.2
Cu	0.6	0.9	614.0	144.5	3.7	4.3	618.0	149.7
Hg	0.5	0.5	4.4	2.7	5.0	5.5	10.0	8.7
Mn	–	–	0.1	0.5	0.2	–	0.3	0.5
Ni	–	0.9	497.0	72.3	0.9	–	498.0	73.2
Pb	91.6	35.5	652.0	144.5	252.0	273.8	995.0	453.8
Sb	3.5	3.6	1.2	4.9	8.5	9.4	13.2	17.9
Se	–	1.8	2.9	4.9	0.1	7.2	3.0	13.9
Zn	7.6	3.6	249.0	96.3	972.0	1056.1	1228.0	1156.0

Sources: S & D = Skeaff and Dubreuil (1997).
P & P = Pacyna and Pacyna (2001)

Evidently copper and zinc are very low on the monetary scale compared to lead, arsenic and cadmium. Damages due to airborne lead emissions, via all exposure routes, range from \$440/kg to \$44,000/kg (Chernick 1992). For purposes of his own analysis Newell uses the geometric mean between \$440 and \$44,000, viz. \$4400/kg. This is the number used to calculate the damage estimates for other metals given in *Table 4.7*. *Table 4.7* also display comparative human toxicity potentials for these metals.

The fragility of these monetary damage estimates is hard to overstate. For example, several runs of a sophisticated air quality model led to estimates ranging from \$605/kg to \$7060/kg, for lead, depending on population density and local meteorological conditions (Rowe et al. 1995a; Newell 1998). a different approach is to equate damage costs to marginal costs of emissions control at lead smelters. This makes sense if (and only if) smelters abate to allowable US emissions levels for lead, and if these levels are socially optimal. In such a case marginal abatement costs would equal marginal abatement benefits. The benefits of abatement, which came out in one study to an (intermediate) value of \$1160/kg, can then be equated to the social costs of non-abatement (Zuckerman and Ackerman 1995).

4.8. Other byproduct metals

A number of other metals are obtained partly or largely from copper, lead and/or zinc mining. *Table 4.8* shows amounts and percentages for these in the US in 1968 and 1982. More recent data has been suppressed by the US government as a result of consolidations in the mining industry which reduce the number of competitors. Some of these metals (e.g. gold and silver) are

Table 4.7: Human toxicity potentials (HTP) and monetary damage estimates for metals emissions in the US c 1996

	Hertwich: Human toxicity potentials		Newell: Monetary damage estimates			
	Cancer HTP (relative to benzene equivalent emissions to air=1)	Noncancer HTP (relative to toluene equivalent emissions to air=1)	US emissions metric tons/year	Emissions data source	Damage cost estimate $/kg	Calculated damage million $/year
To air						
Arsenic	3 300	220 000	12000	(USEPA 1996a)	$2 758	$34 000
Cadmium	83	5 000 000	720	(USEPA 1996a)	$1 510	$1 100
Chromium	170	6 400	270	(USEPA 1996a)	$32 200	$8 600
Copper		28 000	2500	(USEPA 1996b)	$23	$56
Lead	35	1 500 000	4500	(CEQ 1996)	$4 400	$20 000
Mercury		19 000 000	220	(USEPA 1996a)	$1 904	$430
Nickel	3.6	8 300	4900	(USEPA 1996a)	$644	$3 100
Thallium		32 000 000	0.13	(USEPA 1996b)	$27 800	$3.70
Zinc ore		500	3000	(USEPA 1996b)	$0.04	$0.12
To water						
Arsenic	820	52 000			$3.35	
Cadmium	$<10^{-20}$	360 000	0.91	(USEPA 1996b)	$8.54	$0.008
Chromium	$<10^{-20}$	690	82	(USEPA 1996b)	$0.61	$0.05
Copper		17 000	64	(USEPA 1996b)	$0.07	$0.045
Lead	2.6	110 000	30	(USEPA 1996b)	$17.10	$0.51
Mercury		19 000 000	0.15	(USEPA 1996b)	$1 182.00	$0.17
Nickel	$<10^{-20}$	67	45	(USEPA 1996b)	$4.14	$0.18
Thallium		7 100 000				
Zinc		36	680	(USEPA 1996b)	$0.004	$0.003

Sources: Extracted from Hertwich et al. (2001) Appendix 2 and Newell (1998) Table 4.18.

Table 4.8: Byproduct relationships in metals mining in the United States in 1968 and 1982

	unit	Quantity 1968	Quantity 1982	% of domestic production 1968	% of domestic production 1982
Copper Mines					
Copper	mt	1 062 000	1 114 901	98.1	97.2
Lead	mt	44	191	<0.1%	<0.1%
Zinc	mt	8 100	*W*	1.8	*W*
Arsenic	mt	2 820	*W*	100	100
Molybdenum	mt	11 100	12 527	25.5	35.5
Nickel	mt	1 820	*W*	13.4	*W*
Rhenium	mt	2.2	*W*	100	100
Selenium	mt	*W*	243	100	100
Sulphur	mt	520	614	5.4	6.3
Tellurium	mt		*W*	*W*	100
Silver	1000 oz	9 551	9 566	29.2	23.8
Gold	1000 oz	406	235	27.5	16
Palladium	1000 oz	*W*	7	*W*	100
Platinum	1000 oz	5	1	45.9	100
Lead Mines					
Copper	mt	7 150	7 941	0.7	0.7
Lead	mt	253 000	474 554	78.8	92.6
Zinc	mt	50 800	63 680	10.7	21
Antimony	mt	172	163	54.3	41.6
Bismuth	mt	312	*W*	100	100
Sulphur	mt	*W*	101 568	*W*	1
Tellurium	mt	16	*W*	29.7	*W*
Silver	1000 oz	5 573	2 245	17	5.8
Gold	1000 oz	75	*W*	5.1	*W*
Zinc Mines					
Lead	mt		1 065	19.2	0.2
Cadmium	mt	1 715	*W*	100	*W*
Germanium	mt	22	*W*	100	*W*
Manganese	mt	7 257	*W*	16.7	*W*
Sulphur	mt	290 000	*W*	3.0	*W*
Thallium	mt	1.2	*W*	100	*W*
Silver	1000 oz	4 136	*W*	12.6	*W*
Gold	1000 oz	21	*W*	1.4	*W*
Indium	1000 oz	230	*W*	100	*W*
Silver Mines					
Copper	mt		20 616		1.8
Lead	mt		20 001		3.9
Gold-Silver Mines					
Copper	mt		644		0.1
Lead	mt		248		<0.1

Source: USBuMines annual (1968 and 1982).

economically very important. Others are important because of their toxicity (e.g. arsenic, cadmium, thallium) and still others (e.g. antimony, bismuth, selenium) have small but significant current market niches or (like bismuth, germanium, indium, rhenium, tellurium, thallium) they may have future importance in specialized electronic applications.

It is worth noting, once more, that consumption is a potentially misleading term, since all forms of consumption result in emissions, when the products in question finally become wastes. It is of some possible interest to display consumption-related emissions factors for different metals associated with various uses. An emissions factor, for our purposes, is the fraction of the metal in use that is lost to the environment over a ten-year period. Emissions factors for many of the heavy non-ferrous metals, compiled by one group, are shown in *Table 4.9.*

4.8.1. *Antimony (Kesler 1994)*

Antimony is classed as a chemical metal. It exists in nature as the simple sulfide (stibnite) and as complex copper-lead-antimony sulfides (tetrahedrite and jamesonite), generally precipitated from brines. It is found in a few deposits with arsenic or mercury but is produced mainly as a byproduct of lead-zinc-silver smelting. Annual production is around 60 kMT. It was formerly used as an alloying agent with lead in automotive batteries, but this use has declined markedly in recent years. The major current use is in chemical (trioxide Sb_2O_3) form, as a fire retardant additive for plastics and textiles, primarily for children's clothing. It is more toxic than arsenic, although its current use has not apparently caused problems. There is some atmospheric

Table 4.9: Estimated emission factors[a] arising from different types of metals consumption

	Ag	As	Cd	Cr	Cu	Hg	Pb	Zn
Metallic use	0.001	0.001	0.001	0.001	0.005	0.05	0.005	0.001
Plating, coating	0.02	0	0.15	0.02	0	0.05	0	0.02
Paint, pigments	0.5	0.5	0.5	0.5	1.0	0.8	0.5	0.5
Electrical tubes, batteries	0.01	0.01	0.02	na	na	0.2	0.01	0.01
Other electrical equipment	0.01	na	na	na	0.1	na	na	na
Chemical uses in final product	0.4	0.05	0.15	0.05	0.05	na	0.75	0.15
Chemical uses not in final product	1	na	1	1	1	1	1	1
Agricultural biocides	na	0.5	na	na	0.05	0.8	0.05	0.05
Non-agricultural biocides	na	0.8	na	1	1	0.9	0.1	0.1
Medical, dental	0.5	0.8	na	0.8	na	0.2	na	0.8
Miscellaneous	0.15	0.15	0.15	0.15	0.15	0.5	0.15	0.15

Source: Karlsson et al. (1997) Table 7.2, p. 46.

[a] The emission factor defined here is the loss of the metal to the environment over a period of ten years.

pollution resulting from fossil fuel combustion (coal) and some water pollution has been observed near lead-zinc-silver mines.

4.8.2. *Arsenic (Kesler 1994; Loebenstein 1994; Sancha 2000)*

Arsenic is a white transition (semi-conducting) metal with a low boiling point, and high vapor pressure. It is classed as a chemical metal, since most of its uses are as chemicals. It has no known biological function, although small amounts have been added to chicken feed. Most of its compounds are toxic, and most arsenic uses are based on its extreme toxicity.

Arsenic has been recovered primarily as arsenic trioxide (As_2O_3), a byproduct of copper smelting, starting in 1901. In the US arsenic averages 6.5 kg per metric ton of copper. The US was, for a long time, the primary global producer, with peak production of 24.88 kMT in 1944, from six producers. However by 1960 only two producers were left, Anaconda (Montana) and ASARCO in Tacoma, Washington. The latter was closed in 1985, mainly due to the high cost of bringing the plant into compliance with state and federal government regulations.

Arsenic (as trioxide) is recovered from copper, lead, gold and silver ores, mainly from smelter flue dusts. The arsenic is concentrated in flue dust because of its high volatility. Purification is subsequently accomplished by stepwise distillation. Arsenic is not recovered from mill tailings or slag. Global production of the trioxide in 1999 was 38.8 kMT, increasing to 40 kMT (est.) in 2000 according to the USGS. China is the biggest producer (16 kMT) followed by Chile (8 kMT) and Ghana (5 kMT).

The US is by far the largest global consumer of arsenic: 30 kMT in 1998, 2 kMT in 1999 and 34 kMT (est. in 2000). All arsenic consumed in the US today is entirely imported, mainly from China (49 percent), Chile (30 percent), and Mexico (7 percent).

Historical uses of arsenic are quite diverse, including some medical and cosmetic uses. However the major use prior to the 1960s was in insecticides, notably lead arsenate (used against pests of vegetables and fruits, especially apple orchards), calcium arsenate (against the cotton boll weevil), copper aceto-arsenate ('Paris green', also once used as a dye) and sodium arsenite. Agricultural uses accounted for 90 percent of arsenic use in the 1940s and continued strong until the mid-60s. The use of arsenic acid for desiccating cotton leaves (to facilitate mechanical cotton picking) began in 1965 and continued until 1992. Three other arsenical herbicides, cacodylic acid, MSMA and DSMA, became popular for a time, but were phased out by the early 1990s.

Of the 22,400 t. of imports in 1989, herbicides and other desiccants (mainly for cotton) accounted for 4900 t. in 1989, sharply down from a peak of 20,600 t. in 1974. Meanwhile non-agricultural uses, mainly for wood preservatives, rose sharply, especially after 1980. The sum total of non-agricultural arsenic containing products consumed from 1969 through 1989 (mainly wood preservatives) accounted for 175,000 t. In 1989 wood preservatives (known as chromated-copper-arsenic or CCA) accounted for 70 percent of US consumption. As of 2000 this use accounted for more than 90 percent of arsenic consumption. Since 1989 agricultural uses in the US have essentially disappeared as seen in *Figure 4.7*.

Arsenic metal imports for consumption in 1999 were 1.3 kMT, and 1.0 kMT in 2000. The metal is used as an alloying additive (for non-ferrous alloys, mainly for lead-acid batteries). Former uses in glass-making and ceramic glazes are mostly phased out.

Atmospheric emissions to the air in the early part of the 20th century were very large. In 1917 alone over 10 kMT (arsenic content) were emitted to the air from smelting ores from just two sites, Butte Montana (Anaconda) and Tintic district (Utah). Meanwhile around 10 kMT per year are discarded into the environment from copper processing, although atmospheric emissions from US smelters were only around 0.8 kMT in 1989. As noted in the copper chapter, about 2.4 kMT were discarded in copper concentrator tailings, and 5.3 kMT were discarded in smelter slag. Another 1.4 kMT was embodied in exported concentrates.

Apart from copper processing and consumption-related wastes, significant amounts of arsenic are emitted by coal-burning and volcanos. A global summary of arsenic emissions for 1983 and 1995 is shown in *Table 4.10* (Nriagu and Pacyna 1988; Pacyna and Pacyna 2001). For a breakdown of atmospheric arsenic emissions for Europe in 1992–1993 see *Table 4.11*

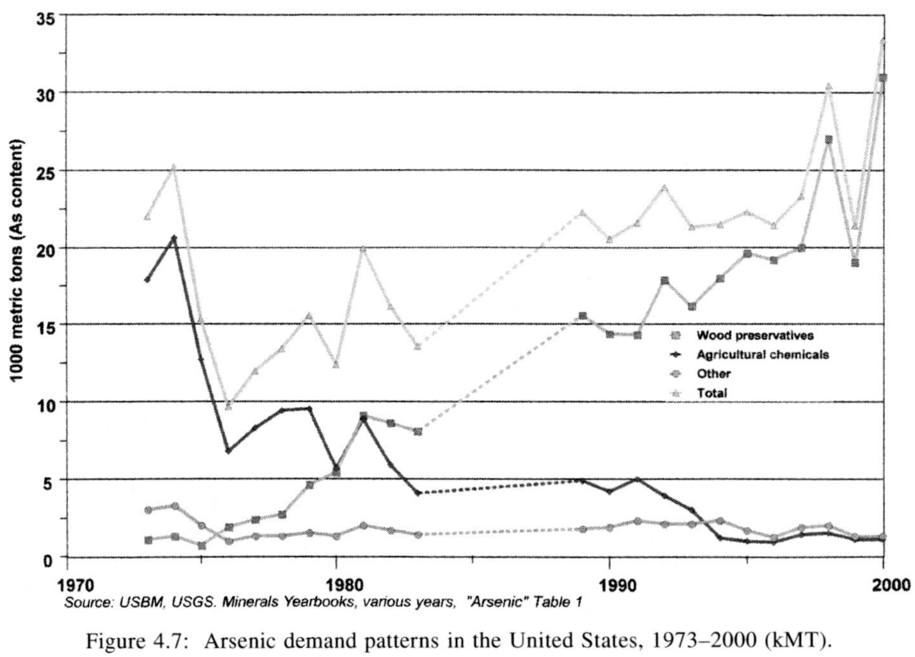

Figure 4.7: Arsenic demand patterns in the United States, 1973–2000 (kMT).

Table 4.10: Global arsenic emissions estimates 1983 and mid-1990 (metric tons)

Source	1983 to air	1983 to water	1983 to soil	mid90s to air
Coal combustion: electric utilities	232–1550		6.7–37	809
Coal combustion: domestic + industry	198–1980	2.4–14		
Oil combustion: electric utilities	5.8–29			
Oil combustion: domestic + industry	7.2–72			
Wood combustion	60–300			
Metals mining	40–80	0–0.75	7.2–11	3457
Copper and nickel production	8500–12750	1.0–13	4.5–9	
Lead production	780–1560			
Zinc and cadmium production	230–690			
Iron and steel production	355–2480	–		353
Secondary non-ferrous production	–			
Cement	178–890			268
Other manufacturing	1250–2800	1.2–12.8	0.01–0.21	
Municipal waste incineration	154–392			
Sewage sludge incineration	15–60			
Phosphate fertilizers & peat for agriculture	–		0.04–0.5	
Municipal waste and other urban			0.09–0.7	124
Sewage sludge & organic wastes		0.4–6.7	0.01–0.49	
Atmospheric deposition		3.6–7.7	8.4–18	
Domestic wastewater		3.0–15.3		
Agriculture and food wastes			0–0.6	
Animal wastes, manure			1.2–4.4	
Logging and other wood wastes			0–3.3	
Misc. wastage of commercial products			36–41	
Total emissions	12000–25630	12–70	64–132	5011
Median emissions	18820	41	98	

Sources: 1983 data adapted from Nriagu and Pacyna (1988).
1990s from Pacyna and Pacyna (2001).

Table 4.11: Atmospheric emissions of arsenic and cadmium in Europe, 1992–1993 (metric tons per year)

Source	Arsenic	Cadmium
Stationary fuel combustion	857.1	278.5
Non-ferrous metallurgy	1307.5	390.0
Iron and steel production	158.7	83.8
Waste disposal (incineration)	4.5	33.2
Other sources	253.2	110.0
Total	2581.0	895.5

Source: Pacyna (1998).

(Pacyna 1998). Other estimates of global anthropogenic releases into the atmosphere range from 24 kMT to 124 kMT, as compared to volcanic emissions estimated at 2.8 to 8.0 kMT (Loebenstein 1994).

Arsenic emissions have been particularly serious in parts of Northern Chile, where the copper ore has a high arsenic content. Three major copper mines are Potrerillos, Chuquicamata and La Negra, emitting – pretreatment – 8.2, 5.4 and 1.3 metric tons of arsenic into the air per day, respectively, or 5.4 kMT (Sancha 2000 Table 2). These three mines alone have a total capacity of 2.28 MMT, which is a substantial fraction of the total for Chile. Populations exposed, directly and indirectly amount to 195,000 people (ibid.). Abatement measures introduced prior to 1996 had cut arsenic emissions 12 percent at Potrerillos, 85 percent at Chuquicamata and 40 percent at La Negra. Further planned investments were expected to cut arsenic emissions by an additional total of 72 percent at Potrerillos and La Negra, and 87 percent at Chuquicamata (ibid., Table 11). Yet, even after these improvements, arsenic emissions to the atmosphere (from just those three mines) would still amount to 0.84 metric tons/day, or over 0.3 kMT. Presumably most of the arsenic is to be recovered for sale (mainly to the US).

Two rivers in Northern Chile, Rio San Salvador and Rio Loa have arsenic concentrations in excess of 1.5 mg/liter, while the Rio Vilama and Rio Toconce exceed 0.6 mg/l and half a dozen other rivers exceed 0.2 mg/l (ibid., Table 3). For the most part, this arsenic is probably leached from mine wastes. The safe limit, recommended by the World Health Organization is 0.01 mg/l. The Chilean government has invested significantly in water treatment facilities to improve the situation, but much remains to be done.

Arsenic emissions to the ground water are a newer and perhaps more serious problem in certain regions, especially in Bangladesh where many rural people now obtain water from tube wells. It is likely but not absolutely certain that this arsenic is of natural origin. There is growing epidemiological evidence that arsenic in water causes skin, lung, liver and bladder cancers. Citing evidence from a number of studies, including a long-term study of 40,000 villagers in southwestern Taiwan, the United States Environmental Protection Agency (USEPA) acted in March 2001 to reduce the acceptable standard from 50 micrograms per liter to 10 micrograms per liter. A 1999 report of the US National Academy of Sciences said that daily ingestion of water containing 50 micrograms per liter of arsenic would add 1 percent to the odds of dying of cancer (*Scientific American*, June 2001, p. 10). It is not certain that even 10 micrograms per liter is safe. Some experts want the standard reduced to 5 micrograms. On the other hand the newly appointed Administrator of USEPA, Christina Todd Whitman, has initiated a study to review the findings of the NAS study above cited, with a reported objective of weakening the standard to 20 micrograms per liter, because of the high cost of compliance with the proposed standard.

While not all arsenic pollution is associated with arsenic presence of copper mining, some is, notably in Arizona and northern Chile (where the copper ores have a high arsenic content). In a high-arsenic region of northern Chile it was found that 7 percent of all deaths were arsenic-related (Scientific American, June 2001, p. 11).

Overall, anthropogenic mobilization as of 1990 or so was already of the order of 10 times natural mobilization. If copper mine production and byproduct arsenic consumption continues to increase exponentially for the next half century or so, as the discussion in Chapter 2 suggests, the arsenic buildup – and mobilization – in the future will be much greater. This is a very serious concern.

4.8.3. *Bismuth (Kesler 1994)*

Bismuth is mainly associated with lead, molybdenum and tungsten. It has a very low melting temperature and a low boiling point, whence it is easily volatilized. It is recovered primarily from lead smelters. World production is about 3.2 kMT, which is worth $20 million. Recoverable reserves are estimated at 110 kMT. It has some pharmaceutical uses, notably (as carbonate) for soothing upset stomach. It also adds luster to cosmetics. Its main metallurgical use is as an alloying agent for lead (in batteries) and for castings that must fit very tightly at high temperatures (because it expands when it crystallizes from a melt). It may find a future use in lead-free solder. It is not generally regarded as environmentally hazardous, but there have been suspicions of a relationship with brain disease leading to a ban in France (1978) and a ban on cosmetic use in Austria (1986).

4.8.4. *Cadmium (Kesler 1994; Llewellyn 1994; ICDA 2001)*

Cadmium is a soft, malleable, ductile metal that is chemically similar to zinc (and generally found with zinc minerals), but is much more toxic. It is one of the most volatile of all metals (along with antimony, arsenic, lead, selenium, tellurium, tin, and zinc; only mercury is significantly more volatile). Its sulfides are brightly colored, which accounts for its major early usage for yellow and orange pigments. It averages between 0.1 ppm and 0.2 ppm in the earth's crust, but 200 ppm (up to 500 ppm) in zinc ore. Greenockite (CdS), the main cadmium mineral, is practically always associated with the zinc mineral sphalerite (ZnS), even when the latter is also associated with copper or lead. About 3 kg of cadmium are produced per metric ton of zinc.

Cadmium was first used in Germany (as yellow-orange pigment) in 1850 and first produced as a metal, also in Germany in 1890. It has been recovered as a by-product of zinc, lead or copper smelting in the US since 1907.

By 1917 there were six cadmium producers in the US. Peak domestic output occurred in 1967 at 5736 metric tons from 12 plants. Domestic production declined rapidly thereafter, especially after 1992, due to the closure of two domestic primary producers motivated partly by environmental concerns. US production in 2000 was 1890 metric tons, much of it secondary. Imports took up the slack. Worldwide production and consumption of cadmium increased until the mid 90s, mainly due to increasing demand for nickel-cadmium batteries, but has flattened since then. However, total apparent consumption of cadmium in the US now remains significantly lower than it was in the 1960s.

The cadmium content of zinc concentrates typically ranges from 0.3 percent to 0.5 percent. Very little of the cadmium is lost to the tailings. It is recovered in a cadmium-rich zinc-cadmium precipitate from the hydro-metallurgical process, and as zinc-cadmium flue dust from the retorting process. The next stage of recovery is to dissolve these materials in sulfuric acid, with further filtration and chemical separations, resulting in a cadmium sulfate solution. The cadmium is then precipitated by zinc dust to form a cadmium sponge, which is subsequently briquetted, remelted and cast into ingots. Recovery (in the US) is now estimated to be 95 percent. There are essentially no wastes at this stage, since all residues are recycled to the zinc smelters. (Some cadmium ends up in the slag.)

Cadmium is also present in trace amounts in iron ore, limestone and coal (up to 170 ppm in mid-continental US coals). In consequence, cadmium is also released into the atmosphere at iron and steel operations. Cadmium averages 500 ppm in electric arc furnace (EAF) bag-house dust. EAF bag-house dust has a high zinc content – around 18 percent – and can be sent to zinc smelters, where the cadmium can be recovered, in principle. As of 1985, however, only 15 percent of the EAF dust was reprocessed for zinc and cadmium recovery, while 75 percent was sent to landfills. By 1989 the recovery rate had increased to 20 percent, and the projection to 2000 was for 90 percent recovery, due to economic benefits and regulatory pressures. It has been estimated that iron and steel operations account for 16.7 percent of human exposure to cadmium, as compared to only 2.5 percent for cadmium production and the use and disposal of cadmium products (Van Assche 1998).

Cadmium is also released from coal burning power plants (especially in the ash). A mass balance for a modern Austrian power plant indicated that 96.3 percent of the cadmium in the coal was deposited on fly ash particles, which were mostly (up to 99 percent) captured by electrostatic precipitators and deposited in landfills. A small amount (0.4 percent) was captured by the flue gas desulfurization waste. However a significant fraction (4.2 percent) was emitted as vapor in the stack gases (Pacyna 1998, p. 47).[4] Cadmium is a contaminant of phosphate rock (5–15 ppm for North Carolina and Florida phosphate rock; 30–300 ppm for Idaho phosphate). Cadmium in phosphate fertilizer ranges from 2 ppm to 20 ppm. It is accumulating in topsoils, although no immediate hazard has been identified from this buildup. Fuel combustion

---

4 Unfortunately the numbers do not add up to 100 percent. However I would assume that the percentage allocated to stack gases is accurate and that the fraction remaining in fly ash is a calculated residual, in which case it should be 95.4 percent.

and fertilizer use have been estimated to account for 22 percent and 41.3 percent respectively, of human exposure (Van Assche 1998).

Cadmium release has been associated in the past with impure zinc used in galvanized iron, and in rubber tires (where it is a contaminant in the zinc oxide). It is released when these products wear. At one time tire wear was a significant source of cadmium in the environment (Davidson et al. 1976). However the cadmium levels in zinc used for galvanizing and zinc oxide pigment have declined considerably in the last decade (ICDA 2001). These sources of cadmium in the environment are no longer regarded as significant.

Consumption patterns have changed dramatically. As noted above, early uses (up to 1920) were primarily as pigments. Another early use was for dental fillings (cadmium 26 percent, mercury 74 percent), although this use was always very minor. Cadmium was used as a plating element after 1920 and as a substitute for tin during World War I. Cadmium anti-corrosion plating for steel was popular in the auto industry and became the main usage for many years. It accounted for 70 percent of consumption in 1945. This use continues on a small scale, though other plating materials such as zinc-nickel and zinc-cobalt have replaced some (but not all) cadmium plating. In 1980 plating still accounted for 34 percent of total Cd consumption; in 1989 plating was down to 30 percent, and 26 percent a year later. It is currently around 10 percent.

Meanwhile demand for pigments was declining more slowly. As recently as 1979 pigments accounted for 27 percent of cadmium consumption. It is now (2000) about 12 percent. Stabilizers and other additives for plastics (e.g. PVC) and synthetic products were at 12 percent in 1979; the current figure is 4 percent and such uses are being phased out. Use in special alloys also took only 1 percent in 2000, down from 4 percent in 1979. (A small amount of cadmium (0.5 percent–1.2 percent is a useful hardening agent for copper wire, since it does not adversely affect its conductivity. Cadmium is also used in some solders.) Small amounts of cadmium and cadmium compounds are also used in phosphors for TV screens, fluorescent lights, solar cells, catalysts and nuclear reactor controls. Another use, since the 1950s, was cadmium laureates and stearates as UV stabilizers for the plastic PVC. This use is being phased out due to regulatory concerns and the availability of satisfactory substitutes, such as Ba-Zn, Ca-Zn and organo-tins.

Rechargeable nickel cadmium batteries were discovered by Jungner in Sweden and Edison in the US shortly after 1900. They were introduced in some industrial applications in the 1920s. Sintered NiCad batteries were developed in Germany in the 1930s and used by German military aircraft during the war. The industrial applications dominated until two decades ago. Consumer applications in small appliances took off in the 1980s. Batteries accounted for only 1 percent of consumption in 1940–1945. This had risen to 16 percent by 1980, to 35 percent by 1989, 40 percent by 1990 and 75 percent by 1999 thanks to the popularity of portable laptop personal computers. Today consumer applications account for 80 percent of the battery business.

In view of the importance of rechargeable batteries to the cadmium market, a few comments on future battery technologies may be appropriate. Possible competition for nickel cadmium rechargeable batteries will arise from nickel metal hydride, lithium ion and lithium polymer chemistries. All three of these are under rapid development; experts now believe that the lithium batteries have the edge in future applications. There is only an outside chance that nickel zinc batteries will find greater use.

Cadmium is a component of one of the most promising candidates for high efficiency thin-film photovoltaic cells, namely the cadmium telluride (CdTe) cell. These cells are also amenable to low cost, high volume manufacturing. Systems for nearly complete recycling of these cells have been developed. The economics of recycling are favorable because of the high price of tellurium. Nevertheless, tellurium scarcity is likely to limit the maximum future use of such cells in relation to the availability of cadmium (Andersson 2000).

A detailed account of cadmium flows in the Federal Republic of Germany was carried out for the years 1980, 1986 and 1989 for the Enquete Commission of the German Bundestag on the "Protection of Humanity and the Environment" (Bätcher and Böhm 1994). The Enquete Commission report document is widely available, and need not be summarized in detail. The main points of importance are that cadmium production and consumption both declined during the 1980s, as some major uses were phased out. Domestic consumption fell from 1787 metric tons in 1980 to 886 metric tons in 1989, while recovery from scrap also declined from 270 metric tons to 120 metric tons. The amount of cadmium stored in various forms in Western Germany were estimated at between 9 kMT and 16 kMT, while 1.5 kMT/y goes to landfills. An interesting sidelight uncovered by the German study is that the cadmium entering the environment as an impurity in zinc (metal or oxide) declined during the period from 55 metric tons to only 1 metric ton. However, it should be noted that the German case is not necessarily representative, because of several special situations.

On the other hand, in 1991 25 metric tons of cadmium entered German topsoils as a contaminant in phosphate fertilizers, while a further 43 metric tons entered agriculture via other routes, including sewage sludge (2.4 metric tons) but primarily via atmospheric deposition. Studies have shown that one third of the cadmium remains in the soil (bound to clay particles) while the other two thirds are removed by leaching and erosion, ending up in ground water or rivers. Uptake into plants has not been shown to be a threat to human health, however. (See *Appendix B*)

Cadmium in the environment has attracted a great deal of attention in recent years. A summary of global cadmium emissions to the atmosphere for 1983 and 1995 is shown in *Table 4.12* (Nriagu and Pacyna 1988; Pacyna and Pacyna 2001). Compare with *Table 4.11*, which gave only atmospheric emissions for

Table 4.12: Global cadmium emissions estimates 1983 and mid-1990s (metric tons)

Source	1983 to air	1983 to water	1983 to soil	mid90s
Coal combustion: electric utilities	77–387		1.5–13.0	
Coal combustion: domestic + industry	99–495	2.4–14.0		691
Oil combustion: electric utilities	25–174			
Oil combustion: domestic + industry	18–72			
Wood combustion	60–180			
Metals mining	0.6–3.0	0–0.3	2.7–4.1	
Copper and nickel production	1700–3400			
Lead production	39–195		1.9–3.2	2171
Zinc and cadmium production	920–4600			
Iron and steel production	28–284			64
Secondary non-ferrous production	2.3–3.6	0.01–3.6		
Cement	8.9–534			17
Other manufacturing		0.6–4.3	0.08	
Municipal waste incineration	56–1400			
Sewage sludge incineration	3–36			
Phosphate fertilizers & peat for agriculture	68–274		0–0.041	
Municipal waste and other urban			0.88–7.5	40
Sewage sludge & organic wastes		0.08–1.3	0.02–0.35	
Atmospheric deposition		0.9–3.6	2.2–8.4	
Domestic wastewater		0.48–3.0		
Agriculture and food wastes			0–3.0	
Animal wastes, manure			0.2–1.2	
Logging and other wood wastes			0–2.2	
Misc. wastage of commercial products			0.78–1.6	
Total emissions	3100–12040	2.1–17	9.9–45	2983
Median emissions	7570	9.4	27.5	

Sources: 1983 data adapted from Nriagu and Pacyna (1988).
1990s from Pacyna and Pacyna (2001).

Europe. According to a more recent IIASA study of 26 European countries, cadmium emissions to the atmosphere peaked around 1960 at just over 2.6 kMT, declining slowly to 2.227 kMT in 1970, 1.948 kMT in 1975, 1.318 kMT in 1982 and just under 1 kMT in 1987. The big emitters by far were Poland and the western USSR, largely from poorly controlled soft coal combustion (Olendrzynski et al. 1995 Table 2).

A considerably lower figure for European cadmium emissions in 1990 (612 metric tons) was given by the European Emissions Inventory for 1990 (Berdowski 1998, p. 90). More recent and more detailed data for 1996 have been provided for both emissions and depositions, in 'extended Europe' (38 countries) as shown in *Table 4.13*. It is noteworthy that only 4 percent of the cadmium emitted in the region was deposited outside the region, but 16 percent was deposited on water surfaces. The latest estimate for Europe (c. 1995) is 362 metric tons (Pacyna and Pacyna 2001).

Table 4.13: European cadmium deposition in 1996 (metric tons)

	Deposition Emissions	Total	Own
Russia	109	103	81
Poland	92	65	53
Italy	60	28	27
Ukraine	54	42	30
Spain	37	19	18
Germany	31	25	14
United Kingdom	25	10	8
Romania	22	19	14
Czech Republic	12	10	5
Slovakia	10	8	4
Belgium	10	3	2
Macedonia FYR	9	6	5
Yugoslavia	8	7	5
Bulgaria	8	7	4
Belarus	7	12	3
Hungary	5	6	2
Austria	5	5	3
Finland	4	4	2
Switzerland	4	3	2
Estonia	4	2	1
Greece	4	4	1
Lithuania	3	4	1
Portugal	3	2	1
Latvia	3	3	1
Croatia	3	4	2
Ireland	2	1	0
Moldavia	2	1	0
Sweden	2	5	1
Denmark	2	2	0
Netherlands	2	2	0
Norway	2	3	1
Albania	1	2	0
Slovenia	1	2	1
Luxembourg	1	0	0
Iceland	0	0	0
Cyprus	0	0	0
Bosnia Herzegovina	0	2	0
France	na	na	na

Source: Gusev (1998, p. 116).

4.8.5. *Germanium (Kesler 1994)*

Germanium is less scarce than the other byproduct metals (1.4 ppm in the continental crust), but it rarely forms its own minerals. Known reserves are associated mainly with zinc deposits of the Mississippi Valley type (20 ppm in zinc ore), and some copper ores. One old copper mine (the Apex in Utah) has been converted into a germanium mine. Mostly it is recovered as a byproduct of zinc smelting. World production averages 60–80 tpy, worth $100 million. Prices have been rising. US recoverable reserves are estimated at 450 metric tons. No figures for world reserves have been reported. It is the original semiconductor element, although germanium transistors have long since been displaced by silicon. The main current use is in opto-electronics, especially fiber-optic components. There is a large potential for increased demand in amorphous silicon (aSi) photovoltaic cells.

4.8.6. *Gold (Kesler 1994)*

Gold is extremely scarce, with a crustal abundance of 0.0025 ppm. Nevertheless it is mostly mined for itself, both from placer mines and underground deposits. Gold ores average 5 ppm, though 1 ppm is not uncommon.[5] However gold is also a significant byproduct of copper, lead and zinc mining and refining. Gold is recovered from the residues of electrolytic copper refineries. In 1982 16 percent of US output came from copper smelting, down from 27.5 percent in 1968. Gold is found in porphyry copper deposits at about 0.5 ppm, but the volumes are large enough to justify recovery. A single US mine (Bingham, Utah) recovers 16 tpy gold. There are several other copper mines in the world with very significant gold reserves. For example, P.T. Freeport Ertzburg copper mine in Irian Jaya, Indonesia, is said to be the largest single gold reserve in the world (UNEP/ICME 1996, p. 8). Gold is also recovered from lead and zinc smelters, though to a lesser extent. In 1968 – the last year for which data are available – lead smelters provided 5.1 percent of (US) gold, while zinc smelters yielded 1.4 percent (USBuMines 1975; USBuMines 1985)

Gold was once used widely for coinage (as well as jewelry) and later became an official backstop for paper currency. The formal connection was ended in 1971 when the US dollar was delinked from the price of gold (i.e. gold was permitted to float). Gold is still widely regarded as a "stock of value" and much of the world supply is kept in central banks for reserve purposes, although some central banks have begun to sell gold gradually. This has put downward pressure on prices.

Another large fraction of the global gold stock is kept by individuals, espe-

---

[5] Gold mines often contain significant amounts of copper, lead or zinc byproducts, which can be environmental hazards if they are not present in quantities large enough to justify recovery for their own sake. An example is the Hope Brook gold mine in Newfoundland, which generates tailings from its main cyanide recovery facility that are subsequently subjected to a new flotation process commissioned in 1993. In full operation the flotation plant accepts an input feed with 0.12 percent Cu and 0.55 g/t of Au. It recovers 70 percent of the copper (in a concentrate with 20 percent Cu) and 35 percent of the gold, resulting in overall recovery of 86 percent (Au) and 70 percent reduction of copper in the final mill tailings (UNEP/ICME 1996, p. 28).

cially in India, where women's dowries are conventionally paid (and retained) as gold bracelets. In fact, India was a major importer of gold from Europe – ultimately from the Americas – in the 18th and 19th centuries (in exchange for other commodities, such as calicoes, muslins, and tea). The mining of gold is still a major source of income for several countries, notably South Africa and Ghana – that are heavily dependent on gold mining as a source of export earnings and employment.

Apart from monetary and related uses, and jewelry, gold is increasingly important as an industrial metal, especially in the electronics industry. Gold is a good electrical conductor and has the further advantage of being virtually impervious to corrosion.

4.8.7. *Indium (Kesler 1994)*

Indium is a malleable metal with interesting properties. It has a very low melting point (156.6 deg. C.), which makes it very volatile. It is extremely scarce, with a crustal abundance of only 0.05 ppm; known reserves are associated mainly with zinc ores (4 ppm). It is mainly found as 10–20 ppm in sphalerite, although it also occurs as an atomic substitute in tin and tungsten minerals. As of 1982 100 percent of US production came from zinc smelting. Most production comes from electrolytic zinc refining, where it is left in the 'slime' residues. Annual production is about 120 tpy, worth $28 million, but recoverable reserves are estimated at only 1670 metric tons, which is very low in relation to current output. It is alloyed with bismuth, cadmium, lead and tin for various purposes. Uses include indium-tin oxide protective coatings for automotive windscreens, semiconductors – especially for thin-film copper-indium-gallium diselenide (CIGS) photovoltaic cells – breathalyzers and dental crowns. It is toxic only at high concentrations. There are no known environmental hazards.

4.8.8. *Rhenium (Kesler 1994)*

Rhenium is a very scarce metal (1 ppb in the earth's crust). It is found mainly as an atomic substitute for molybdenum, which in turn is often (but not always) associated with copper. Curiously, rhenium in porphyry molybdenite (not discussed in this report) is not highly concentrated (10–100 ppm), but when molybdenite is found together with porphyry copper deposits, the rhenium concentration is much higher (2000 ppm), which makes it worth recovering. World recoverable reserves are estimated at 2500 metric tons. Annual production is around 30 tpy (worth $40 million), but demand is increasing rapidly. It has two main uses: (1) in platinum-rhenium reforming catalysts for petroleum refining and (2) in nickel-rhenium alloys used in jet engines.

4.8.9. *Selenium (Kesler 1994)*

Selenium is quite scarce in the earth's crust (0.12 ppm). Known reserves are mainly found in copper ores (4 ppm) and current output are almost entirely produced as a byproduct of copper smelting. Annual production is about 1.8 kMT. It has very few industrial uses, primarily as a photo-receptor for xerography. This market is now declining (because of competition from organic photo-receptors). Other minor uses are in dandruff-preventing shampoos, as a decolorant for glass and an alloying agent for ferrous and non-ferrous alloys to improve metal-working properties. An important future use could be in the thin-film copper-indium-gallium-diselenide (CIGS) photovoltaic cell, although indium availability is likely to be the limiting factor in this case.

Selenium is highly toxic in excess amounts but it is also a necessary trace element for plants and animals. Thus there are both excess and deficiency diseases. Deformities appear to be associated with the former and the incidence of stroke may be a symptom of the latter. Pollution does not appear to be a problem (at least it appears to be controlled), although wastewater from some oil refineries is enriched in the element.

4.8.10. *Silver (Kesler 1994; USGS annual,1999, p. 62.11)*

Silver has been valued as a precious metal since pre-history. Silver is mined for its own sake, and it is a co-product of some copper, lead, zinc and gold mines (e.g. in Mexico and Peru). In 1982, the last year for which data are available, 23.8 percent of US silver production was a byproduct of copper smelting and refining (down from 29 percent in 1968); lead smelting and refining provided 5.8 percent (down from 17 percent in 1968). Zinc smelting and refining provided 12.6 percent of US silver output in 1968, but 1982 data are unavailable. US industrial demand for silver in 1999 was 7 kMT, as compared to mine production of 1951 metric tonnes. The remainder was imported, recycled or from stock.

Silver has many industrial uses, apart from the historical ones (as coinage and jewelry). Its high electrical and thermal conductivity make it particularly valuable in micro-electronics. However its major industrial use has been in photography, especially in color film. This is also the major source of silver dissipation and pollution, from film processing wastes.

4.8.11. *Sulfur (Ayres and Ayres 1996; Ayres and Ayres 1998; USGS annual, 1999)*

Sulfur is a non-metallic element that has been known since prehistoric times (as brimstone, meaning burning stone, because of its combustibility). It is biologically and environmentally important because it can exist in valence states from –2 (e.g. hydrogen sulfide, H_2S) to +6 (sulfuric acid, H_2SO_4). Like nitrogen and phosphorus, it is essential to living organisms. It is a component of three of the 20 amino acids that, are the "building blocks" of all proteins, viz. cystine, cysteine and methionine. It seems to play a special part in cross-linking amino acid chains to give them their characteristic three-dimensional shapes (especially important for enzymes). On the other hand, a number of sulfur compounds are toxic, including hydrogen sulfide, sulfur dioxide and the mercaptans.

Sulfur is not intrinsically scarce; developed sources amount to at least 5 billion metric tons as a byproduct of petroleum refining and gas processing, as well as in deposits of sulfide ores, pyrites and elemental sulfur. Most elemental sulfur used by industry today comes from oil refineries and/or natural gas processing (66 percent), and non-ferrous metals smelting (19 percent). Other sources – Frasch, pyrites or native sulfur – account for about 15 percent (mostly from China and Poland). An insignificant amount is recovered from coal, lignite or gypsum. It is much easier, hence cheaper, to obtain sulfur in forms useful to industry from reduced sources such as H_2S or S than it is to recover it from coal, or from oxides.

Non-ferrous metals are almost exclusively obtained from sulfide ores. For instance, copper ore consists mostly of chalcopyrite ($CuFeS_2$) – which is the most common – followed by bornite (Cu_5FeS_4) and chalcocite (Cu_2S). Zinc and lead are obtained respectively from the minerals sphalerite (ZnS) and galena (PbS). In chalcopyrite ore (the most common) the sulfur/copper ratio is about 1:1, whence every ton of copper produced from mines results in the mobilization of a ton of sulfur. The corresponding ratios for lead and zinc are 0.15 (for galena) and 0.49 (for sphalerite).

Sulfur recovery from copper and lead smelters was still relatively incomplete until quite recently. It was originally motivated mainly by environmental concerns. As recently as 1974 the US Bureau of Mines estimated a 30 percent sulfur recovery rate for copper smelting and 43 percent for lead smelting (USBuMines 1975). However, the above estimate for copper assumes a sulfur content of copper ores equivalent to 1.47 tons of sulfur per ton of copper, which is inconsistent with the known composition of copper minerals and smelter feeds. For this reason we suspect that the United States Bureau of Mines recovery estimate for copper in that year was slightly too low. Using the ratio for chalcopyrite figure for sulfur content the sulfur recovery rate for US copper smelters in 1974 would also have been 43 percent. This is consistent with the fact that about 60 percent recovery was easily achievable at the convertor stage, but the remainder (40 percent or so) requires scrubbers or the equivalent. It is estimated that sulfur recovery from copper smelters in the US has now reached 95 percent (USBuMines monthly, 1994).

By contrast, complete sulfur recovery from zinc smelters in the US was much easier to achieve (81 percent by 1974). This was apparently because the zinc retorting process itself makes sulfur recovery straightforward and economical. If all by-product sulfur recovery from copper and zinc smelters in 1911 (the first year for which we have data) is attributed to zinc smelting alone, the implied recovery rate would be 79 percent – close to the 1974 level.

Byproduct sulfur from US non-ferrous smelters amounted to 1.4 MMT in 1995, and increased to 1.61 MMT in 1998, but fell to 1.32 MMT in 1998 due to the closure of three copper smelters (out of seven). Because of increasing use of sulfuric acid for the SX-EW process, however, US copper producers as a group switched from net sellers of sulfur to net importers, in 1999.

Globally, sulfur recovery from non-ferrous smelters was at least 8.22 MMT in 1995, out of a global total of 54 MMT, with unspecified sources accounting for 3.92 MMT (USGS annual, 1999). The metallurgical share increased to at least 10.30 MMT in 1998 and 10.20 MMT in 1999, out of global totals of 56.7 MMT and 57.1 MMT, respectively, of which 4.3 and 4.5 MMT respectively was from unidentified sources, mostly oil and gas or metallurgy. In short the metallurgical share of total sulfur production (and consumption) is now between 19 percent and 20 percent and growing. Meanwhile the US contribution to the global total is declining, even in absolute terms, since 1998.

The efficiency of sulfur recovery (as opposed to emissions control) from non-ferrous metal smelters outside the US is not easy to determine. One study (Chapman 1989 Table 3) estimated the sulfur control indices for a number of exporting countries for 1985 as follows: Chile (14 percent), Peru (0 percent), South Africa (39 percent), Zaire (76 percent), Zambia (70 percent), Philippines (0 percent), Australia (0 percent) and Canada (26 percent). It estimated the average control level for copper exporters as 32 percent, compared to an average of 89 percent for the major copper importing countries. A more recent study of 8 US and 29 non-US copper smelters around the world, commissioned by the US Bureau of Mines, concluded that sulfur control (i.e. recovery) at US smelters for the year 1988 was 90.4 percent. but that only 55 percent of the available sulfur was recovered by the non-US group (Towle 1993). However, of the non-US smelters surveyed, three were located in Western Europe (Ronnskar, Sweden, Hamburg, Germany and Huelva, Spain). The first two, at least, are at the 90 percent recovery level. There is strong anecdotal and indirect evidence, however, that sulfur recovery (due to environmental regulations) is increasing rapidly in all copper exporting countries.

Most industrial sulfur is converted initially to its most oxidized form, sulfuric acid. Sulfuric acid, because of low cost and its variety of uses as a powerful oxidizer and pH control in the chemical industry and other industries, is one of the most reliable indicators of general industrial activity. However, currently, the biggest and fastest growing single use of sulfuric acid, worldwide, is a specialized one: converting phosphate rock to phosphoric acid for fertilizer. In this case, the sulfur is combined with calcium as insoluble calcium sulfate (gypsum).

Outside the mining sector, sulfuric acid is the starting point for most sulfur based chemicals, as well as many processes where the acid is ultimately neutralized by an alkali metal and discarded as a waste. Sodium sulfate has significant industrial uses. Ammonium sulfate is a fertilizer. Calcium sulfate – the result of neutralizing sulfuric acid with lime – is potentially useful as a substitute for natural gypsum to manufacture plaster-of-Paris for wallboard in the building industry, although in practice most of it is discarded.

The biggest single use of sulfuric acid is in phosphate rock processing. In the US crude phosphate ore produced in 1993 (after beneficiation by flotation) was 106.8 MMT to which 24.2 MMT of sulfuric acid was added during treatment. This accounted for 7.9 MMT (S) in 1993, or 54 percent of the total supply of acid. The sulfur used in this process is embodied directly in the concentration waste stream (as phospho-gypsum). The quantity of fertilizer grade phosphoric acid was 35.5 MMT. The difference (95.5 MMT) was concentration waste, none of which is currently utilized for any purpose.

Another big inorganic carrier, but not consumer, of sulfur was synthetic sodium sulfate. It has many uses. For instance, cellulosic fibers such as rayon, consumed 51 kMT (S), or 157 kMT of sulfuric acid in 1993 (USBuMines annual, 1996 "Sulfur" Table 7). Evidently 157 kMT of sulfuric acid inputs, neutralized by 129 kMT of sodium hydroxide, would necessarily yield 228 kMT of sodium sulfate.

Aluminum sulfate or "alum" was once mined (for use as a fixative for dyes). It is now essentially all synthetic, produced from bauxite and sulfuric acid. Total production of aluminum sulfate ($Al_2(SO_4)_3$) in 1993 was 1.15 MMT (17 percent Al_2O_3) with a sulfur content of 278 kMT (S) (USBuMines annual, 1996 "Aluminum" Table 18). Based on the chemistry, the process probably required 0.75 MMT of bauxite and 0.85 MMT of sulfuric acid (100 percent) and generated about 0.45 MMT of waste similar to "red mud" from bauxite processing.

The majority of chemical processes end up with sulfur (from the acid) combined with calcium (from lime) as calcium or sodium sulfate (from caustic soda). In the latter case, the material is soluble and biologically available. However it is reasonable to assume that almost all industrial sulfur ends up (perhaps after several intermediate lives) as insoluble sulfates, notably calcium sulfate. The major source, of course, is phosphate rock processing. Sulfuric acid used in the SX-EW copper process (the third largest use of sulfuric acid today), remains in the ground, probably in the form of light metal sulfates.

These sulfates are unreactive and non-toxic and are not known to be harmful as such. However insofar as the acid combines with alkaline buffers they do contribute to the overall acidification of the earth by reducing buffering capacity.

The major source of sulfur dioxide emissions to the atmosphere is fossil fuel combustion, mainly coal and lignite. However, despite increasingly efficient sulfur recovery, non-ferrous metal smelters continue to be significant (but declining) sources. In roasters and smelters sulfur is oxidized to SO_2 or SO_3. Some of this, especially from the older reverberatory furnaces, is too low in concentration to be recoverable for purposes of making sulfuric acid.

In the air, SO_2 and associated droplets of acid are irritating and harmful to the nasal passages and lungs. Sulfur dioxide was actually the first air pollutant to be subject to regulation, although its sharp and unpleasant odor may account for this early attention. The unreacted sulfur oxides remaining in the atmosphere are the major cause of a haze which reduces visibility and increases the reflectivity (albedo) of the earth to sunlight. Thus sulfur oxide emissions, other factors remaining equal, tend to cool the earth slightly.

The sulfur oxides are eventually deposited as dry sulfite/sulfate particulates, or dissolved in water droplets (producing sulfurous or sulfuric acid), or they react with ammonia, producing ammonium sulfite or sulfate. In any case, the sulfur is eventually deposited on land (or water) as sulfurous or sulfuric acid, or as sulfates. On the ground, ammonium sulfate is a fertilizer. Sulfurous and sulfuric acids are also biologically active. In effect, the natural "sulfur cycle" has been significantly accelerated by industrial activity. One consequence is widespread acidification of forest soils, especially granitic soils in central and northern Europe. On the other hand, since many soils are sulfur deficient, there is also a general fertilization ("eutrophication") effect in some locations, especially in calcareous soils.

The major source of reduced sulfur into the atmosphere is from volcanic eruptions, fumaroles, anaerobic decay in swamps, and vented natural gas.

4.8.12. *Tellurium (Andersson 2000)*

Tellurium is the scarcest of all the byproduct metals, except for gold. Crustal abundance is 0.005 ppm. It is mainly recovered from copper ores (1.5–3 ppm), where it is considerably enriched. There are two known deposits where tellurium is found at much higher concentrations, one in Mexico (0.2 percent) and one in China. Tellurium has a significant potential use in thin-film cadmium telluride (CdTe) photovoltaic cells.

4.8.13. *Thallium (Kesler 1994)*

Thallium is an extremely rare metal, found mainly with zinc minerals, where it has an abundance of 10–40 ppm in sphalerite. It is recovered from zinc smelter flue dust, but very few zinc smelters bother. Recovery is entirely from European countries and Japan. Production is about 16 metric tons, worth $3 million. World recoverable resources are estimated as 377 metric tons. The metal is used in electronics in special solders and other low-temperature melting applications. It is extremely toxic to animals (more toxic than arsenic, cadmium or lead), due to interference with potassium metabolism. There are a few pharmaceutical uses, e.g. for treatment of ringworm, dysentery and TB.

CHAPTER 5

THE FUTURE OF RECYCLING

5.1. Background

It is well-known that the most (only) dynamic and profitable sector of the US steel industry is the so-called 'mini-mills' – exemplified by Nucor. These companies are essentially scrap processors and recyclers. At first they produced mainly rather low grades of steel products from iron and steel scrap. It is authoritatively estimated that 95 percent of the iron and steel embodied in products eventually returns as old scrap (Sibley et al. 1995). In recent years the recycling technology has improved significantly, and with it the quality of the products and the range of products in which recycled steel can be used. In the long run, as steel recycling technology improves further and global demand approaches saturation, it would seem that the so-called 'integrated' iron and steel producers, who start from iron ore, are facing gradual extinction, at least in the industrialized countries.

This long range projection assumes that future steel recycling technology is able to remove so-called 'tramp' metal contaminants, such as aluminum and copper, from scrap iron and steel. No such technology has been developed to date, Undoubtedly one of the reasons is that these elements are not normally found in virgin iron ore from natural sources. The problem of contamination is discussed in more detail later in this chapter. However it is important to emphasize the key point: that 'mining' wastes of a certain 'grade' is not necessarily as easy as mining natural ores of the same grade.

Perhaps this is why, the increasing economic success – approaching dominance – of scrap reprocessors over integrated producers in the iron and steel sector has not (yet) been recapitulated in the non-ferrous metal sector, except possibly for lead. In fact, the trend seems to be the other way. In 1976 the USEPA identified 31 secondary copper producers (USEPA 1983, Table III-1, p. 624). In 1985 there were eight secondary copper smelters and refiners operating in the US, and by year end 1998 only two remained, one of which was expecting to be closed early in 2000 (Edelstein 1999). Recycling of old scrap has not accounted for an increasing share of the global copper supply, as noted in the previous chapter. In fact the ratio of old scrap to new scrap in the US secondary industry has declined from over 80 percent in 1990–1991 to 45 percent in 1998 (Jolly 1999, Figure 11).

In 1976 there were 69 secondary lead recyclers in the US (mainly processing discarded batteries from scrap vehicles) (USEPA 1983, Table III-2, p. 897). In 1998 there were 28 lead recovery plants active (Smith 1999). Recycling constitutes an increasing share of the lead supply, as noted in *Chapter 3*.

In 1979–1980 there were at least 16 active secondary zinc producers, although six plants had recently closed (McElroy and Shobe 1980, Tables 4-1, 4-2). In 1999 three primary smelters consumed scrap zinc and twelve small companies produced zinc chemicals from scrap (Plachy 1999). None produced zinc metal from scrap. There is no evidence of a long-term increase in the old scrap recycling rate, at least in the US.

Declining metal prices and increasingly tight regulation of secondary metallurgical (and other manufacturing) operations have apparently combined to drive the secondary copper sector offshore to countries with lower labor costs and no domestic resources. Russia and the US are the biggest exporters (Germany also exports a lot, but it imports even more.) China is by far the biggest (net) importer of copper scrap (951 kMT in 1998).[1] Other factors may be at work, of course, but the trends so far seem well-marked. We don't have trade data for lead and zinc scrap, but the pattern is probably similar, at least for zinc.

As explained previously, the *recycling rate* is defined as the ratio of old scrap consumption to total (apparent) consumption of the metal. The *recycling efficiency* is defined as the ratio of old scrap consumption to old scrap generated domestically plus old scrap imported (net of exports). Not surprisingly, platinum leads the list in recovery efficiency (98 percent), with iron and steel, vanadium and nickel following close behind at 95 percent. Gold is recovered less efficiently (90 percent) because of dissipative losses in dentistry and electronic uses. A comparison for the US in 1993 is shown in *Table 5.1* for those metals for which data are available (Sibley et al. 1995).

Table 5.1: US old scrap recycling performance, 1993

	Recycling rate	Recycling efficiency	Availability (rate/efficiency)
Copper	22	30	73
Lead	57	64	89
Zinc	12	21	57
Gold	18	90	20
Platinum	38	98	39
Antimony	35	43	78
Molybdenum	13	7	186

Rate = old scrap consumption / total (apparent) consumption.
Efficiency = old scrap consumption / old scrap generated + net imports.
Availability = old scrap generated (domestic + net imports) / total (apparent) consumption.
Source: Sibley et al. (1995).

1 Trade statistics are quite imperfect. According to the International Copper Study Group, total world exports in 1998 added up to 1.791 MMT while total world imports came to nearly double that amount, 3.275 MMT [Jolly 1999, Tables 3 and 4]. Reported scrap imports consistently exceed exports by large margins.

Finally, the *availability* of recoverable secondary metal is defined as the ratio of old scrap generated (domestic plus net imports) to total apparent consumption. The latter measure is the ratio of the first two, viz. *rate* divided by *efficiency*. Evidently the recycling rate will be lower (and the availability of secondary material higher) if the consumption of the metal is rising rapidly, and conversely. On the other hand, the availability also reflects the length of life of the applications of the metal. Low availability (other factors remaining equal) implies long-lived applications and conversely.

A few interpretive comments on these ratios may be helpful. Molybdenum (in 1993) was produced mainly from ore – as usual – but the recycling rate was higher than the recycling efficiency. That means that more scrap molybdenum was being used in that year than was being generated. This can only mean that stocks of old scrap were depleted and/or that uses were changing. In this case it does not necessarily follow that the uses of molybdenum are short-lived.

In the case of gold, the recycling efficiency was quite high (90 percent) but only 18 percent of the gold refined in 1993 was from scrap. This implies that most gold is used in very long-lived applications, which is obviously consistent with ordinary observation. For platinum the recycling efficiency is even higher, but some of the uses (in catalytic convertors) are not particularly long-lived, hence the higher availability of secondary metal.

Lead is also anomalous: it exhibits a surprisingly low recycling efficiency (64 percent) in view of its dominant use (in automotive storage batteries), which are easily recovered. In both cases, the problem may be related to the use of these two metals in electronics and (in the case of lead), in glass.

As noted in a previous section, zinc is also anomalous, insofar as the available data are inconsistent. About 30 percent of current refinery output is recycled, but only 12 percent (in the US) is from old scrap, according to *Table 5.1*. But whereas the US Bureau of Mines estimates the recycling efficiency for zinc at 21 percent, the International Lead and Zinc Study Group (ILZSG 2000b) sets it at 80 percent, which is higher than lead and exceeded only by gold and platinum. The latter figure is, frankly, implausible and we think it is erroneous.

Copper, our main concern, is peculiar. It is dominated by long-lived applications, and yet the recovery efficiency is still surprisingly low, at least according to the authors of *Table 5.1* (Sibley et al. 1995), as well as an earlier USBM calculation based on data from the 1960s and early 1970s (Carrillo et al. 1974). These studies arrived at a recycling efficiency for the US of only 30 percent, implying that 70 percent of obsolete copper products were not recovered. Admittedly, as noted in *Chapter 3*, other authors have arrived at considerably higher estimates; the calculations depend on assumed lifetimes-in-use. There is considerable uncertainty on this point. In an ideal world, however, the pattern for copper would be more like the pattern for platinum or gold.

Given current (increasing) trends in demand, patterns of use and recycling efficiency, the world will either run out of recoverable primary copper or will be mining copper with an extremely low grade, long before the end of the next century. As we have said, we believe that global peak production is likely to occur within two decades, even though demand is likely to grow for several more decades if economic growth continues, especially in Asia. A five or ten-fold increase in the rate of open-pit surface mining to extract ores of extremely low grades is almost certainly not feasible for environmental reasons, if for no others. The only possible 'way out' of this dilemma must be to recycle copper now in use far more efficiently than is now the practice.

Nevertheless, it may also be necessary in the future to 'mine' copper from old mine waste deposits, or even possibly from landfills. *Table 5.2* provides some relevant background on the copper content of waste streams in Europe (15 EU countries plus Poland). The important point to bear in mind is that copper constitutes less than 0.1 percent of the total weight of every waste stream except electrical and electronic equipment (7.8 percent) and end-of-life vehicles (0.9 percent). These two waste streams account for 72 percent of the copper in the wastes; the other 28 percent is divided among five significant waste categories (industrial waste, construction and demolition waste, municipal solid waste, hazardous waste and sewage sludge) in which the copper content ranges from 0.02 percent to 0.07 percent, all far below the grade of ores currently mined.

To be sure, specialized concentration techniques may be available for some of these waste categories that are more efficient than concentration techniques used in the mining industry. For instance, the recovery of copper (wire and pipe) from construction and demolition wastes is much easier than the low 'grade' would suggest. In fact, most of this copper is actually recovered

Table 5.2: Sources of copper waste for 15 EU countries and Poland, 1994

Source	Total waste		Copper content		
	kg per capita per annum	%	kg per capita per annum	% of total waste	% of copper waste
Industrial waste	1000	46	0.2	0.02	6.7
Construction and demolition waste	530	24	0.37	0.07	12.3
Municipal solid waste	475	21	0.24	0.05	8
Hazardous waste	86	4	0.043	0.05	1.4
End of life vehicles	75	3	0.67	0.9	22.3
Electrical and electronic equipment	19	0.8	1.5	7.8	50
Sewage sludge	22	1.0	0.008	0.037	0.3
	2200	100	3	0.14	100

Data source: Eurostat.

manually without great difficulty. But even if all of this copper is recovered (which is not the case) at least 16 percent of the copper in wastes is in concentrations below 0.05 percent. Completely new technologies will be needed to recover this metal.

5.2. Recovery and recycling of copper from old scrap

The recoverability of copper from old scrap depends very much on its origin and composition. Secondary copper is classified initially according to source. Direct or 'home scrap' from within the smelter/refinery has the highest purity; it is normally recycled immediately within the plant. The next category consists of 'new scrap' from downstream metal fabricators, such as trimmings, borings and croppings. It, too, is normally returned to the refinery. Much of this flow is not accounted for in official statistics. Data for the US show that new scrap recovered increased from 723 kMT in 1992 to 950 kMT in 1999, roughly in proportion to overall demand.

The critical problem for copper is 'old scrap' from discarded products and appliances. Here the data are telling a different story. Although overall global demand for copper has increased by 30 percent in the 8 year period of economic boom (from 3,030 kMT in 1992 to 4,080 kMT in 1999), the quantity recovered from old scrap actually declined from 555 kMT in 1992 to 381 kMT in 1999. The decrease was nearly monotonic, with only a single upward 'blip' (1997), and an accelerated decrease from 1997 to 1999.

According to the International Copper Study Group (ICSG), world production of secondary copper declined in 2000, for the second successive year. In 2000 secondary metal accounted for only 13 percent of total refined copper. The global decline of 100 kMT, year to year, was virtually equal to the US decline. Whereas in 1985 there were eight secondary copper smelters in the US, only two remained at the end of 1998 and one of those announced impending closure in early 2000, citing high costs of meeting environmental standards (USGS annual 1999, p. 62.44). Globally the picture is similar: the scrap share in 1970 was over 41 percent; it has fluctuated since then, with lows around 32 percent in 1978–1979 and again in 1984 and 1986 and most recently in 1998. The highs since 1970 were in 1974, 1980 and most recently 1994–1995. Nevertheless the overall trend appears to be down (Dewison 1999, Figure 13).

The decreasing rate of recovery of copper from old scrap in the US probably has several explanations. One factor is increased exports (e.g. to Taiwan, which relies exclusively on secondary copper, and to China, which also relies heavily on imported scrap.) Another factor is that there was a change in copper consumption patterns in the early 1980s, with more going into long-lived infrastructure uses and less to shorter-lived industrial machinery and equipment uses. This would reduce the amount of post consumer scrap becoming available in the past few years. A third factor is almost certainly due to the increasing complexity and geographic dispersion of copper-containing products, especially appliances and electronic equipment. The more complex the product, and the more dispersed its consumption, the more difficult it is to collect and recycle. On the other hand quite a bit of the copper associated with electronic toys, games and the like is actually imported with the product and does not appear as part of the official US consumption statistics.

Whatever the explanation, the trend has two important implications. One is that environmental emissions associated with copper mining and smelting (and also due to copper accumulation in the environment) are on the increase, relative to previous expectations. The second implication is that copper mines will be forced to use lower grade ores sooner than would otherwise be necessary. This may be 'good news' for the mining companies, in the short run, but it is bad news for our industrial society in the longer term.

In the US there are three grades of post consumer copper scrap. The highest grade (Number 1) scrap consists of wire, cable and copper tubing. It is sorted, copper wire is chopped into short pieces for easy disposal of insulation,[2] and finally melted in a reverberatory furnace and cast. The second grade (Number 2 scrap) consists of unalloyed scrap with a nominal 96 percent (minimum 94 percent) copper content. After sorting and preparation it is sent to a reverberatory furnace and then subsequently fire and/or electrolytically refined.

The lowest grade of commercial scrap consisting of impure, contaminated or even unknown materials, with a nominal copper content of 50 percent or less. This scrap is processed like concentrate from ore, except that it is usually smelted in a blast furnace (cupola type), using coke as a fuel, rather than a reverberatory furnace. Brass and bronze scrap are generally remelted without refining, provided they can be sorted into composition categories.

Melting of high grade (No. 1) scrap – mainly wire – in a reverberatory furnace takes 4.45 GJ/t of refined billets or cakes. Of this, 95 percent is process energy, the remainder being for air pollution control. However for No. 2 scrap the process energy requirement is 18.4 GJ/t of wirebar, plus 1.75 GJ/t for space heating and pollution control. This is because of the refining steps but still considerably less than copper from ore, due to the fact that mining and milling (flotation) are avoided. For low-grade scrap (25–35 percent Cu) the process energy requirement jumps to 46 GJ/t plus 3.16 GJ/t for ancillaries. Remelting brass and bronze costs about 6.7 GJ per tonne of alloy in process energy, plus 1.43 GJ/t in ancillaries (ibid). Clearly there is a high incentive for improved methods of sorting and preprocessing, to minimize the need for re-refining.

In Japan, where the recycling sector has been studied intensively, electric power, telecommunications and railways contribute roughly 50 percent of

---

2 The disposal can be a problem, depending on the nature of the insulation. The most common insulating material is polyethylene, which can be burned off quite easily. However PVC is used for some kinds of wire and both the chlorine and some of the additives can be pollutants.

old copper scrap, wiring and pipes recovered from old buildings about 20 percent, industrial machinery 13 percent, automobiles 10 percent and electrical appliances about 7 percent (Sekiguchi and Ichiyama 2000). See also *Figure 3.4*.

Recovery of copper from end-uses varies greatly from product to product. Copper wire from electric power, telecommunications and railways is essentially 100 percent recovered. On the other hand, in Japan copper from small electrical and electronic appliances is only about 20 percent recovered. In the case of automobiles the recovery fraction is 48 percent; for industrial machinery it is a little over 81 percent; for buildings it is a little over 68 percent (Simada et al. 1999, Table 5). Recovery in many other countries is less efficient.

Some scrap is melted for brass without the necessity of re-refining. In 1998 this amounted to 3194 kMT, globally, as compared to 1353 kMT of copper scrap that was refined and total refinery output of 13,315 kMT. Thus, only 10 percent of refined copper was recovered from scrap in 1998 (down from 17 percent in 1995), although the overall ratio of scrap consumption to total consumption was 27.5 percent in that year (actually down from 31.6 percent during the period 1990–1993 inclusive.

The fraction of refined copper recovered from scrap varies widely from country to country. For the year 1995 the International Copper Study Group (ICSG) estimated 6 percent for Korea and 11 percent for India, 100 percent for Taiwan (which has no primary copper refinery), 40 percent for China, 12 percent for Japan, 30 percent for the US and 48 percent for Europe (ICSG 1997). On the other hand, the data from Donald Rogich (personal communication) gives 1 percent for Korea, 15 percent for India, 9 percent for Japan, 42 percent for China and 15.5 percent for the US. The discrepancies are not easy to explain.

The wide variability among countries only partly reflects the efficiency of the collection/recovery system in each country. It also partly reflects differences in historical usage. Some countries use more copper than others in applications (such as pipe and roofing) where the grade of the product is not especially critical. In Japan, little copper is used for such purposes. Copper scrap from the west coast of the US is generally exported because there are no major copper-using industries located there, and it is cheaper to ship scrap by sea than overland by rail or truck. Finally, the fact that secondary copper refineries are relatively dirty and therefore undesirable (unless there are heavy investments in control technology) also accounts for some of the international variability.

As noted there is no existing commercial demand for copper scrap with a copper content of less than 30 percent in the US and Europe (50 percent in Japan). There is a lot of such scrap, especially from automobiles and trucks, and household appliances, not to mention electronic devices. As a percentage of the total, the low grade fraction is probably increasing. The non-ferrous metal (Al, Cu, Pb, Zn) content of an automobile is about 7.5 percent (of which only about 10 kg is copper); for a clothes washer about 5 percent and for a dishwasher it is 3 percent (Poetzschke 1991). The picture for Europe is summarized in *Table 5.3* (WRF 1996). It is noteworthy that the non-ferrous metal content for video and sound equipment, household appliances and office equipment is below 6 percent.

The recovery efficiency in general is proportional to the percentage of copper in the various waste streams, because it is difficult to upgrade the non ferrous metal content to the 30–50 percent level by means of technologies developed for the mining sector. The technology of old scrap recovery from electrical and electronic products, automobiles, and appliances is still comparatively primitive, and the most effective method is hand sorting. But it is not practical to disassemble most discarded items in industrial countries, due to high labor costs. This is unfortunate, in light of the fact that, in principle, many of the motors (where much of the copper is embodied) could be re-used, or remanufactured.

The current technology of low grade scrap recycling consists of four stages (Simada et al. 1999). The first stage is shredding. After the shredding process, ferrous fragments can be largely removed by magnets. However the magnetic separation process is not completely efficient, and some non-ferrous metals accompany the iron and steel scrap. This is a problem for recycling the latter, because the non-ferrous metals are undesirable contaminants in the secondary steel. Magnetic separation leaves a residue that may have a copper content as low as 0.3 percent. Organic materials may be eliminated by combustion or pyrolysis, reducing the mass and raising the copper content of the residue to 0.6 percent or more. Combustion is problematic because temperatures must be very high (over 1,000 °C) to prevent dioxin formation. Even so, the combustion products require emissions treatment. Pyrolysis takes place at lower temperatures. The pyrolysis product – rich in CO – may be recovered as fuel, unless it contains halogens from chlorinated plastics like PVC. However, pyrolysis systems must be isolated and halogen removal is still difficult.

Table 5.3: Material distribution by percent of electrical and electronic waste: predictions for Europe 1998

Categories	Ferrous metals	Non-ferrous metals	Glass	Plastics	Other
Data processing equipment	35.0%	19.0%	19.0%	22.0%	5.0%
Offices and services	62.3%	26.7%	0.3%	10.0%	0.7%
Telecommunications	28.0%	15.0%	0.0%	55.0%	2.0%
Video and sound equipment	20.0%	6.0%	30.0%	25.0%	19.0%
Household appliances	52.8%	5.9%	2.4%	18.6%	20.3%
Equipment; hotels, restaurants	77.0%	5.7%	1.6%	9.9%	5.8%
Cables	4.0%	53.0%	0.0%	38.0%	5.0%
Lamps	4.5%	5.9%	86.4%	1.4%	1.8%

Source: WRF (1996).

The solid residues from the above are then upgraded by either wet (flotation) or dry (gravitic) techniques to around 30 percent Cu content. Some of this material can be fed into a primary copper smelter. The smelting and refining techniques used thereafter depend upon whether the scrap contains precious metals or not.

5.3. Recovery and recycling of electronic scrap

Significant quantities of copper, lead, zinc and other rare byproduct metals are embodied in electrical and electronic products. In Europe scrap electrical and electronic equipment averages nearly 8 percent copper and accounts for 50 percent of all copper in wastes (*Table 5.2*). Recovery and recycling in most cases is theoretically possible but not much practiced. Electronic scrap (including PC boards) is a special problem, inasmuch as the copper content is very low (*Table 5.4*) and the quantities are rapidly increasing. According to one source, a metric ton of circuit board scrap (even after the equipment in which the PCs are embodied has been dismantled) contains less than 100 kg (10 percent) copper (Arpaci and Vendura 1992).

A recent survey indicates that 25 kMT of electronic scrap was recycled in the US during 1997–1998 (Amore undated, p. 16). See *Table 5.5*. The recycling rate for personal computers was only 11 percent (as compared to 70 percent for large appliances such as refrigerators, washing machines and air conditioners). In fact the recycling rate went down between 1997 and 1998 and 20 recyclers went out of business (ibid). Most of the recycled PCs were collected from large-scale users. TV sets were even less efficiently recycled: in 1998 24.7 million TVs were manufactured and only 19,000 units were recycled. (ibid).

Table 5.4: Average metal content of 1 metric ton (1000 kg) of PCB scrap (kg)

Metal	According to	
	Arpaci and Vendura (1992)	Poetzschke (1991)
Copper	< 100	200
Iron	na	80
Nickel	< 12	20
Tin	20	40
Lead	na	20
Aluminum	12	20
Zinc	na	10
Antimony	na	4
Silver	6	1
Gold	1	0.5
Palladium	na	0.05

Table 5.5: US electronic product recycling; 1997–1998

Product type	Quantity recycled (metric tons)		Number recycled (> 000s)	
	1997	1998	1997	1998
Desktop PCs	26800	26300	2362	2308
Mainframes	25400	24400	56	54
Workstations	4500	5400	342	413
CRT monitors	20900	23100	1307	1453
CRT TVs	317	454	14	19
Notebook PCs	1360	1497	301	327
Peripherals	33100	33100	2915	2933
Telecom	9500	9980	2122	2197

Source: Amore (updated) *Table 1*, p. 68.

The problem with TVs and PCs is that effective recycling (at present) requires prior dismantling. In the case of a TV there are nine main components: cables, capacitors, cathode ray tubes (CRTs), copper wire, aluminum, steel, printed circuit boards (PCBs), plastics and wood. The CRT typically weighs half of the total; for recycling it must be further separated into the mask (metal), the front glass – from which the fluorescent coating must be removed prior to remelting – and the tapered back glass, which may be up to 33 percent lead (WRF 1996). Cables are copper wire with insulation, typically PVC or PE, which must also be removed prior to remelting.

To illustrate the issues involved, we refer again to the two recent studies that show the metals content of printed circuit board (PCB) scrap (see *Table 5.4* above). The discrepancies between the two lists is a sufficient indication of the uncertainties of the data. Nevertheless it is clear that significant quantities of a number of rare metals are being lost. These metals would be worth recovering by existing methods if the quantities were large enough and if the scrap were uniform enough in quality, but neither is true. However, what is clearly true is that the recovery process must be designed in an integrated fashion to take into account the entire range of components.

A simple but somewhat inefficient solution is to use the cyanide process for dissolving the gold and silver (as in gold mining), leaving the rest for landfill disposal. Obviously, the residue then has significantly less recovery value, not to mention the fact that cyanide wastes and the lead content makes this residue very undesirable. (Landfill disposal of electronic waste has already been prohibited in a number of US states because of lead content.)[3]

---

3 Since there is, as yet, no effective recycling system for these wastes, they are simply accumulating in informal storage areas. The magnitude of the eventual electronic waste treatment/recycling problem is growing all the time. Industry will inevitably be forced to play a more important role, possibly by the enforcement of 'take back' legislation along the lines of the German scheme.

Fortunately, it seems that electronic (solid) waste can be fed directly into a copper smelter or convertor, as long as the quantities involved are not too great in relation to the other feeds. The copper is recovered automatically, while the lead, zinc, nickel, tin and other low value byproduct metals can be discarded in the slag or recovered, as prices dictate. The valuable silver and gold, can be recovered (90 percent) from the anode copper, via existing refining processes. Dioxin formation is not a problem. The Noranda facility in Ontario is now profitably recycling electronic waste commercially, by bundling it into other scrap. Rare metals are recovered from anode slime. Emissions are taken care of by existing pollution control equipment. What Noranda is doing ought to be feasible for many other smelter-refiners.

There is no commercially viable technology for recovering copper from dilute aqueous waste streams. However, in the electronics industry, where such wastes are increasingly important, copper recovery may be necessary to reduce the copper concentration of wastewater below acceptable legal limits. As mentioned previously, copper metalization is replacing tungsten and aluminum metalization for semiconductor devices. This development involves three new copper technologies: copper seed and/or fill, copper chemical-mechanical polishing (CMP) and copper wafer reclaim (Mendicino and Brown 1998). The first of these requires vapor deposition or plating, either electro- or electrodeless. Both plating processes involve a concentrated bath and rinse water. In the case of CMP, again, there is a waste stream containing the copper removed from the wafer. Wafer reclaim is needed because fabricators generate large numbers of test wafers. New reclamation processes are needed; the possibilities include nitric acid, sulfuric acid/hydrogen peroxide or ammonium persulfate/sulfuric acid. The latter appears preferable, since it is also compatible with electroplating waste (ibid).

There are a number of treatment possibilities, none of which is yet standardized. Most of the work so far has focused on the CMP effluent, which averages 2.5–5 ppm copper, almost entirely in soluble form with a neutral pH of 7–7.5 (Mendicino and Brown 1998). The copper is typically 95 percent removed by electro-winning, which means it can theoretically be recycled without refining (ibid).

In this connection, the approach pioneered by Noranda, discussed above, is worthy of further development and wider use, inasmuch as most of the rare metals in the scrap can be recovered economically from electrolytic refinery slimes.

5.4. Copper as a contaminant of recycled steel

Unfortunately it seems to be easier to separate iron from copper scrap (or ore) than vice versa. When the scrap metal is primarily steel (e.g. from automobiles) destined for recycling by electric furnaces in the secondary steel sector, there is a serious problem of copper contamination in the steel. To be specific, when copper contamination exceeds approximately 0.1 percent, the workability of EF steel is degraded somewhat [Katayama and Mizukami 1996]. The upper limit of copper contamination for steel used in the machinery sector (including automobile manufacturing), where a high degree of workability is required, seems to be around 0.3 percent, although the upper limit may actually decrease in future as the need to reduce weight induces steel producers to roll steel thinner and thinner. Thus, primary (convertor) steel from iron ore is needed for the machinery sector and secondary steel is almost exclusively destined for use in the construction sector, where workability requirements are less severe.

A model of the steel sector in Japan suggests that in coming decades secondary EF steel will be oversupplied for the construction sector, depending on the level of allowed copper contamination. The higher the allowed contamination level in EF steel, the sooner the oversupply point will be reached: for 0.4 percent contamination limit the oversupply year appears to be around 2032; for 0.3 percent contamination it occurs in 2040, and for 0.2 percent the crossover year is delayed until 2090 (Kakadate et al. 2000). The actual level of copper contamination in Japanese EF steels in 1995 was estimated as 0.4 percent (Toi et al. 1997; Kakadate et al. 2000).

The problem of achieving these lower limits remains unsolved, however. Whereas iron is effectively removed from copper concentrates during the smelting and refining sequence, the removal of copper traces from secondary steel seems to be extremely difficult. Up to now the only demonstrated method seems to be vacuum distillation, which is far too costly for practical application. As a consequence, the global stock of steel is being gradually contaminated by copper buildup, which also removes copper from the anthropogenic copper inventory.

It has been estimated by Japanese researchers that the output of EF steel in Japan reached about 40 MMT by 2000 (Kakadate et al. 2000, Figure 2). Assuming an average contamination rate of 0.3 percent, this would imply a diversion of 120 kMT of copper out of the copper cycle into the steel cycle in that single year in Japan alone. On a worldwide basis this would imply a loss closer to 500 kMT, or 4 percent of annual copper production, which seems unlikely at first glance, but not impossible. If these figures are anywhere near accurate the accumulation of copper in the global steel inventory of at least 10,000 MMT, of which at least 25 percent is of secondary origin, must be in the neighborhood of 5 MMT (assuming 0.2 percent copper contamination), and could be twice that or even more. A more detailed analysis of this question is clearly needed.

The percentage of copper in secondary steel is far too low to be interesting as a future source of copper, but it constitutes a growing barrier to recycling steel scrap with significant copper content. The obvious solution is to remove the copper before recycling the steel, but this is more easily said than done. (A similar problem exists with regard to copper contamination of aluminum scrap.)

A possible future process which has been explored by Daimler-Chrysler in conjunction with the steel producer Voest-Alpine is direct melting of end-of-life vehicles (ELVs) in a melt-reactor without prior dismantling. The organics contribute heat for melting and carbon for the steel fraction. It is alleged that non-ferrous metals can be separated from the melt, and most hazardous materials can be left in the slag, although the problem of copper contamination remains unsolved and the quality of the metal product is still uncertain. No commercialization is envisaged as yet (Naegele et al. 1996; Zoboli et al. 2000].

5.5. Copper recycling wastes and emissions

In general, recycling consumes less energy (exergy) from fuels and generates far less wastes than primary production, although precise data are very case specific. For instance chopping copper wire before melting costs about 2.05 GJ/t (including ancillaries) of which 1.2 GJ is for the process *per se*. If the wire is insulated (but not with PVC) the incineration stage – to burn off the insulation – takes about 1.95 GJ/t. If PVC is involved the combustion gases must be cleaned by wet scrubbers with subsequent effluent treatment (Jolly 2000, p. 26). Mechanical means of removing the insulation are increasingly favored because the burning process generates significant particulate emissions. But in this case the waste material ('fluff') must be utilized or disposed of. In the past some insulating plastics contained lead compounds as stabilizers; in this case the fluff has to be buried in a landfill.

Arsenic and sulfur, present in ore, are not problems for secondary copper smelting. However, volatile alloying metals including lead, zinc, cadmium and mercury can be emitted and must be captured by electrostatic precipitators (ESPs), baghouse filters or wet scrubbers. Baghouse dusts, in particular, are rich in recoverable metals, but they are also often classed as hazardous materials, which increases handling costs (ibid).

Table 5.6 (Bruch et al. 1995), based on three secondary copper mills in Germany, gives energy consumption, wastes and emissions resulting from the conversion of 1.04 t of scrap and 0.58 t of other secondary raw materials into 1 t of cathode copper. An idea of the amount of variation due to differing material inputs can be gained by comparing this table with *Table 2.4* (German production of 1 metric ton of copper from concentrate and scrap) and *Figure 2.1* (theoretical production of 1 metric ton of copper from raw ore only).

Globally, secondary copper refining in 1995 generated 317 metric tons of zinc, 160 metric tons of copper, 62 metric tons of lead, 5 metric tons of cadmium and antimony, 3 metric tons of arsenic and 2 metric tons of nickel, as particulates (Pacyna and Pacyna 2001, Table 5).

Table 5.6: Energy consumption, emissions, wastes and byproducts from the production of 1 metric ton of copper from secondary materials,[a] Germany 1992

	Total	Production only	Electricity and other
Primary energy (GJ/t)	*20.55*	*9.21*	*11.34*
Emissions			
Fumes (Nm^3)	32000	30440	1560
Dust (kg)	0.31	0.21	0.10
SO_2 (kg)	4.32	3.92	0.40
NO_X (kg)	3.20	1.96	1.24
CO (kg)	3.52	3.08	0.44
HC (kg)	0.23		0.23
CO_2 (kg)	1.96	1.23	0.73
Solid wastes (kg/t)			
Ash	< 20		< 20
Gypsum	< 10		< 10
Waste water (m^3)	*< 1*	*< 1*	
Marketable byproducts (kg/t)			
Slag	530	530	
Fly ash	120	120	
Anode mud	10	10	
$NiSO_4$	20	20	

[a] Major inputs are 1040 kg scrap and 580 kg other secondary raw materials.
Source: Bruch et al. (1995) Table 9, p. 8.

5.6. Recovery and recycling of lead (USGS annual, 1999, p. 627 and p. 62.13).

In the case of lead, secondary production is becoming more and more important. In 1960–1962 only about 22 percent of refined lead was from secondary sources; by 1986–1988 the recycled component had increased to 36 percent. By 1996 it was already about 50 percent globally. The recycled fraction of the current US supply has hovered between 61 percent and 65 percent since 1995 (USGS annual, 1999 Table 1). In 1999 about 76 percent of refined lead produced in the US was recycled from scrap, largely (93 percent) from automotive starting-lighting-ignition (SLI) batteries, although other lead-containing materials such as cable sheathing were also recovered.

As noted, a fairly high percentage of the global lead supply is now recycled from old scrap. This reflects the fact that the stock of lead in the technosphere of the industrialized countries is no longer growing rapidly. (This is obviously not true of developing countries, however.) In the past recycling was generally a small local business. The combination of tighter regulation and more sophisticated capital-intensive technology has driven out many of the small operators, especially those performing hand breaking and sorting operations.

Nevertheless, in 1998 there were 28 recovery facilities belonging to 20 companies.

The trend is toward more automated operations, based on shredders, flotation techniques and even whole battery furnace charging. The disadvantages of these methods are that battery acid and plastics are more likely to be discarded, with associated environmental costs. However in the long run the distinction between primary and secondary processors will become blurred as more and more refiners are able to handle both concentrate and scrap in a range of combinations.

Recycling of lead from products other than batteries remains problematic. Lead in motor fuel and lead pellets in shotgun ammunition are totally dissipative. The same is largely true for solder. The US National Safety Council estimated that 20.6 million PCs were discarded in 1998 alone; allowing for a recycling efficiency of 11 percent, and a lead content of 4–5 kg – mainly in the monitor glass – annual discards of PCs in the US (c. 1998) contained 70–90 kMT of lead, of which two-thirds to three-quarters was embodied in the glass.[4] Assuming the discarded PCs were 3 years old and that sales have been growing ever since at double digit rates, the annual usage of lead in new ophthalmic quality glass for TVs and PCs today (2002) must therefore be significantly larger than this, perhaps as much as 150 kMT/year in the US alone. Globally, the figure is at least double. Virtually none is recycled.

It must be recognized that the recycling system is still very 'leaky'. There are losses to the environment at every stage, from collection to refining, as well as in use. The lead cycle has been studied in some detail for Sweden for the year 1989. In that year approximately 20 kMT of lead, mostly from batteries, was recycled, corresponding to a recycling *rate* of about 57 percent. (The US recycling rate in 1993 was exactly the same.) Sweden was one of the first European countries to ban the use of lead as a gasoline additive, so dissipative losses from this use were insignificant. Emissions to air and water registered by local authorities amounted to about 0.5 kMT (about 0.25 percent of the recycled flow or 0.15 percent of the total flow) while another 3 kMT of lead in other products – 8.5 percent of total throughput – went into landfills [Karlsson et al, 1997, p. 49].

Globally, it has been estimated that lead recycling operations in 1995 generated particulates containing 143 metric tons of lead, 46 metric tons of zinc, 6 metric tons of cadmium, 2 metric tons of copper and 1 metric ton of arsenic (Pacyna and Pacyna 2001, Table 6).

---

4 See 'Please dispose of properly' by David Pescovitz in *Scientific American*, February 2000; also reply to a letter to the editor in the June 2000 issue of the same magazine by Robert Knowles (of Technology Recycling Inc.), who was quoted by Pescovitz,

5.7. Recovery and recycling of zinc

In the US the zinc recycling rate for old scrap in 1993 was 12 percent (Sibley et al. 1995). However from 1995 through 1999 the total recycling rate ranged between 24.2 percent (1995) to 27.5 percent (1998) (USGS annual 1999, Table 1). The difference is that the total recycling data refer mostly to new scrap. In fact, only about one quarter (6 percent overall) of the 1995–1999 recycling rate was old scrap, which is sharply down from 1993. The decline may be due to increased exports of scrap, although we have no data to confirm that.

Recovery of zinc from recycled galvanized iron and steel is becoming increasingly important, but is still small. The remainder was mostly from brass products, flue dust, old die castings and old rolled zinc products.

Baghouse dust from electric arc steel furnaces in the US contain about 18 percent zinc (and a little over 2 percent lead, plus some cadmium), on average, mostly from galvanized iron scrap (Llewellyn 1994). It was estimated several decades ago that between 200,000 and 250,000 tons of zinc were potentially available from electric furnace stack dusts in the US (Higley and Fine 1977). This zinc is economically recoverable by the so-called Waelz process, although only 20 percent of EAF dust was recycled for mineral recovery as late as 1990 (Llewellyn 1994, p. 13). Current research is aimed at stripping the zinc from galvanized surfaces prior to melting.

The major problem for recyclers is that secondary zinc is almost invariably associated with other metals, including copper, lead, arsenic, cadmium, and chromium. Baghouse dusts, for instance, are generally classed as hazardous materials, because of the impurities present, and this complicates disposal or transfer to other locations for processing. Moreover, zinc-containing materials are extremely variable. They include automobile scrap, brass scrap, flux skimming dross from the wet galvanizing process, zinc ash from the dry process, arc furnace dusts, and so on. Compositions vary widely, and recovery processes would have to be equally variable.

There are indications that hydro-metallurgical techniques may be adaptable to this situation, although the research is still at an early stage (Jha et al. 2001). In the medium to long term, it seems plausible that instead of specialized processes adapted to utilize very well-defined types of scrap for particular users, it may be feasible to design a more flexible batch-type system in which the most suitable lixiviant (either a strong acid or alkali) is determined for the specific mix of materials in the batch, with subsequent extraction by precipitation (cementation), solvent extraction, ion-exchange or electro-winning. The latter technology is already fairly well advanced for copper, of course. Undesired materials can be left in a slag, or simply recycled with other batches for recovery of other metals.

5.8. Recovery and recycling of byproduct metals

5.8.1. *Antimony (USGS monthly May 2002, pp. 22–23)*

We have found no data on amounts. Some antimony in the US is recovered from secondary lead smelters as antimonal lead (alloyed with lead in batteries). The value of antimony recovered in 2001 was $55 million, a decline from prior years due to declining use of lead in batteries.

5.8.2. *Arsenic (Loebenstein 1994, p. 8)*

Only 300–400 metric tons of post-consumer arsenic were recycled in the US in 1989 (alloyed with lead in lead-acid batteries). Process water from copper-chrome-arsenic (CCA) wood preservative operations is recycled internally as 'new scrap'. The same is true of arsenic wastes from gallium arsenide semiconductor manufacturing.

5.8.3. *Cadmium (Llewellyn 1994; ICDA 2001)*

NiCad batteries can be recycled, quite easily, but as of 1989 the actual recycling rate for consumer products was negligible. There were recovery operations in France, Japan, Sweden and Korea (since closed), but only for large industrial batteries. Other countries, notably Denmark, Germany, the Netherlands, UK and the US were trying to create necessary infrastructure for collection and sorting, to facilitate future recycling activities. It was expected by some optimists, in the late 1980s, that the recycling rate would approach 90 percent by 2000. However, this has not occurred.

The recycling efficiency for large industrial NiCad batteries is currently estimated to be 80 percent (ICDA 2001; USGS annual). For consumer NiCad batteries it is currently from 30 percent to 40 percent in the OECD countries, and may be as high as 60 percent in some countries (ibid). Recovered cadmium from batteries (as of 2000) was about 2.5 kMT, which amounts to about 17 percent of battery consumption and 12.5 percent of total cadmium consumption (ibid). About 10 percent of the global cadmium consumption was secondary in 1999, mostly from recycled batteries (USGS annual, p. 62.2). Secondary output in the US was only about 0.5 kMT/y, from one recovery plant, which started cadmium recovery in 1996. The recovery process for large batteries (> 2 kg) involves dismantling by hand. Detached cadmium plates are then remelted. Smaller batteries are treated by burning off the casings and separators prior to remelting (ibid).

As noted in connection with zinc recycling, significant quantities of cadmium are contained in EAF flue dust. The recovery rate from this source was about 20 percent in 1990, although the recovery rate was expected to rise rapidly due to the hazardous nature of the material (and the increasing demand for cadmium). Scrap aluminum remelting is another potential source.

However, low prices for cadmium in recent years have inhibited commercial recovery of cadmium from baghouse dusts.

Recycling would undoubtedly increase if prices were higher, which would follow from more accurate public perceptions of the real hazards vs the real benefits of cadmium. While it is true that cadmium is highly toxic, most human exposure arises from cadmium that is disseminated into the environment from fossil fuel combustion, phosphate fertilizer, ferrous metallurgical operations and incineration of wastes, such as PVC containing cadmium stabilizers or printed paper with cadmium-based inks. In short, from processes involving materials where cadmium is a trace contaminant. There is actually very little hazard from metallic uses of cadmium as such.

As a matter of fact, wider use and higher prices for cadmium could result in substantially reduced human exposure if it resulted in the development of practical technologies for the recovery of cadmium from phosphate fertilizers (prior to use), and iron ore. This possibility deserves some attention from researchers.

5.8.4. *Germanium (USGS monthly May 2002, pp. 70–71)*

In the US 75 percent of germanium consumption (29 metric tons Ge content) is used in manufacturing electronic and optical systems. Of this more than 50 percent is recycled as new scrap; that is, more that 11 metric tons (Ge) is recycled. Around 25 percent of world consumption, or 17 metric tons (Ge) is produced from recycled materials. Essentially no old scrap is recycled.

5.8.5. *Gold (USGS annual, p. 62.4)*

Old scrap from jewelry and dental materials accounts for 13–25 percent of US gold supply. Domestic old scrap recovered in 1999 was 73 metric tons. Data for the rest of the world is non-existent. New scrap (from processors and manufacturers) is mostly recovered. Domestic consumption of new scrap in 1999 was 100.5 metric tons. The US is also a net exporter of gold scrap.

5.8.6. *Indium (USGS annual, p. 62.5)*

Indium is not recycled to a significant degree under normal circumstances; however the year 1996 was exceptional, due to unusually high prices. In that year much of the US domestic supply was from recycled scrap, while imports declined by 50 percent in that year. In Japan in 1999 imports were 91 metric tons, ingot production of primary indium was 20 metric tons, while recycling accounted for 55 metric tons.

5.8.7. *Selenium (USGS annual, p. 62.11)*

Scrap from plain paper copier drums is efficiently recycled in Japan, Europe, Canada and the Philippines. There is no secondary recovery in the US; all selenium-bearing scrap is exported.

5.8.8. *Silver (USGS annual, p. 62.11)*

Secondary silver recovery in the US in 1999 was 1,801 metric tons, of which 1,300 was from photographic film processing laboratories. The rest was from jewelers sweepings, spent catalysts, electronic scrap, and other miscellaneous sources.

5.8.9. *Tellurium (USGS monthly May 2002 , pp. 168–169)*

No quantitative data on recovery or recycling has been found. Small amounts of tellurium and recovered from selenium-based photreceptor scrap exported from the US, but there is no US secondary production of tellurium.

5.9. Further comments on losses and potential recoverability

A point that is implicit in much of the preceding discussion, but which deserves emphasis, is that the potential for recycling is strongly dependent on the nature of the use. At one extreme, metal objects of known composition, such as steel structures or copper wiring, are easy to recycle. It is this which accounts for the very high recycling efficiency of iron and steel, for instance (subject to the growing problem of contamination, discussed above.) At the other extreme, metals used in chemical form are very difficult or impossible to recycle. In between these cases is a wide range of intermediate situations. *Table 4.9* is a rough indication of the range of uses and losses associated with those uses (Karlsson et al. 1997).

In general, very small metallic household objects such as fasteners, bottle caps, light bulbs, batteries and small appliances (such as radios and TVs) end up in landfills. Recoverability of metals from landfills in the distant future is unlikely but barely possible in some cases, as where the landfill site occupies valuable land and/or must be cleared for other reasons. The enormous Fresh Kills landfill in Staten Island (New York City) is a possible candidate for future 'waste mining' if (and only if) a very clean and efficient technology for doing so can be developed and tested in less contentious locations.

Yet there are policies, and technologies, that can sharply reduce the loss of scarce metals (such as copper and its co-products and byproducts) into landfills. Separation of waste streams at the point of collection is already becoming standard practice in Europe and some US cities. Separate collection of aluminum cans, paper and glass bottles considerably reduces the bulk of the remaining wastes and correspondingly increases the concentration of metals in the remainder.

High temperature incineration with energy recovery (as practiced currently in a number of cities) also yields bottom ash and flue gas residues that are much 'richer' in these metals. Moreover, the combustion process itself partitions the metallic components by separating the metals with high melting points and low vapor pressure (like copper) from the more volatile metals like mercury, cadmium and zinc. (Improved air pollution control technologies now available are able to increase the transfer of volatile metals from vapor into solid residues from 20 percent to 99 percent for cadmium and from 8 percent to over 98 percent for mercury (Rechberger and Brunner personal communication 2001).

The same point can be made with regard to liquid process wastes from many cleaning, electro-plating, and other operations in the manufacturing and service sectors. Again, treatment to allow recycling of the process water – already becoming scarce in many locations – yields sludges that incorporate greater concentrations of these metals. While these sludges are not currently rich enough to 'mine' they must be stored somewhere and these storage sites may be targets of future 'cleanup' campaigns. The toxic heavy metals concentrations that make these sludges hazardous may also make them future potential 'ores'. Indeed, based on 1986 data on industrial hazardous wastes, it appears that 85–92 percent of the roughly 95 kMT of copper in US hazardous waste streams and 96–98 percent of the 250 kMT of zinc in identified hazardous waste streams in the US is potentially recoverable (Allen and Behmanesh 1994). Yet, in 1986 only 10 percent of the copper and 13 percent of the zinc was actually recovered from these streams. We have no data on recent developments in this regard, if any. As regards the other metals of interest, Allen and Behmanesh (ibid) estimated that 84–95 percent of the lead was recoverable, while only 56 percent was recovered. For arsenic the corresponding figures were 98–99 percent recoverable and 3 percent recovered; for cadmium 82–97 percent recoverable, 7 percent recovered; for selenium 93–95 percent was recoverable while 16 percent was recovered; and for silver 99–100 percent was recoverable in principle, yet only 1 percent was actually recycled from waste streams.[5]

[5] The silver case illustrates some of the practical problems. Most of the silver wastes consist of silver halides from photographic processing. However, the photographic processors were mostly small, generating only a few gallons of spent developer solutions per month. There was no economical way to recycle these wastes on site, nor any infrastructure for collecting them and processing them off-site. Regulatory restraints made the problem worse (Allen 2002).

CHAPTER 6

CONCLUSIONS AND QUESTIONS

6.1. Introduction

In this section we note the major conclusions and implications of the study, without attempting to summarize each chapter in detail. We consider long term supply, recovery technology, usage (and possible substitutes), recycling and environmental impact, in that order.

6.2. Copper availability

Copper, lead and zinc are so-called exhaustible resources. The question arises: is there enough economically recoverable ore in the earth's crust to support projected rates of economic growth throughout the next century?

The major cause for concern is that the three metals are currently being mined from mineral ores at grades far higher than the average in the earth's crust. These high grade ore bodies exist because of natural geochemical concentration processes, some of which are still only imperfectly understood. However, it is quite unlikely that these ores are distributed like the more common metals, i.e. in such a way that the quantity of ore increases monotonically as the grade decreases, down to the crustal average. It is far more likely that copper (and lead and zinc) are characterized by double-peaked (or, conceivably, multiple peaked) quantity-grade distribution functions, as was seen in *Figure 1.7.*

In other words, it is probable that a relatively small fraction of the total crustal copper is in relatively high grade mineral ores (mainly sulfides) while most of the remainder is dispersed more or less uniformly at extremely low concentrations (a few parts per million) in so-called atomic substitution sites in ordinary rock. In effect the two peaks of the are separated by a 'mineralogical barrier'.

Over a very long time horizon, then, there are two (or more) distinct parts of the distribution curve. These correspond to phases of the extraction history. The first phase of mining history (not yet completed) can be characterized as 'climbing up' the quantity-grade distribution function, toward the (unknown) peak, where the quantity of ore at the given grade is maximum. As a rule the highest grade ores are mined first, but there is not much available of the very richest ores. As mining continues, during this phase of the cycle, the known and identified reserves tend to grow as mining firms continue to explore and new deposits are discovered. In fact, the current rate of decline appears to have slowed down while the identified reserves appear to be growing, but at a very slow rate.

Published 'official' reserves must not be confused with actual mineable resources in the ground. The reasons are twofold. First, and most important, it is not economically worthwhile for mining companies (who do most of the exploration) to undertake detailed geological mapping to discover resources that will not be needed for more than two decades in the future. The second reason why published reserves are unrealistically low is that new technology makes exploration and extraction more efficient, thus reducing the costs of extraction for a given grade of ore or reducing the ore grade that can be economically utilized at a given point in time.

However, at some point in time, the rate of discovery will slow down, the peak of the quantity-grade distribution will be reached, the decline in ore grade being mined will accelerate, and the stockpile of known reserves will also begin to fall. These events will mark the beginning of the second phase of extraction history. Unfortunately there is no means of forecasting this turn-around point exactly, in the absence of complete data on underground resource quantities as a function of grade. It is thought by some experts that this point may occur in the case of copper when the lowest ore grade being mined falls to around 0.1 percent or so. In the second phase of copper mining the energy requirements, and materials handling costs of mining and concentration will begin to increase sharply. Prices would probably begin to rise fairly sharply around that time, if not sooner.

Copper ores (and lead-zinc ores) can be, and are being, extracted and concentrated by technologies that exploit the physical and chemical properties of the ore minerals, notably the sulfur content of the ores. However no current or projected technology can recover copper economically from sea-water or from atomic substitution sites in ordinary crustal rock, such as granite or feldspar. In either case, the amount of raw material and energy required for processing would be hundreds or thousands of times greater than current mining practice. Inevitably there will come a time in the long run when recycling from anthropogenic stocks will be the dominant source, while recovery and recycling of these metals from landfills or polluted industrial sites will be less costly than extraction from crustal rock or sea water.

All of the mine production scenarios we have investigated predict peaks. In four scenarios (2, 4, 6, 8) the peaks occur between 2055 and 2060. In the other four scenarios (1, 3, 5, 7) the peaks are later, between 2065 and 2085. The lowest peak production rate (scenario 7, is just under 50 MMT/y (around 2065–2070) while the highest peak (scenario 1) is close to 70 MMT/y (around 2080–2085).

On the other hand, currently known global copper reserves of 310 MMT – or perhaps a little larger (see *Table 2.1*) are sufficient for less than 30 years production at current levels and only 5 or 6 years at projected peak production levels. The known reserve stock, including copper already mined and the so-called 'reserve base', allowing for steadily increasing demand, will be exhausted in all scenarios by 2025 (see *Figure 3.21*). In fact, depending on the scenario, new reserves amounting to 4–5 times the total of all copper already mined, or now counted as reserves would have to be found to support projected consumption over the coming century. Given the known distribution of copper ores around the globe, it is implausible that such a lot of new copper reserves remain to be discovered.

To be sure, existing reserve stocks will probably increase for some time to come, and ores can be profitably extracted from porphyry-type ores of much lower grade than the current world average. Moreover, the average ore grade mined in the US has not declined recently. It has remained close to 0.5 percent for the past decade or so, despite increasing production levels (*Figure 2.8*). The existence of surplus production capacity in the world is further indicated by the continuing price declines of recent years (*Figure 3.6*) which has actually prompted the closure of several US mines and smelters.

Nevertheless, we suspect that these indicators are somewhat misleading. The price decline of recent years probably mainly reflects two things. One of them is sharply reduced current consumption in the former USSR since 1990 (accompanied by large exports of copper scrap), plus the 1997 economic collapse in East Asia. The other is rapid introduction of the low cost SX-EW process, first in the US and increasingly elsewhere. Both cost-depressing factors will have run their course in another ten years or so. Meanwhile low prices depress exploration and development of new resources, so even if there are significant copper resources left to be discovered, the search will slow down near-term.

For these reasons we think peak global copper mine production – and the transition from phase 1 to phase 2 – will probably occur in less than 15 years at levels not much more than double global current production, and possibly less. The peak may not be followed by an immediate decline in reserves and grades. The transition period (during which the US will probably cease to be a major primary producer) is likely to be somewhat extended.

However by mid century, whatever happens, the total stock of copper in use and the total quantity of copper in gangue, slag heaps, other waste disposal sites, and dispersed into the terrestrial environment will be several times the present level. Taking into account future leaching of existing waste deposits, copper concentrations in topsoil and surface waters are likely to be much higher than at present. However, recovery of copper from topsoil or sediments is not likely to be feasible under any circumstances.

6.3. Copper demand: the continuing electrification of the global energy system

Economists tend to argue that there are substitutes for every substance, depending only on price. This may be true in some ultimate sense. However copper has unique physical characteristics – notably electrical and thermal conductivity – that, until recently, made it appear essentially non-substitutable in its major electrical applications. While it is true that aluminum is also a reasonably good electrical conductor, it is not really a practical alternative except in very high voltage transmission lines (where it is already used). For applications at lower power levels, the use of aluminum wire in place of copper wire would entail significantly greater thermal losses (i.e. lower efficiencies). Moreover, copper has other physical properties that make it easier to work with than aluminum.

But the world economy is steadily electrifying, in the sense that an increasing fraction of total energy from all sources is converted first to electricity and utilized in that form. Light was the first application to be electrified, followed by stationary motive power for factories, refrigeration, air conditioning, ventilation, cooking and water heating. Telecommunication and information processing is inherently dependent on electric power, notwithstanding the substitution of glass fibers for copper wires or microwaves as information carriers. Electric power is increasingly important in metallurgical and chemical applications (including both aluminum reduction, and copper electro-winning and electrolytic refining).

The two major energy 'holdouts' against electrification, so to speak, are residential and commercial space heating and transportation. But electric heat pumps and co-generation will become increasingly important for space heating, while hybrid or fuel-cell-battery-motor combinations will probably eventually displace IC engines with variable speed mechanical transmissions in motor vehicles. These developments will require ever more copper.

Non-electrical uses of copper (especially as brass) will doubtless continue to be important for several decades to come, but the electrical applications of copper are already dominant and will eventually displace most other uses. A scarcity of copper a few decades from now could become a significant drag on economic growth, especially in developing countries lacking a large stockpile of copper in use.

The light at the end of the tunnel, so to speak, may be the commercialization of high temperature superconductors (HST) starting with underground cables but eventually for local distribution in cities and even applications in large buildings. Until 1986 all known superconductors required refrigeration at temperatures a few degrees above absolute zero. A new class of ceramic superconductors made it possible to operate at the temperature of liquid nitrogen, which is commercially available at very low prices. Moreover, new insulating materials appeared on the scene, facilitating the possibility of refrigerated cables. None of this is commercially available now, but rapid commercialization may begin within a few years. This could release quite a lot of copper from existing underground copper cables, and possibly heavy-duty generating equipment, over the coming half century.

What HTS will not do, of course, is to replace the copper in household wiring, most industrial machinery, vehicles, and the vast majority of electrical and electronic appliances, where refrigeration is not really an option. On the other hand, a breakthrough in organic superconductors, which has been an object of serious investigation for a number of years, might do just that. Hold the phone.

6.4. Lead, zinc and byproduct metals availability and uses

In the case of lead, peak production has probably already occurred, and future demand will be increasingly limited to the two major existing markets, namely lead-acid automotive storage batteries and ophthalmic quality glass used in TV screens and computer monitors. The latter market is much the smaller of the two, although it is growing rapidly. On the other hand, large flat screen units, soon to replace large numbers of TVs, will not need leaded glass.

The lead-acid battery market is also still growing, as the world automotive vehicle fleet keeps growing, but recycling can be extremely efficient and – before any critical shortage of lead occurs – a new and more efficient type of battery (most likely nickel zinc or nickel-metal hydride) will start to replace lead in this application. Thus the decline in mine production of lead is likely to continue indefinitely (although short term increases may occur).

Zinc is a very different situation. It is less scarce than copper, though by no means common. It is still being mined from much higher grade ores. Its uses are very diverse, but the major one by far is as an anti-corrosion coating for steel. That use will probably continue to increase for several decades, inasmuch as steel is not likely to be replaced for construction purposes in the foreseeable future. On the other hand, the largest user of galvanized steel at present is the motor vehicle industry, where lighter alternative materials (aluminum and engineering plastics) will continue to reduce the need for steel.

Many of the other uses of zinc are probably substitutable, over time. We anticipate no supply crisis, but rather an indefinite continuation of mine output and demand at levels not very different from those of today, until prices begin to rise significantly. Increased demand (if it occurs) can be accommodated for quite some time by increased recycling rates, especially from galvanized steel.

Some byproduct metals like indium, rhenium, selenium and tellurium have properties that make it likely that they will find valuable new uses of an environmentally acceptable nature. In these cases, the amounts available will never be enough to satisfy potential demand. Platinum, silver and gold are already very valuable for their special properties, so much so that they are extensively mined for themselves. For environmental reasons such mining – especially for gold – is likely to be more and more discouraged. Moreover a number of countries have large stockpiles they would like to sell off, and gold mining tends to depress the price. All of this will eventually tend to drive prices up, relative to other commodities, although not soon enough to make the gold-bugs happy. As things stand, the additional values obtained from separating these byproduct metals from copper, lead and zinc smelters is already important. It will constitute an increasing proportion of the revenues and profits of copper, lead and zinc mining.

On the other hand certain other byproducts, notably arsenic and cadmium, and possibly antimony, bismuth and thallium, will be available in quantities much greater than anticipated acceptable uses. We discuss this problem below.

6.5. Concentration, reduction and refining technology

Three major near-term trends are worth mentioning, although already well-known to the mining industry. First, hydro-metallurgical methods (for copper), exemplified by the SX-EW process, are now practical and economically advantageous in dealing with low grade oxide ores, including some wastes, although the recovery rate is currently lower than it is for traditional concentration and pyro-metallurgical processes. These processes are also applicable to lead and zinc mining, and further development seems to be well under way.

Second, in pyro-metallurgy there is a clear trend away from multi-step batch operations toward continuous one-step operation from concentrate to blister. This has not yet been fully achieved, but several different technologies introduced since the 1970s are promising. The old reverberatory furnace-based batch technology has definitely been superceded by flash furnaces (e.g. Outokumpu and INCO) or Noranda furnaces. There are strong indications that incremental improvements will eventually enable fully continuous operation, as pioneered by Mitsubishi.

Third, the economics of pyro-metallurgical operations are already (and will increasingly be) driven by sulfur recovery (see also the next section). The reduction of sulfides ores is potentially exothermic, which means that in an ideal process no other fuels or electricity would be needed. In effect, all sulfur recovered has a treble value: first, the byproduct acid has value (it is needed for the SX-EW process); second, the reaction heat eliminates the need for supplementary fossil fuel energy; and third, efficient use of sulfur as fuel (later recovered as acid) reduces both sulfur dioxide and carbon dioxide emissions.

6.6. Sulfur recovery and acidification of the environment

Sulfur recovery is the antidote to acidification, at least in the mining regions. By-product sulfur – mainly from copper smelters – is already a significant share of global elemental sulfur supply (around 19 percent) and will continue to be important for the next few decades. However thanks to the solvent-extraction electro-winning technology (SX-EW) the copper industry in the US has already switched from being a net seller of sulfur to being a net buyer. In the long run we suspect that the pyro-metallurgical and hydro-metallurgical branches of the industry will – for good economic reasons – gradually approach approximately a zero sulfur balance: that is, the hydro-metallurgical branch will utilize the surplus sulfur from the pyro-metallurgical branch.

Even so, the non-ferrous metals industry is likely to be a significant polluter in the more remote developing countries like Zaire, Zimbabwe, Papua New Guinea and Indonesia due to corruption, lax environmental standards, lax enforcement or all three. This, in turn, contributes to regional (and global) acidification as well as more localized pollution and environmental health problems.

Despite increasingly effective sulfur recovery, not all the sulfur is recovered from the ore itself, and even more remains with the gangue and overburden. This constitutes a potential long-term problem, known as acid rock drainage (ARD). The problem is associated mainly with closed and abandoned mines, rather than active ones. It arises in the first instance from the exposure of sulfide-bearing waste materials, mainly iron pyrites, to air and water. The exposed sulfides are gradually oxidized, yielding sulfurous and sulfuric acid which reacts with minerals in the rock surface and causes further cracking or spalling. The rate of abiotic oxidation is fairly slow, but in certain circumstances the presence of the bacteria *Thiobacillus ferrooxidans* can catalyze the process and increase the oxidation/acidification rate radically. Once this happens, the acidification proceeds auto-catalytically, and is very hard to control. Of course, in a highly acid environment many trace metals, including arsenic, cadmium, copper, lead and zinc, are mobilized. However, concentrations are much too low for recovery via electrowinning, for instance. These contaminants may be precipitated naturally onto iron oxyhydroxides (ochers) or aluminum hydroxides, depending on pH and other factors. But under worst case conditions they may subsequently leach into groundwater or surface waters. Unusual events, such as heavy storms, can also cause the acidic waters to escape from ponds or other containment facilities and pollute downstream areas, as has happened several times.

The ARD problem has not been addressed sufficiently by the mining industry in the past, and the detailed mechanisms involved are not yet sufficiently well-understood. In principle the problem can be prevented before it starts. But prevention requires prediction (as well as long-term investment) and the state of the science is still not what it should be. It is still much too easy for mine owners to walk away from an exhausted deposit, leaving problems that may continue for decades or generations if not addressed adequately from the beginning.

6.7. Copper, lead and zinc recycling

The irreversible trend toward global electrification together with limited availability of copper ore means that copper prices will inevitably start to rise and the recycling rate for copper will necessarily approach 100 percent, probably before the end of the present century. Whether the peak in copper mine output occurs within the next two or three decades, or not, there is no doubt that recycling must and will become the dominant source of supply for copper in the fairly near future (as it is already for lead), and that zinc recycling cannot be far behind.

Unfortunately the non-ferrous metal recycling system, as a whole, is disorganized, technologically backward and lacking in economies of scale at present. It also suffers from being essentially in competition with the mining sector, which has inhibited its growth. For these reasons, the recycling efficiency – i.e. the fraction of the metal in use that is eventually recovered – is significantly lower than it should be in all three cases, especially for copper and zinc. To say it in another way, too much of these metals is currently being dissipated and 'lost' with adverse environmental consequences as well as unfortunate long-term implications for supply. This is perhaps not the place to introduce new (to the mining industry) ideas, but it must be said that a vertical integration of the primary and secondary sectors would be beneficial in terms of promoting a more comprehensive approach.

Best of all, we think, would be an evolutionary transformation of the primary copper-lead-zinc miners from an 'extract, refine and sell' industry, to a true 'service' industry which treats each of the metals as a capital asset rather than as a commodity. This shift would trigger changes at every stage in the life cycle, from 'cradle to grave' as the phrase has it. Most important, probably, would be changes in design philosophy to facilitate product remanufacturing and metal recovery, and the development of an efficient 'reverse logistics' system for collection and return. Considering the fact that goods transport at present tends to be uni-directional, with substantial excess capacity in the reverse direction, the problem is essentially one of organization and coordination.

The details of such a fundamental restructuring cannot be discussed here, but it would mean that copper and the other metals would be (in effect) rented to users and taken back at the end of use. Of course this 'take back' could occur at any stage along the product chain. For instance, it seems likely that many electric motors could be refurbished and resold as such, rather than being recycled to the component metals.

From a technological point of view, there is much to be done in terms of developing practical and cost-effective ways of recycling very complex products such as TVs and PCs. The most promising approach for the latter seems to be to incorporate them with ore concentrates and other scrap in Noranda-type furnaces, where the scrap would actually enrich the byproduct stream. It is worth noting that this option is not available in a 'pure' recycling society.

6.8. Emissions and accumulation of copper and zinc in agricultural soils; probably a non-problem

Emissions from mine operations and smelters are increasingly under control. Apart from health-related environmental concerns, the economic benefit of sulfur recovery, not to mention the recovery of byproduct metals like silver and gold, is driving the mining industry toward more and more efficient materials and waste management.

On the other hand, many copper mines are located in dry areas and desert areas are problematic from another perspective, viz. it is more difficult to control acid rock drainage (ARD) in normally dry areas subject to only occasional inundations. (For example, the west coast of South America, where a significant fraction of the world's known copper reserves are located, is subject to occasional El Niño events, accompanied by heavy rainfall but at intervals of years or even decades.

The buildup of heavy metals in topsoil has attracted much attention. However the potential harm is probably much less than originally feared. As noted at the beginning of *Chapter 4*, copper and zinc (and some of the other byproduct metals, such as selenium) are essential for plant growth, although they can also be toxic in high ionic concentrations. Each of these metals is actually concentrated by a large factor in plants. As long as the metals are deposited on the soil surface gradually, whether by atmospheric deposition or as trace elements in fertilizer or manure, they are quickly adsorbed by clay particles or chelated (i.e. bonded to organic materials as ligands) and subsequently, over a longer time, converted to insoluble compounds that are permanently immobilized. In short, there is no correlation between the concentration of these metals in soil and the uptake by plants or the ecotoxity of the metal. What matters is the concentration of ionized metal compounds, which is normally very small for the reasons noted.

This does not rule out localized pollution problems, e.g. where the surface runoff from a large copper roof or a zinc galvanizing operation is channeled into a small stream, where it can cause devastation for some distance downstream. However, the best available evidence suggests that the capacity of most agricultural soils to contain copper and zinc is much higher than the actual levels observed now or likely to be reached even a hundred years from now. This issue is discussed in more detail from a technical perspective in *Appendix B*.

6.9. Accumulation of arsenic, cadmium and other toxic byproduct metals in the terrestrial environment: A real problem

As noted above, there are certain metals for which the byproduct supply from mine production will probably exceed any environmentally acceptable use.

Arsenic is the prime example, though others may also pose problems of this kind. We have noted already the likely oversupply of antimony, bismuth, cadmium, and thallium. Arsenic is mostly a byproduct of copper mining. Although there is no fixed relationship between arsenic content and copper content of ores, the average arsenic content of US ores is 6.5 kg per metric ton of copper (*Chapter 4, section 4.8.2*).

A simple mass balance calculation makes the point. Assuming this proportional relationship between arsenic and copper holds for the rest of the world, the amount of arsenic associated with and mobilized by copper mining, worldwide (12 MMT) would be 78 kMT for the year 2000. Of this amount – that which is originally in the ore – only 24 kMT was actually produced and utilized (in the form of 40 kMT of arsenic trioxide) as noted in *Chapter 3, section 3.8.2*. This leaves about 54 kMT of arsenic unaccounted for. Obviously this figure is very uncertain, but the error could be in either direction. In other words, the amount of missing arsenic could be greater as well as smaller.

Much of the missing arsenic presumably remains in slag or other wastes left at mine or smelter site, depending on the efficiency of the air pollution control equipment installed. Whatever is not captured is presumably released in one form or another. Since arsenic is comparatively volatile, it is likely that a few percent, at least, escapes into the atmosphere. The escaping fraction depends on the technology of the smelter-refinery.

Most past uses of arsenic, as rodenticides, pesticides, herbicides and so on depend on its toxicity. Of the arsenic that is still utilized in the economy (mainly in the US), almost all of it is now in the form of a wood preservative known as chromated copper arsenate, or CCA, which inhibits the growth of decay organisms that cause wood to rot. Most other uses of arsenic in industrialized countries have been phased out because of its toxicity. However the CCA that is impregnated into wood products extends the natural life of the wood by as much as a factor of 5, but it is not permanently immobilized. Wood that is exposed to sunlight and rain does eventually break down – after as few as ten years in some circumstances – and the arsenic (as well as the chromic acid and copper) does eventually leach out.

This is no longer just a theoretical possibility. A recent issue (May 2002) of *Time* magazine carried an article entitled 'Toxic playgrounds' calling attention to the fact that 98 percent of the wood sold for outdoor use in the US (including children's playgrounds) is treated with CCA, and that in the state of Florida alone, nearly 30,000 tons (27,000 metric tons) of arsenic have accumulated in this form. There is, so far, no convincing proof that people have been harmed, but it takes little imagination to see that a few decades from now there could be serious problems of arsenic buildup in soil or sediments. Disposal of end-of-life wood products, whether by burning, mulching or deposition in landfills, will eventually release the arsenic. Arsenic buildup in soil does not automatically imply uptake by plants or any associated threat to humans. (See *Appendix B*). However if the arsenic leaches into aquifers, the threat would be direct and potentially serious.[1]

It would be helpful if a truly environmentally benign use of arsenic could be found, such as the semiconductor compound gallium arsenide. However only 15 metric tons of gallium arsenide were produced in 1998, and there seems little prospect of an increase in demand large enough to affect the arsenic supply.

A superficially similar problem arises with cadmium, a byproduct of copper, lead and zinc mining, as well as a contaminant of phosphate rock and some coals. Most of the cadmium used in commerce is recovered from zinc (on average 3 kg/t). Assuming the US cadmium-zinc ratio holds for the rest of the world, the cadmium mobilized by zinc mining alone (7.2 MMT in 1996) would be around 20 kMT, and that does not include the other potential sources of cadmium, such as phosphate rock, that are not currently recovered. As it happens this is fairly close to world refinery production of cadmium. This is understandable in view of the fact that cadmium currently enjoys a very strong market for use in rechargeable nickel-cadmium batteries. NiCad batteries are used in a variety of small appliances, including video cameras, electric toothbrushes, power tools, flashlights and so forth, but the big and growing market is for laptop and notebook personal computers.

On the other hand, there is no guarantee that this market will continue growing indefinitely. There is increasing evidence that lithium batteries will eventually displace NiCad batteries for many uses. Some countries in Europe are considering a ban on cadmium, based largely on the fact that nickel cadmium batteries after use are mostly discarded and taken to landfills, where the cadmium can eventually leach into the groundwater. Actually it seems likely that cadmium in batteries could be recycled without too much difficulty by means of a deposit-return system, although both merchants and customers are likely to resist.

A point that is often overlooked but that deserves emphasis in this context is that the cadmium pollution in the environment is almost totally disconnected from the industrial use of cadmium, especially as batteries. For the most part it is associated with and released by coal burning, iron and steel operations and phosphate fertilizer use. A minor contribution may arise from incinerators. However, while the final disposition of nickel-cadmium batteries should certainly be controlled better than it is now, it seems likely that the use of this metal in a long-lived and easily recycled product is probably safer from an environmental standpoint than any attempt to 'ban' cadmium usage, given that cadmium is going to be produced as long as zinc is mined.

---

[1] The widespread use of tubewells in Bangladesh has had tragic results for many rural families, due to arsenic contamination in the water (probably from natural sources, in this case.) However, many people in the US and Europe also depend on water from wells.

6.10. The threat of 'toxic time bombs'

Even if recycling becomes extremely efficient in the future – far more so than at present – there remains the problem of toxic heavy metals – other than copper and zinc – building up in river, lake and estuarine sediments and in agricultural soils, where they can be taken up into the human food chain. The sort of mechanisms by which this can happen were well illustrated by the Minimata experience in Japan back in the early 1960s. In that case there was a discharge of inorganic mercury oxides (used as catalysts) from a chemical plant. The concentration of mercury in the discharge was not high, and it was greatly diluted by the waters of the Minimata Bay. Nevertheless, the mercury compounds – being heavy and insoluble – sank to the bottom and accumulated in the anaerobic sediments under the bay. Under such conditions (it is now known) certain anaerobic bacteria can convert the inorganic mercury oxides to methyl mercury, meanwhile utilizing the oxygen for their own metabolisms. Methyl mercury was then taken up by benthic organisms on the sea bottom, and thence by shellfish, bottom dwelling fish, then other fish, and finally humans (as well as sea birds, cats and so on.) Several hundred people were poisoned by the mercury and a number of them suffered irreversible central nervous system damage.

The same sort of problem could conceivably arise from other toxic metals under certain circumstances. As pointed out several times in the foregoing, many heavy metals tend to bond with clay particles or organic materials as ligands, under conditions of moderate to high pH (alkalinity). As long as the environment is not too acid, the metals are 'permanently' immobilized and harmless. However, increased acidity increases ionization. Even moderate changes in the ionic concentration of some toxic metals will result in much higher uptake rates into plants, with potentially deadly consequences, either to the plants or the plant-eaters.

As noted previously in connection with ARD, oxidation of sulfur (or even organic carbon) will increase environmental acidity. There are possible circumstances, such as floods, that could suddenly expose large amounts of previously anaerobic river sediment to oxygenated water. That would result in more or less instant mobilization of the heavy metals previously 'locked up' by clay particles. As far as agricultural soils are concerned, liming is the key. Any large scale conversion of previously limed agricultural soils to tree crops, for instance, could also remobilize toxic heavy metals.

6.11. The long-term prospect

Copper, and its co-products are scarce metals with valuable physical properties. We are currently exploiting natural stockpiles of these metals. The global demand for copper (and zinc) is still growing and will certainly increase several-fold in the coming century. The exact magnitude of the increase is uncertain, of course, but what is of primary importance is that toxic byproduct metals such as arsenic and cadmium will also be entering the economy – and the environment – in increasing quantities. The temptation to let these byproduct metals build up in mine wastes or use them in dissipative ways (e.g. arsenic as a wood preservative) will be hard to resist.

Meanwhile the global resources of copper and its co-products will be effectively exhausted within a century and perhaps significantly earlier. Purely dissipative chemical uses are gradually declining in importance, but with a few exceptions (e.g. tetraethyl lead) they are not being actively phased out.

More serious, from a future resource availability standpoint, is the unintentional dissipation that results from the use of scarce metals in complex electronic and other products that are discarded and lost because they are too complex to recover and recycle economically. Copper roofing, copper pipes and copper telephone wires or underground cables are easily recovered and recycled. Unfortunately this is not true for TVs, PCs, kitchen appliances, electric toys, or automobiles where the copper (or much of it) is combined with much larger quantities of other materials, making recovery and recycling far more difficult if not impossible. The recycling efficiency for copper and zinc is actually declining, not increasing. Very little is being done to reverse this trend.

A related difficulty is that copper is becoming an increasingly problematic contaminant of recycled iron and steel, and aluminum. Not only is the contaminant copper unavailable, but the quality of secondary (recycled) steel is also being adversely affected. A practical means of recovering copper and other scarce metals from secondary steel and secondary aluminum is becoming an urgent need.

In a few generations, at most, mankind will be forced to recycle copper, lead, zinc and the other byproduct metals with very high efficiency, comparable to the efficiency with which gold is recycled today. The alternative will be to live without these metals (except insofar as they are undesirable contaminants of sediments, topsoil or other materials). True, substitutes may be feasible for most uses of copper, lead and zinc. But in the case of copper and zinc, it will be a long time before substitutes are likely to be available for the most important uses. Neither aluminum nor any high temperature superconductors known today can replace copper in household wiring, vehicles, electrical appliances or electronics. Even if practical alternatives were available, substitution in most products would take several decades and, in buildings, considerably longer.

In short, there is a lot to do, and not much time to do it. The dynamics of the market are not helping. Market forces would call forth better recycling technologies and substitutes for some uses, if prices were rising and if a rising price trend could be forecast with some confidence into the future. But prices have recently been falling, not rising. This promotes even more dissipative uses, and discourages recovery and recycling.

Yet short-term trends (like market bubbles) tend to be self-sustaining – for a while. Then changes of direction can occur very suddenly, and often when they are least expected.

The copper-lead-zinc industry, comprising mines, smelter-refiners, and fabricators, needs to change, too, in anticipation of the day when the market suddenly realizes that copper and the other metals are scarce. The industry must soon begin to think in system terms. The metals are a depreciable asset. Rather than selling the product, with no thought for recovery and even encouraging dissipative use, the time will come, inevitably, when producers will be reluctant to sell outright, but will prefer to rent or lease the metals to major fabricators. They, in turn, will be forced to rent or lease, rather than sell, to downstream users. Such a system would have powerful built-in incentives not only for conservation but for recovery and recycling.

The details of an organized system to accomplish this radical, but necessary change are far beyond the scope of a book such as this. In any case, only the firms involved in the sector can take the next steps – with help from government regulators and academics. But in the long run such a change is inevitable and it is not a moment too early to begin planning for the future.

REFERENCES

Ahrendts, J., 1980. 'Reference states.' *Energy* **5**(5): 667–677.

Allen, David T., 2002. 'Wastes as raw materials.' In: Robert U. Ayres and Leslie W. Ayres, *Handbook of Industrial Ecology*. Cheltenham UK and Lyme MA: Edward Elgar: 405–420.

Allen, David T. and Nasrin Behmanesh, 1994. 'Wastes as raw materials.' In: Braden R. Allenby and Deanna J. Richards, *The Greening of Industrial Ecosystems*. New York: National Academy Press.

Amore, Dawn, undated. *Electronic product recovery and recycling baseline report*. Washington DC: National Safety Council, Environmental Health Center: Warmer Bulletin: 68.

Andersson, Bjorn A., 2000. 'Materials availability for large-scale, thin film photovoltaics.' *Progress in Photovoltaics Research and Applications* **8**: 61–76.

Arndt, Richard A. and L. David Roper, 1976. *The metals and mineral fuels crisis: Facts and predictions*. Blacksburg VA: University Publications.

Arpaci, E. and Th. Vendura, 1992. *Recycling of industrial copper materials*. Berlin: Association of German Engineers (VDI): VDI report: 917.

Ayres, Robert U., 1988. 'Optimal growth paths with exhaustible resources: An information-based model.' *Journal of Environmental Economics and Management*.

Ayres, Robert U., 1995. 'Life cycle analysis: A critique.' *Resources, Conservation and Recycling* **14**: 199–223.

Ayres, Robert U., 1995. 'Thermodynamics and process analysis for future economic scenarios.' *Journal of Environmental and Resource Economics* **6**: 207–230.

Ayres, Robert U. and Leslie W. Ayres, 1996. *Industrial ecology: Closing the materials cycle*. Cheltenham, UK and Lyme MA: Edward Elgar.

Ayres, Robert U. and Leslie W. Ayres, 1998. *Accounting for resources 1: Economy-wide applications of mass-balance principles to materials and waste*. Cheltenham, UK and Lyme MA: Edward Elgar.

Ayres, Robert U. and Leslie W. Ayres, 1999. *Accounting for resources 2: The life cycle of materials*. Cheltenham, UK and Lyme MA: Edward Elgar.

Ayres, Robert U. and Katalin Martinás, 1992. 'Learning from experience and the life cycle: Some analytic implications.' *Technovation* **12**(7): 465–486.

Ayres, Robert U. and Katalin Martinás, 1995. 'Waste potential entropy: The ultimate ecotoxic?' *Economie Applique* **XLVIII**(2): 95–120.

Ayres, Robert U. and Katalin Martinás, 1996. 'Wealth accumulation and economic progress.' *Evolutionary Economics* **6**(4).

Ayres, Robert U. and Richard A. McKenna, 1972. *Alternatives to the internal combustion engine*. Baltimore MD: Johns Hopkins University Press.

Ayres, Robert U., Leslie W. Ayres, James McCurley, Mitchell Small, Joel A. Tarr and Rolande C. Widgery, 1988. *An historical reconstruction of major pollutant levels in the Hudson-Raritan basin 1800–1980*. Rockville MD: National Oceanic and Atmospheric Administration: October. Technical Memorandum: NOS OMA 43.

Ayres, Robert U., Katalin Martinás and Leslie W. Ayres, 1998. 'Exergy, waste accounting and life cycle analysis.' *Energy* **23**(5): 355–363.

Azar, Christian, John Holmberg and Kristian Lindgren, 1996. 'Socio-ecological indicators for sustainability: methodological and ideological options.' *Ecological Economics* **18**: 89–112.

Bachmann, H. G., 1960. 'The origin of ores.' *Scientific American*: 146–156.

Bätcher, K. and E. Böhm, 1994. *Zusammenfassende Darstellung des Kenntnisstandes zu*

Cadmium-Stoffstroemen in der Bundesrepublik. Bonn, Germany: Enquete Commission of the German Bundestag.

Ballance, Robert and Helmut Forstner, 1990. *International competitiveness and environmental policy: the case of the lead industry*. Undated draft report for UNIDO: unpublished.

Bare, J. C., P. Hofstetter, D.W. Pennington and Helias A. Udo de Haes, 2002. 'Life cycle impact assessment workshop summary, midpoints vs. endpoints: the sacrifices and benefits.' *International Journal of LCA* **5**(6): 319–326.

Berdowski, Jan, 1998. *The European emission inventory of heavy metals for 1990*. Proceedings of the workshop on the atmospheric transport and fate of metals in the environment, Antwerp, Belgium, The International Council on Metals and the Environment.

Biswas, A. K. and W. G. Davenport, 1976. *Extractive metallurgy of copper*. Oxford, UK: Pergamon Press.

Boustead, Ian and W. T. Dove, 1999. *Ecoprofile of the European zinc industry: Primary zinc production*. Brussels, Belgium: International Zinc Association: September.

Boutron, Claude F., 1998. *Evidence in ice core research*. Proceedings of the workshop on the atmospheric transport and fate of metals in the environment, Antwerp, Belgium, The International Council on Metals and the Environment.

Bruch, H., D. Gohlke, C. Kruger, M. Reuter, I. V. Roepenack, E. Rombach, G. Rombach and P. Winkler, 1995. *LCI of copper production and processing*. Dusseldorf, Germany: German Copper Institute (DKI): special printing: s198.

Carrillo, Fred V., Mark H. Hibbshman and Rodney D. Rosenkranz, 1974. *Recovery of secondary copper and zinc in the United States*. Washington DC: United States Bureau of Mines: Information Circular: 8622.

Council on Environmental Quality, The, 1996. *Environmental Quality, 25th Annual Report*. Washington DC: United States Government Printing Office.

Chapman, Duane, 1989. *Environmental standards and international trade in automobiles and copper: The case for a social tariff*. Ithaca NY: Department of Agricultural Economics, Cornell University: February.

Chernick, Paul, 1992. *Direct computation of certain damage costs of lead from power plants*. Boston MA: Testimony for Massachusetts Department of Public Utilities.

Cleveland, Cutler J. and Matthias Ruth, 1997. 'When, where and by how much do biophysical limits constrain the economic process? A survey of Nicholas Georgescu-Roegen's contribution to ecological economics.' *Ecological Economics* **22**(3): 203–224.

Coates, D., J. Rotondo and A. Charkey, 1999. 'Nickel-zinc: a new/old technology.' In: *Electric and Hybrid '99*. Dorking, Surrey, UK: Vehicle Technology International.

Committee on Mineral Resources and the Environment, 1975. *Mineral resources and the environment*. Washington DC: National Academy of Sciences – National Research Council.

Coppa, Louis V., 1984. *Copper, lead, zinc, gold and silver waste disposal activities and practices in the United States*. Washington DC: United States Bureau of Mines: Information Circular: IC.

Crowson, P., 1992. *Minerals handbook 1992–93: statistics and analyses of the world's mineral industry*. New York: Stockton Press.

D'Arge, Ralph C. and K. C. Kogiku, 1973. 'Economic growth and the environment.' *Review of Economic Studies* **XL**(1): 61–77.

Davenport, W. G., 1999. *Copper extraction from the 60's into the 21st century*. Plenary lectures/Movement of copper and industry outlook/Copper applications and fabrication, Warrendale, PA: The Minerals, Metals and Materials Society (MMMS).

Davidson, Cliff I., 1976. *Mass balances for pollutants in urban regions*. California: W. M. Keck Laboratories, California Institute of Technology: November. Special Report: to EPA.

Demetrio, Sergio, M. A. Duran, R. Alvarado, L. Contreras, J. Ahumade, E. Mast, J. Sanhueza and E. Morales, 1999. *The Chilean copper smelter management way*. Plenary lectures/Movement of copper and industry outlook/Copper applications and fabrication, Warrendale, PA: The Minerals, Metals and Materials Society (MMMS).

Dewison, Paul, 1999. *Markets for copper into the millennium: a new approach to copper market forecasting*. Plenary lectures/Movement of copper and industry outlook/Copper applications and fabrication, Warrendale, PA: The Minerals, Metals and Materials Society (MMMS).

German Copper Institute (DKI), 1984. *What is copper?* Berlin: German Copper Institute (DKI): information printing: i14.

German Copper Institute (DKI), 1990. *Recycling von Kupferwerkstoffen*. Berlin: German Copper Institute (DKI).

Dorfman, Robert, 1969. 'An economic interpretation of optimal control theory.' *American Economic Review* **59**(part 2): 817–831.

Edelstein, Daniel (coordinator), 1999. 'Froth flotation in the United States, 1997.' In: *Mineral Industry Surveys*. Washington DC: United States Bureau of Mines.

Research Institute for Energy Economics (FfE), 1999. *LCA of basic and finished materials: part III metals*. Munich, Germany: Research Institute for Energy Economics (FfE).

Fitzgerald, William F., 1970. 'A study of certain trace metals in seawater using anoidic stripping voltametry.' PhD. Woods Hole Oceanographic Institute, Massachusetts Institute of Technology, Boston.

Flemming, C. A. and J. T. Trevors, 1989. 'Copper toxicity and chemistry in the environment: A review.' *Water, Air and Soil Pollution* **44**: 144–158.

Forrest, David and Julian Szekely, 1991. 'Global warming and the primary metals industry.' *Journal of Metallurgy* **43**: 23–30.

Frear, D. R., J. W. Jang, J. K. Lin and C. Zhang, 2001. 'A metallurgical study of Pb-free solders for flip-chip interconnects.' *JOM* **53**(6): 28–32.

Freeman, A. Myrick III, 1993. *The measurement of environmental and resource values: Theory and methods*. Washington DC: Resources for the Future.

Frei, M., A. Kueng and B. Schuler, 1993. *Kupfer-Flüsse in der Automobilbranche*. Zurich: Vertiefungsblock-Arbeit ETHZ.

Gaines, Linda L., 1980. *Energy and material flows in the copper industry*. Argonne IL: Argonne National Laboratory.

Gilbert, T., G. C. McLeod, R. Maekel, K. V. Ladd, A. Clay and A. Barker, 1972. *Trace metal analysis of Boston Harbor water and sediments*. Boston: New England Aquarium.

Giurco, D. P., M. Stewart and J. G. Petrie, 2001. *Decision making to support sustainability in the copper industry: technology selection*. Sydney, Australia: Department of Chemical Engineering, University of Sydney.

Gordon, R. B., J. J. Koopmans, William B. Nordhaus and Brian J. Skinner, 1987. *Towards a new iron age*. Cambridge MA: Harvard University Press.

Gössling, Stefan, 2001. 'Entropy production as a measure for resource use.' PhD. Institute for Experimental Physics, University of Hamburg, Hamburg, Germany.

Guinee, J. B. et al., 1996. *LCA impact assessment of toxic releases*. Bilthoven, The Netherlands: CML/RIVM.

Gusev, Alexei V., 1998. *Air transport models*. Proceedings of the workshop on the atmospheric transport and fate of metals in the environment, Antwerp, Belgium, The International Council on Metals and the Environment.

Gyftopoulos, E. P., L. J. Lazaridis and T. F. Widmer, 1974. *Potential fuel effectiveness in industry*. Cambridge MA: Ballinger Publishing Company.

Heijungs, R. et al., 1992. *Environmental life-cycle assessment of products: Manual and background*. Leiden, The Netherlands: CML.

Henstock, Michael, undated. *The potential and the limitations of copper recycling*. Kettering, UK: Consultant.

Herfindahl, O., 1967. *Depletion and economic theory. Extractive Resources and Taxation*. M. Gaffney. Madison WI: University of Wisconsin Press.

Hertwich, Edgar G., William Pease and Catherine P. Koshland, 1997. 'Evaluating the environmental impact of products and production processes; A comparison of six methods.' *The Science of the Total Environment* **196**: 12–29.

Hertwich, Edgar G., Sarah F. Mateles, William S. Pease and Thomas E. McKone, 2001. 'Human toxicity potentials for life-cycle assessment and toxics release inventory risk screening.' *Environmental Toxicology and Chemistry* **20**(4): 928–939.

Higley, Leo W. and M. M. Fine, 1977. *Electric furnace steelmaking dusts: a zinc raw material.* Washington DC: United States Bureau of Mines: Higley and Fine 77: 8209.

Hong, Sungmin, J.-P. Candelone, Clair C. Patterson and Claude F. Boutron, 1994. 'Greenland ice evidence of hemispheric lead pollution two millennia ago by Greek and Roman civilizations.' *Science* **265**: 1841–1843.

Hubbert, M. King, 1956. 'Nuclear energy and the fossil fuels: Drilling and production practice.' *American Petroleum Institute*: 7–25.

Hubbert, M. King, 1962. *Energy resources.* Washington DC: National Research Council/National Academy of Sciences: Publication: 1000-D.

Hubbert, M. King, 1973. 'Survey of world energy resources.' *The Canadian Mining and Metallurgical Bulletin* **66**(735): 37–54.

Huibregts, Mark J. A., 2000. *Priority assessment of toxic substances in the frame of LCA; time horizon dependency of toxicity potentials cacluated with the multi-media Fate, Exposure and Effects Model (USES-LCA).* Amsterdam, The Netherlands: University of Amsterdam Institute for Biodiversity and Ecosystem Dynamics.

Huijbregts, Mark J. A., 1999. 'Priority assessment of toxic substances in LCA: Development and application of the multimedia fate, exposures and effect model USES-LCA.' PhD. IVAM Environmental Research, University of Amsterdam, Amsterdam, the Netherlands.

Hunt, R. G., W. E. Franklin, R. O. Welch, J. A. Cross and A. E. Woodall, 1974. *Untitled report.* Washington DC: US Environmental Protection Agency: 1974.EPA/530/SW-91c.

International Copper Development Association, 2001. *Brochure.* International Copper Development Association.

International Copper Study Group (ICSG), 1997. *ICSG Copper Bulletin.* International Copper Study Group (ICSG): ICSG Copper Bulletin: 4(1).

Illman, D. L., 1994. 'Unknown.' *Chemical Engineering News* **72**(31): 8.

International Lead and Zinc Study Group, 1999. *Study on the image of lead.* London: International Lead and Zinc Study Group: March.

International Lead and Zinc Study Group, 2000. *Study on the economic and environmental role of zinc.* London: International Lead and Zinc Study Group: April.

International Lead and Zinc Study Group, 2001. 'Lead and Zinc Statistics.' *Monthly Bulletin of the International Lead and Zinc Study Group* **41**(7).

INFORM, 1995. *Toxics Watch 1995.* New York: INFORM Incorporated: 521.

Intergovernmental Panel on Climate Change (IPCC), 2000. *Special Report on emissions scenarios.* Cambridge, UK: Cambridge University Press.

International Zinc Association (IZA), 2001. *Zinc aspects of 'the life cycle of copper, its co-products and byproducts'; Comments for MMSD.* Brussels: International Zinc Association: 30 September.

International Zinc Association, 2000. *Pocket guide to world zinc.* Brussels and London: International Zinc Association and International Lead and Zinc Study Group.

Jha, M. K., V. Kumar and R. J. Singh, 2001. 'Review of hydrometallurgical recovery of zinc from industrial wastes.' *Resources, Conservation and Recycling* **33**(1): 1–22.

Johansson, Hank, 1987. 'A new age for accounting (interview).' *CIME.*

Jolly, James H., 1992. *Materials flow of zinc in the United States 1850–1990.* Washington DC: United States Bureau of Mines: April. Open File Report: OFR 72–92.

Jolly, Janice L. W., 1999. *The US copper-base scrap industry and its byproducts: An overview.* Washington DC: United States Geological Survey: July 23. unpublished.

Jolly, Janice L. W., 2000. *The US copper-base scrap industry and its byproducts: An overview.* New York: Copper Development Association Inc.: July 31. Technical Report: A1309-00/00.

Kakudate, Keiji, Yoshihiro Adachi and Toshio Suzuki, 2000. *A macro model for steel usage pattern in Japan using a population balance model*. Second International Conference on Processing Materials for Properties, Tokyo, The Minerals, Metals and Materials Society.

Karlsson, Sten, Christian Azar, Goran Berndes, John Holmberg and Kristian Lindgren, 1997. *Man and materials flows: Towards sustainable materials management*. Uppsala, Sweden: The Baltic University Programme, Uppsala University.

Katayama, H. and Y. Mizukami, 1996. 'Recycling of ferrous scrap.' *Materia Japan* **35**(12): 1285.

Kaufman, T. D. and R. C. Drury, 1987. 'Aluminum and copper are no longer substitutes.' *Natural Resources Forum* **11**: 373–377.

Kennedy, Paul, 1989. *The rise and fall of great powers*. New York: Vintage Books.

Kesler, Stephen E., 1994. *Mineral resources, economics and the environment*. New York: Macmillan Co.

Landner, Lars and Lennart Lindeström, 1999. *Copper in society and in the environment; an account of the facts on fluxes, amounts and effects of copper in Sweden*. Västerås, Sweden: Swedish Environmental Research Group (MFG).

Lave, Lester B., Chris T. Hendrickson and Francis C. McMichael, 1995. 'Environmental implications of electric cars.' *Science* **268**: 993–995.

Leygraf, C. and I. Odnevall Wallinder, 1997. 'A study of copper runoff in an urban atmosphere.' *Corrosion Science* **39**(12).

Llewellyn, Thomas O., 1994. *Cadmium (materials flow)*. Washington DC: United States Bureau of Mines: Information Circular: IC9380.

Loebenstein, J. Roger, 1994. *Materials flow of arsenic in the United States*. Washington DC: United States Bureau of Mines: Information Circular: IC9382.

Lovins, Amory B., 1973. *Open pit mining*. London, Earth Island.

Lovins, Amory B., 2001. *Unpublished*. London: International Institute for Environmental Development (IIED).

Mackay, Donald M. and S. Patterson, 1991. 'Evaluating the multimedia fate of organic chemicals: A level III fugacity model.' *Environmental Science and Technology* **25**: 427–436.

Mackay, Donald M., W. Shui and K. Ma, 1993. *Illustrated handbook of physical-chemical properties and environmental fate for organic chemicals*. Boca Raton FL: Lewis Publishers.

Malenbaum, W., 1978. *World demand for raw materials in 1985 and 2000*. New York: McGraw-Hill.

Masini, Andrea, Leslie W. Ayres and Robert U. Ayres, 2003. 'An application of exergy analysis for five basic metal industries.' In Robert U. Ayres, Armin Von Gleich and Stefan Gössling (Eds.), *Sustainable Metals Management*. Dordrecht, NL, Kluwer Academic Publishers.

McElroy, A. D. and F. D. Shobe, 1980. *Source category survey: Secondary zinc smelting and refining industry*. Kansas City MO: Midwest Research Inst.: May. Technical Report: EPA-450/3-80-012.

McKelvey, V. E., 1960. 'Relation of reserves of the elements to their crustal abundances.' *American Journal of Science* **258**(A): 234–241.

McMahon, Alfred D., 1964. *Copper: a materials survey*. Washington DC: United States Bureau of Mines: Information Circular: 8225.

Mendicino, Laura and Paul Thomas Brown, 1998. 'The environment, health and safety side of copper metalization.' *Semiconductor International*.

Mitchell, Paul, 1999. 'Prediction, prevention, control and treatment of acid rock damage.' In: Alyson Warhurst and Ligia Norohna, *Environmental policy in mining: Corporate strategy and planning for closure*. Boca Raton FL: CRC Press: 118–143.

Morris, D. R., F. R. Steward and P. Evans, 1983. 'Energy efficiency of a lead smelter.' *Energy* **8**(5): 337–349.

Murray, C. J. L. and A. D. Lopez, Eds., 1996. *The global burden of disease*. Global Burden of Disease and Injury vol I. Boston MA: WHO/Harvard School of Public Health/World Bank/Harvard University Press.

Naegele, J. R. (coordinator), P. Canarelli, T. Hernandez, M. Nicolai Amorelli, F. Scapolo and

C. Heller, 1996. *The car of the future, the future of the car*. Seville, Spain: Institute for Prospective Technological Studies – Joint Research Centre: EUR WP:EUR 17277 EN.

NAS/NRC Committee of Resources and Man (Preston Cloud, Chairman), 1969. *Resources and man*. San Francisco: W. H. Freeman.

Newell, S. A., 1998. 'Strategic evaluation of LCA metrics: Making use of LCIs.' PhD Thesis; Massachusetts Institute of Technology, Cambridge MA.

Nilarp, 1994. *Omsättningen av koppar i Sverige i ett bärkraftighetsperspektiv*. Stockholm: Svenska naturvårdsverket.

Nriagu, Jerome O., 1989. 'A global assessment of natural sources of atmospheric trace metals.' *Nature* **338**: 47–49.

Nriagu, Jerome O. and Josef M. Pacyna, 1988. 'Quantitative assessment of worldwide contamination of air, water and soils by trace metals.' *Nature* **33**: 134–139.

Olendrzynski, Krzysztof, Stefan Anderberg, Jerzy Bartnicki, Josef Pacyna and William M. Stigliani, 1995. *Atmospheric emissions and deposition of cadmium, lead and zinc in Europe during the period 1955–1987*. Laxenburg, Austria: International Institute for Applied Systems Analysis: April. Working Paper: WP-95-35.

Pacyna, Josef M., 1998. *Anthropogenic emissions of heavy metals*. Proceedings of the workshop on the atmospheric transport and fate of metals in the environment, Antwerp, Belgium: The International Council on Metals and the Environment.

Pacyna, Josef M. and Elisabeth G. Pacyna, 2001. *An assessment of global and regional emissions of trace metals to the atmosphere from anthropogenic sources worldwide*. Kjeller, Norway: Norwegian Institute for Air Research (NILU).

Patterson, C. C., T. J. Chow and Murozumi, 1970. 'The possibility of measuring variations in the intensity of worldwide lead smelting during medieval and ancient times using lead aerosol deposits in polar snow strata.' In: R. Berger, *Scientific methods in medieval archeology*. Berkeley CA: University of California Press: 339–350.

PEDCo-Environmental, 1980. *Industrial process profiles for environmental use: Chapter 29; primary copper industry*. Cincinnati OH: PEDCo-Environmental: July. NTIS: PB81-164915, EPA-600/2-80-170.

PEDCo-Environmental, 1980. *Industrial process profiles for environmental use: Chapter 27; primary lead industry*. Cincinnati OH: PEDCo-Environmental: July. NTIS: PB81-110926, EPA-600/2-80-168.

PEDCo-Environmental, 1980. *Industrial process profiles for environmental use: Chapter 28; primary zinc industry*. Cincinnati OH: PEDCo-Environmental: July. NTIS: PB80-225717, EPA-600/2-80-169.

Pinhao, C. M., 1996. *Energia e noves materiais: o caso das fibras oticas*. Mestre Programa de Planejamento Energetico, COPPE_UFRJ, Rio de Janeiro, Brazil.

Plachy, 1999. 'Recycling metals.' In: *Metals and Minerals*. Washington DC: United States Geological Survey: I.

Poetzschke, M., 1991. *User goods as complex raw material sources: value preserving material cycles of metals from a raw material producers perspective*. Berlin: Association of German Engineers (VDI): VDI report: 906.

Pontryagin et al., 1962. 'The mathematical theory of optimal processes.' *Interscience*.

Rant, Z., 1956. 'Exergy, a new word for technical available work.' *Forsch. Ing. Wis.* **22**(1): 36–37.

Roberts, M. C., 1996. 'Metal use and the world economy.' *Resources Policy* 22: 183–196.

Raw Materials Group Stockholm and Roskill Information Services Ltd., 1997. *Roskill's metals databook 1997*. London, Roskill Information Services Ltd.

Roskill Information Services Ltd., 1990. *The economics of copper 1990*. London: Roskill Information Services Ltd.

Rowe, Robert D. et al., 1995. *The New York environmental externalities study: Summary of approach and results*. Workshop on the External Costs of Energy, Brussels, Belgium.

Russell, E. W., 1988. *Russell's soil conditions and plant growth*. Bath, UK: Bath Press.

Ruth, Matthias, 1995. 'Thermodynamic implications of natural resource extraction and technical change in US copper mining.' *Environment and Resource Economics* **6**: 187–206.

Sancha, Ana Maria F., 2000. *Gran mineria y medio ambiente: Estudio de caso; contaminacion por arsenico en el norte de Chile y su impacto en el ecosistema y la salud humana.* Chile: University of Chile; Faculty of Physical and Mathematical Sciences.

Schmidt-Bleek, Friedrich B., 1993. 'MIPS – A universal ecological measure?' *Fresenius Environmental Bulletin* **2**(6): 306–311.

Schmidt-Bleek, Friedrich B., 1993. 'MIPS revisited.' *Fresenius Environmental Bulletin* **2**(8): 485–490.

Sekiguchi, Hiroshi and Kazumi Ichiyama, 2000. *The recycling of copper metal in Japan.* Lisbon, Portugal: International Copper Study Group ICSG), Environmental Panel.

Semiconductor Industry Association (SIA), 1997. *National technology roadmap for semiconductors*, 1997 ed. California: Semiconductor Industry Association (SIA).

Sibley, S. F., W. C. Butterman et al., 1995. 'Metals recycling in the United States.' *Resource Conservation Recycling* **15**: 256–267.

Simada, Masanori, Katagiri Nozou and Maeda Sayori, 1999. *Supply and consumption of copper in Japan.* Lisbon, Portugal: International Copper Study Group (ICSG).

Skeaff, J. M. and Alain A. Dubreuil, 1997. 'Calculated 1993 emission factors of trace metals for Canadian nonferrous smelters.' *Atmosphere and Environment* **31**: 1449–1457.

Skinner, Brian J., 1976. *Earth resources.* Englewood Cliffs NJ: Prentice-Hall.

Skinner, Brian J., 1987. 'Supplies of geochemically scarce metals.' In: D. J. Maclaren and Brian J. Skinner, *Resources and world development.* Chichester, UK: John Wiley and Sons: 305–325.

Smith, Gerald R., 1999. *Lead recycling in the United States in 1998.* Washington DC: United States Geological Survey: Flow Studies for Recycling Metal Commodities in the United States.

Spatari, Sabrina, M. Bertram, K. Fuse, Thomas E. Graedel and Heinz Rechberger, 2002. 'The contemporary European copper cycle: One year stocks and flows.' *Ecological Economics* (July).

Spencer, D. W. and P. G. Brewer, 1969. 'The distribution of copper, zinc, and nickel in seawater of the Gulf of Maine and Sargasso Sea.' *Geochimica Cosmochimica Acta* **33**(325).

Steen, Bengt and Sten-Olof Ryding, 1993. *The EPS enviro-accounting method.* Stockholm: Swedish Waste Research Council for a Low-Waste, Ecocyclic Society: May. AFR-Report 11: ISSN 1102–6944.

Stigliani, William M., Peter Doelman, Wim Salomons, Rainer Schulin, Gera R. B. Smidt and Sjoerd E. A. T. M. van der Zee, 1991. 'Chemical times bombs: Predicting the unpredictable.' *Environment* **33**(4): 4–30.

Sundberg, R., 1998. 'The fate of copper released from the Vasa Ship Museum.' *Metall* **52**(4).

Svenska naturvårdsverket, 1990. Rapport 4506 *Metaller, Materialflöden i samhället.* Stockholm: Svenska naturvårdsverket; Redovisning till regeringen del 2.

Szargut, Jan, David R. Morris and Frank R. Steward, 1988. *Exergy analysis of thermal, chemical, and metallurgical processes.* New York: Hemisphere Publishing Corporation.

Tellus Institute, 1992. *CSG/Tellus packaging study: Inventory of material and energy use and air and water emissions from the production of packaging materials.* Boston MA: Tellus Institute: May. Technical Report: 89-024/2.

Tellus Institute, 1992. *Assessing the impacts of production and disposal of packaging and public policy measures to alter its mix.* Boston MA: Tellus Institute: May.

Thomas, R., Ed., 1977. *Operating handbook of mineral processing.* New York: McGraw-Hill.

Thornton, Iain, Ed., 1983. *Applied environmental geochemistry.* Academic Press.

Thornton, Iain, 1995. *Metals in the global environment: Facts and misconceptions.* Ottawa, Ontario, Canada: International Council on Metals and the Environment.

Tilton, John E., Ed., 1990. *World Metal Demand.* Washington DC: Resources for the Future.

Toi, A., J. Sato and T. Kanero, 1997. 'Analysis of tramp element(s) in iron scraps.' *Tetsu-to-Hagane* **83**(12): 852.

Towle, Stuart W., 1989. *World copper smelter sulfur balance 1988*. Washington DC: US Bureau of Mines, Information Circular 9349: Information Circular: IC9349.

Towle, Stewart W., 1993. *World copper smelter sulfur balance 1992*. Washington DC: United States Bureau of Mines: USBM Information Circular.

Udo de Haes, Helias A., 2002. 'Industrial ecology and life cycle assessment.' In: Robert U. Ayres and Leslie W. Ayres, *A Handbook of Industrial Ecology*. Cheltenham, UK and Lyme MA: Edward Elgar. Chapter 12: 138–148.

United Nations Environment Programme and International Council on Metals and the Environment, 1996. *Case studies illustrating environmental practices in mining and metallurgical processes*. New York: United Nations Press.

United Nations Environment Programme Industry and Environment Programme Activity Centre, 1991. *Environmental aspects of selected nonferrous metals (Cu, Ni, Pb, Zn, Au) ore mining: A technical guide*. Paris: United Nations.

United States Bureau of Mines, 1975. *Mineral facts and problems*. Washington DC: United States Government Printing Office.

United States Bureau of Mines, 1985. *Mineral facts and problems*. Washington DC: United States Government Printing Office.

United States Bureau of Mines, 1991. *Minerals yearbook: 1991*. Washington DC: United States Government Printing Office.

United States Bureau of Mines, annual. *Minerals yearbook*. Washington DC: United States Government Printing Office.

United States Bureau of Mines, monthly. *Mineral commodity summaries*. Washington DC: United States Bureau of Mines: through 1996 (United States Geological Survey since 1997).

United States Bureau of the Census, 1975. *Historical statistics of the United States, colonial times to 1970*. Washington DC: United States Government Printing Office.

United States Environmental Protection Agency, 1983. *Development document for effluent limitations guidelines and standards for the nonferrous metals point source category*. Washington DC: United States Environmental Protection Agency: Report:Vol 3, part 1.

United States Environmental Protection Agency, 1996. *National air pollution emission trends 1900-1005*. Washington DC: United States Environmental Protection Agency.

United States Environmental Protection Agency, 1996. *1994 Toxics Release Inventory*. Washington DC: United States Environmental Protection Agency: June. Public Data Release:EPA 745-R-96-002.

United States Geological Survey, 2000. *Minerals Yearbook, 1999: Volume I Metals and Minerals*. Washington DC: United States Government Printing Office.

United States Geological Survey, annual. *Minerals Yearbooks*. Washington DC: United States Geological Survey: annual since 1996.

United States Geological Survey, monthly. *Mineral commodity summaries*. Washington DC: United States Geological Survey: monthly since 1996.

Van Assche, F. J.,1998. *A stepwise model to quantify the relative contribution of different enviromental sources to human cadmium exposure*. NiCAD '98, Prague, Czech Republic.

van Tilborg, Wim J. M., 1998. *Fate of metals in soils: Geochemical transformation and bioavailability of metals: the time factor*. Proceedings of the workshop on the atmospheric transport and fate of metals in the environment, Antwerp, Belgium, The International Council on Metals and the Environment.

Van Vuuren, Detlef P., Bart J. Strengers and Herbert J. M. de Vries, 2000. 'Long-term perspectives on world metal use: a system-dynamics model.' *Resources Policy* **25**: 239–255.

Von Arx, U., 1996. *Kupfer in der Schweiz*. Bern, Switzerland, Bundesamt für Umwelt, Wald und Landschaft (BUWAL).

Waldhauer, Ruth, Albert Matte and Robert E. Tucker, 1975. *Lead and copper in the waters of*

Raritan and Lower New York Bays. Highlands NJ: National Marine Fisheries Sandy Hook Laboratory.

Wall, Goran, 1977. *Exergy: A useful concept within resource accounting*. Goteborg, Sweden: Institute of Theoretical Physics, Chalmers University of Technology and University of Goteborg: Research Report: 77–42.

Wedepohl, K. H., 1995. 'The composition of the Earth's crust.' *Geochimica et Cosmochimica Acta* 59(7): 1217–1232.

WestTech Market Information Services, undated. *Cable technology ready for full-scale demonstrations of commercial HTS technology*. Los Angeles: WestTech Market Information Services.

Woodbury, William D., Daniel Edelstein and Stephen M. Jasinski, 1993. *Lead materials flow in the United States 1940–1988*. Washington DC: United States Bureau of Mines: draft.

World Resources Foundation, 1996. 'Electrical and electronic wastes: information sheet.' *Warmer Bulletin*.

Yen, T. F., 1975. *The role of trace metals in petroleum*. Ann Arbor MI: Ann Arbor Science Publishers.

Yoshiki-Gravelsins, K. S., J. M. Toguri and R. T. C. Choo, 1993. 'Metals production, energy and the environment: Part II; environmental impact.' *Journal of the Metals and Materials Society* **45**: 23–29.

Zeltner, C., H.-P. Bader, R. Scheidigger and P. Baccini, 1999. 'Sustainable metal management exemplified by copper in the USA.' *Regional environmental change* **1**(1): 31–46.

Zoboli, R., G. Barbiroli and F. Leone, 2000. *Regulation and innovation in the area of end-of-life vehicles*. Seville, Spain: Institute for Prospective Technological Studies – Joint Research Council: EUR WP: EUR 19598 EN.

Zuckerman, B. and Frank Ackerman, 1995. *The 1994 update of the Tellus Institute packaging study impact assessment method*. SETAC Impact Assessment Working Group Conference, Washington DC, Society for Environmental Toxicology and Chemistry.

APPENDIX A

THE EXERGY CONCEPT

A1. Definition and description of exergy calculations

The idea of available energy dates back to the last century, when it was first understood by the French engineer Sadi Carnot for the specialized case of heat engines. In the next decades the concept of 'available work' was further developed theoretically, especially by Herman Helmholtz and J. Willard Gibbs. It has been applied to many kinds of processes, for different purposes, under several different names: availability, available work, essergy, physical information. But only recently a standard definition has been formulated and the name exergy definitely adopted (Rant 1956; Gyftopoulos et al. 1974; Wall 1977; Szargut et al. 1988). However, for the purposes of this study it is sufficient to present only the essential features of the theory.

An adequate definition of exergy is the following: "Exergy is the amount of work obtainable when some matter is brought to a state of thermodynamic equilibrium with the common components of the natural surroundings by means of reversible processes, involving interaction only with the above mentioned components of nature" (Szargut et al. 1988). In short, exergy is an extensive non-conservative variable which synthesizes in a concise and useful expression both the first and second law of thermodynamics. It is definable and computable (in principle) for any substance, or system, with respect to the real environment in which the system is located and/or operates.

In principle, four different types of exergy B can be identified. These are denoted, respectively, as kinetic, potential, physical and chemical exergy, viz.

$$B = B_k + B_p + B_{ph} + B_{ch} \tag{A1}$$

Kinetic and potential exergy have the same meaning as the corresponding energy terms. Kinetic exergy is relevant for analyzing a flywheel or turbine. Potential exergy is relevant for electrical or hydraulic systems. But these two terms can safely be neglected for purposes of analyzing most common industrial processes. Physical exergy is "the work obtainable by taking a substance through reversible physical processes from its initial state (temperature T, pressure p) to the state determined by the temperature T_o and the pressure p_o of the environment" (Szargut et al. 1988). Physical exergy assumes an important role for purposes of optimization of thermal and mechanical processes, including heat engines and power plants. But it is of secondary – in fact negligible – importance when attention is focused on very large scale systems, such as chemical and metallurgical processes at the industry level.

In this case chemical exergy plays a major role for purposes of resource accounting and environmental analysis.

Chemical exergy is "the work that can be obtained by a substance having the parameters T_o and p_o to a state of thermodynamic equilibrium with the datum level components of the environment" (Szargut et al. 1988). It has two components: a component associated with chemical reactions occurring in isolation and a component associated with the diffusion of reaction products into the surroundings.

All the foregoing definitions stress the importance of defining a *reference state*, or system, when calculating both physical and chemical exergy. As a matter of fact, the exergy function is a measure of the difference between two states, namely the state of the 'target' system and that of its surroundings (or, more precisely, the ultimate state of the combined system and surroundings, after they have reached mutual equilibrium). In short, exergy cannot be calculated without defining appropriate parameters for the environment where the target system operates, in terms of temperature, pressure and chemical composition.

The importance of defining the parameters for the common environment emerges clearly when we consider the analytical expressions for exergy. They also show that exergy is a measure of the thermodynamic 'distance' of the target system from equilibrium. Another way of saying this is that exergy is a measure of the 'distinguishability' of the target system from its environment. These statements follow from the fact that exergy vanishes when the target system under consideration has the same thermodynamic state as the environment.

In general, for a closed system with temperature T, pressure p, entropy S, and volume V, exergy can be written as:

$$B = S(T - T_o) - V(p - p_o) + \Sigma N_i(\mu - \mu_{io}) \qquad \text{(A2)}$$

where N_i is the number of moles of the ith system and μ_i is its chemical potential. The subscript 'o' refers to the final state of equilibrium of the system plus the environment, combined together. Again, the exergy of a flow crossing the system boundaries of an open system can be written as the sum of three terms:

$$H = H - H_o - T_o(S - S_o) - \Sigma\mu_i(N_i - N_{io}) \qquad \text{(A3)}$$

where the letter H stands for enthalpy. The third term of these expressions takes account of the contribution due to the chemical transformation of the system.

In both of these expressions it is straightforward to recognize how the choice of the reference state affects the value of the function B. For the purpose of calculating physical exergy, this choice does not represent a major problem, as it is relatively easy to define an appropriate level for pressure and temperature of the environment, namely ambient atmospheric temperature and pressure. This is not the case for calculation of chemical exergy. The latter step requires knowledge of the detailed average chemical composition of the reaction products and the environmental sink with which the system interacts.

In this context, considerable efforts have been undertaken by a number of authors. One possible approach would be to assume, as the reference level, the average chemical composition of the earth's crust after reaching an hypothetical (calculated) equilibrium with the atmosphere and oceans (Ahrendts 1980). However, the results vary dramatically according to the depth of the crustal layer that is assumed to be equilibrated. A more practical generic solution has been proposed by (Szargut et al. 1988). This approach recognizes that the three main sinks – atmosphere, oceans and crust – are not in equilibrium with each other, but assumes that the reaction products in any given case must go to one of the three, depending on whether they are volatile (to air), soluble in water (to oceans) or neither (to earth's crust). They calculate standard chemical exergy for a number of chemical compounds and pure elements. The latter procedure has been adopted and extended in several later works (Ayres and Martinás 1995; Ayres and Ayres 1996; Ayres and Martinás 1996; Ayres et al. 1998; Masini et al. 2001] and its results have been also used in the present study.

A2. Exergy as a tool for resource and waste accounting

The intensive use of natural resources for anthropogenic activities is progressively depleting the reservoirs accumulated over the millennia. At the same time the large quantities of waste materials and effluents released to the atmosphere, the oceans, or to the land surface are altering the delicately balanced natural cycles that make the life possible on earth. Therefore minimization of resource use, as well as reduction of dangerous emissions associated to industrial processes, constitute the primary objective of policies to be pursued for sustainable development. For this purpose it is of great importance to develop a general measure capable of accounting both for materials use and waste residuals.

We suggest that exergy is the most suitable indicator for both resource accounting and waste accounting. Nonetheless, up to now it has not been adopted for this role. For historical reasons, resources have been always divided in two categories, namely fuels (measured in energy units) and mineral, agricultural or forest resources (measured in a variety of mass units). This distinction leads to some incongruities and much confusion, as the choice of a different 'currency' for each flux does not enable analysts to evaluate and compare all inputs and outputs on a common basis. In particular, non- fuel flows – such as minerals – are often neglected.

The use of exergy as a general environmental indicator for resource accounting would improve the situation in two important ways. First, an exergy balance automatically combines both mass and energy flows, thus providing a concise representation of the process. This makes processes easier to characterize. Second, the use of exergy enables the analyst to take into account automatically both the first law (energy conservation) and the second (entropy) law of thermodynamics. In addition, by virtue of reflecting second law constraints, exergy analysis – rather than energy analysis – is a suitable tool to identify areas of potential technological improvements. In fact, irreversibilities and process inefficiencies cause exergy losses (i.e. the difference between exergy of inputs and outputs of the process) which reflect increasing entropy. Any exergy loss shows that the system under consideration could be further improved – at least in principle – in order to increase its thermodynamic efficiency and to reduce the use of natural resources. But more important, comparing the relative magnitude of such losses, both within a complex process, and between alternative processes, is a useful guide to identifying the most promising technology choices and targets for R&D.

A third advantage for exergy accounting, in contrast to conventional approaches, is that it opens the possibility of comparative evaluation of different kinds of materials, not only in mass terms but also in terms of available energy 'saved' in the sense of not being required for separation and purification from the average composition of the environment. This is essentially equivalent to the 'energy content' of an ore or mineral. (Here we use the term 'energy' in the familiar but inexact sense of normal language, rather than the language of specialists. In fact, it is exergy we are talking about.) As already remarked, metal ores, minerals and even agricultural products, as well as chemical compounds, are all precious resources for which heat of combustion (which is the usual measure of energy) is zero or irrelevant. These substances can easily be measured and compared in terms of exergy.

Finally, chemical exergy 'content' can be used as a tool for a first-order evaluation of the environmental impact associated with the waste effluents of any industrial process. In fact, exergy is not only a natural measure of the resource inputs, it is also a measure of the material outputs. Indeed, as emphasized in a previous work by one of the authors, the exergy content (or 'potential entropy') of a waste residual can be considered as its potential for doing harm to the environment by driving uncontrolled reactions (Ayres and Martinás 1995; Ayres et al. 1998).

The use of chemical exergy as a tool for evaluating the potential harm of wastes could lead to over-simplification if directly applied to emissions. As a matter of fact, the chemical exergy content of any substance cannot be directly related to its toxicity to humans or other organisms. On the other hand, it represents a measure of how far the substance is from equilibrium with the common state of the environment. In this sense, an high exergy content is a simple indication that the substance under consideration is likely to drive further chemical reactions when it is discharged to the atmosphere, to watercourses, or deposited in landfills. Or, from a different perspective, it suggests that the materials discharged could be further processed to extract potentially useful work.

A3. Composition of mixtures, including fuels

Ores, fuels, intermediate materials and even finished 'pure' metals are, in fact, mixtures. In order to construct balances, either of mass or exergy, it is necessary to assume the average chemical composition of each mixture. For a given mine, metals processing plant or petroleum refinery, relatively precise data might be available; for anything larger, enormous simplifying assumptions must be made. We have done so for the materials involved in this analysis, using data available from authoritative sources. Nevertheless we think that the results can be taken as approximating the US as a whole in 1993. Yet it must be constantly kept in mind that the presentation is exemplary only and does not exactly represent the true situation for any actual time, country, plant or region. Details of the chemical composition and chemical exergy content of all the materials used in our analysis are shown in *Table A1*.

Table A1: Typical mixtures assumed for copper, lead, and zinc flow analysis

Name	Exergy (kJ/g)	Cu	Pb	Zn	Fe	S	Si	Ca	As	Sb	C	H	N	O
Copper, anode	2.1029	99.49%	0.10%			0.01%			0.10%	0.05%				0.09%
Copper, cathode	2.1117	99.99%	0.01%			tr			tr	tr				tr
Copper, concentrate	8.5463	25.75%	0.20%	1.50%	29.85%	30.90%	2.89%	1.00%	0.10%	0.04%				5.47%
Copper, oxide ore	0.6476	0.58%	1.14%	0.66%	4.06%	0.34%	29.28%	5.29%	0.04%	0.42%	0.03%	0.03%		46.52%
Copper, sulfide ore	0.2096	0.63%	0.10%	0.46%	1.52%	1.44%	30.35%	5.33%	0.45%	1.31%				46.00%
Copper, sulfide ore smelting dusts	1.8222	0.51%	0.28%	1.97%	29.24%			41.13%	0.12%	0.02%				25.88%
Leachate (solids), cu oxide ore processing	1.8723	0.09%	0.18%	0.10%	0.63%	21.70%	4.55%	0.82%	0.01%	0.06%	1.74%	2.29%		58.34%
Offgas, copper smelting	1.6116					15.85%					0.10%	tr	67.37%	16.68%
Slag, copper cu sulfide concentrate	0.9774	0.50%	0.01%	0.78%	31.09%	1.50%	18.70%	3.22%	tr	0.04%				38.65%
Slime, copper, refining (solid portion)	1.5488	70.69%	3.61%			3.67%			3.67%	1.82%				10.95%
Solid tailings, cu oxide ore processing	0.0265	0.01%	1.16%	0.67%	4.12%	0.36%	29.71%	5.37%	0.04%	0.42%				46.86%
Solid tailings, cu sulfide ore processing	0.0422	0.09%	0.09%	0.43%	0.91%	0.80%	30.27%	6.11%	0.45%	1.31%	0.26%	0.01%	tr	46.88%
Waste gases, cu oxide ore processing	0.1092										5.04%	tr	81.57%	13.39%
Waste gases, sulfuric acid, cu processing	0.0814					0.08%					0.11%	0.01%	99.63%	0.17%
Lead bullion	1.3056	2.00%	94.35%	tr	0.71%	0.25%			1.00%	1.52%				0.04%
Lead concentrate	3.4041	0.64%	74.80%	0.11%	2.08%	15.10%	1.34%	0.67%	0.01%					3.31%
Lead gangue	0.0111	tr	0.27%	tr	0.01%	0.10%	46.19%	0.43%	tr	0.02%			tr	52.93%
Lead ore	0.2275	0.04%	5.00%	0.01%	0.14%	1.05%	43.56%	0.13%	tr	0.02%				49.89%
Lead sinter	0.4097	1.48%	79.84%	0.13%	2.53%	1.08%		0.67%	1.00%	1.50%				9.87%
Lead, drossed	1.2332	0.01%	96.80%	tr	0.70%			tr	0.80%	1.50%	tr		0.06%	
Lead, refined	1.1238	tr	99.99%	tr	0.01%			tr	tr	tr	tr		tr	
Offgas, lead sintering	0.9807					8.90%					5.26%	0.27%	60.39%	25.19%

Table A1 *(Continued)*

Name	Exergy (kJ/g)	Cu	Pb	Zn	Fe	S	Si	Ca	As	Sb	C	H	N	O
Slag, lead blast furnace	0.8512		45.16%	0.55%	9.72%		1.50%	13.96%	1.16%	1.69%				17.19%
Slag, lead drossing	3.5823	62.27%		0.01%	0.83%	15.85%	0.10%		6.87%	1.65%				12.19%
Solid Waste, lead refining	0.6832	0.10%	tr	0.07%	16.69%			0.01%	19.21%	36.01%				24.80%
Solid Waste, lead sintering	0.9791		40.83%	0.05%	1.19%		1.99%	29.98%	0.01%					24.27%
Waste gases, lead drossing	0.0913										3.57%	0.23%	84.40%	11.70%
Waste gases, lead refining	0.0320										0.02%	tr	99.98%	tr
Waste gases, sulfuric acid, lead processing	1.0390					0.07%					6.96%		79.98%	12.99%
Waste gas, lead blast furnace	0.1543					0.19%					6.71%	0.37%	71.75%	20.97%
Waste gas, lead concentrating	0.1033										3.63%	9.70%		86.67%
Offgas, zinc roasting, purifying	1.5912					15.92%					0.01%	tr	67.40%	16.68%
Sludge, zinc, final refining (solid)	0.7287	0.75%	5.18%	33.10%	1.48%	18.91%		0.03%	0.38%	0.02%				35.95%
Solid Waste, zinc leaching	0.7843	2.87%	7.88%		40.37%	6.66%		4.34%	1.32%	0.71%	0.01%		0.01%	32.06%
Solid Waste, zinc s ore roast/pur	0.3872	6.64%	0.57%	34.95%	23.30%			7.09%	1.61%	1.18%				24.29%
Waste gases, zinc leaching	3.4399					15.33%					3.10%	0.66%	65.61%	15.30%
Waste gases, zinc refining	0.0374										0.01%	3.38%	69.81%	26.80%
Waste gases, zinc sulfide ore concentrating	0.0850										3.14%	9.43%		87.43%
Zinc, leachate	3.2107	3.00%	0.33%	55.50%	7.90%	0.15%	4.39%	3.00%	0.70%	0.50%	1.00%	0.02%		23.34%
Zinc, oxide ore (8.3%)	0.8096	0.16%	6.18%	8.32%	1.35%	3.34%	33.96%	0.13%	0.08%	0.06%	0.21%	0.55%		45.41%
Zinc, prerefined	4.9846	0.25%	1.75%	95.18%	0.50%	0.15%		0.01%	0.13%	0.01%	tr		0.63%	
Zinc, s ore tailings	0.0008	tr	7.66%	1.71%	1.10%	1.97%	41.48%	0.17%	0.06%	0.05%	0.04%		tr	45.50%
Zinc, slab	5.1822	tr	0.07%	99.85%	0.02%			tr	tr	tr	tr		0.02%	
Zinc, sulfide concentrate	6.1896	2.80%	0.85%	47.74%	9.70%	26.38%		2.90%	0.70%	0.48%				7.80%
Zinc, sulfide ore (9.0%)	0.9968	0.45%	6.18%	9.00%	2.42%	5.78%	32.72%	0.35%	0.16%	0.12%	0.14%	0.55%		41.88%

APPENDIX B

THE BEHAVIOR OF COPPER, LEAD AND ZINC IN SOIL

by Benjamin Warr

B1. Metals in soils

Soil is porous medium formed at the land surface by weathering processes and composed of a complex mixture of liquids and gases, organic and inorganic solids (sand, silt and clay).[1] Soil is a spatially and temporally heterogeneous, open and multi component system that exchanges matter and energy with the atmosphere, biosphere and hydrosphere. Soil pollution occurs when the 'normal' level of fluxes into the soil are perturbed by anthropogenic activity. Soil chemistry, a function of soil solid and aqueous phases, defines the fate and mobility of inputs into the soil.

Under natural conditions the main source of heavy metals in soil is the *parent material*, from which the soils are developed. Lead (Pb) is found in naturally high concentrations as galena (PbS), cerussite ($PbCO_3$) and anglesite ($PbSO_4$), each an economically important source, but rarely in it's elemental state. Copper is widely distributed in nature in the elemental state, in sulfides, arsenites, chlorides, and carbonates. Zinc occurs naturally as calamine, smithsonite and wurzite. Other than these concentrated forms the concentration of both zinc and copper in the continental crust tends to be highest in the ferromagnesium minerals, such as the basalts, pyropene and biotite.

Once the soil has formed, the quantities of each metal from anthropogenic sources largely exceed natural emissions and inputs to the soil. Atmospheric emissions are the main route of entry to the terrestrial environment. Mining and metal production are the primary industrial activities causing atmospheric pollution. For example, the total flux of copper to the atmosphere is approximately 75,000 mt per annum of which 85–95 percent is estimated as being deposited on the land by both dry and wet deposition,[2] subsequently entering the hydrological cycle. On a global basis, the atmospheric copper flux from anthropogenic sources is approximately three times higher than its flux from natural sources.

Soils can be considered polluted if the concentration of an element exceeds the natural background concentration. However in practice this definition is difficult to verify and of little significance with regard to potential risks posed

1 Approximate proportions 20 percent, 10 percent, 70 percent respectively.

2 Anthropogenic inputs are therefore both solids and in solution.

to human health, because no reference is made to the form of the element in the soil, or whether it is in a (potentially) mobile and toxic form. Moreover the spatial heterogeneity of the soil confounds the issue. It is for this reason that we do not present typical 'background' or 'polluted' concentration levels, which can be very misleading.[3] Instead we shall briefly describe the concept of bioavailability (see below).

It is often stated that heavy metals are so strongly bound to the soil in non-toxic forms that there is little risk of their moving to groundwater. Indeed, where pollutants exist in very high concentrations disturbing and moving the soil are often considered to pose a greater threat to human health than if the soil is left in its present state. The soil is considered as a convenient 'permanent' storage facility. If the soil were not a dynamic biome this might well be the case. Instead, soil is subject to diurnal, seasonal and long-term trends, that can cause changes to its properties and functions and the possible release of such toxic elements to the hydrosphere (which includes the soil solution) and subsequently to humans, animals and plants. Each copper, lead and zinc species will interact with the dynamic (*soil solution*) and static (*solid phase*) components of the soil in a different manner over time and spatially as the soil properties vary in three-dimensions.

Both copper and zinc are essential for life (*micro-nutrients*). Lead is not. All are damaging to life (both plant and animal) if present in excessive quantities. The most toxic form of each of these metals is the soluble (*mobile*) free ion species. Therefore we shall describe how the anthropogenic emissions of each element, deposited on the soil, may transform in the soil, which properties and processes operating in the soil render these elements immobile and how these properties can be modified, either by natural or anthropogenic cycles and long-term trends.

The diversity and complexity of the soil and the form of pollutants entering the soil means that a discussion that limits itself to the specifics of a single metal or type of pollution will fail to provide an effective overview of the dominant processes specific to each metal. There are many interactions between other elements in the soil. Instead it is far better to describe soil properties in general, then to consider the fate of individual elements, as regards their position in the periodic table and the soil properties at a specific site. The proximity of copper, lead and zinc in the periodic table suggests, as is the case, that they will behave similarly in the soil.

---

[3] Not least because the probability distribution of trace metal concentrations is highly skewed; therefore we would require the mean, variance, skew and kurtosis for a complete description.

B2. Aqueous phase speciation

The relative concentration of elements in solution defines which species are present and therefore what types of interaction occur at the mineral-solution interface. We describe the soil solution as an aqueous *phase*, implying that it has uniform macroscopic properties such as temperature and electrolyte concentration. Be aware though, that spatial variability occurs at micro-scales within the soil and that we are never really able to fully characterize the diversity of chemical environments. Neither can the effects of microbiological activity be ignored.

To simplify matters it is assumed that the soil solution is in equilibrium and that the species are stable. This is often the case in natural soils despite the range of times taken to reach equilibrium (10^{-9}s for the dissociation of $MnSO_4^0$, to 10^6s for the formation of $FeCl^{2+}$). In general oxidation-reduction reactions are fast and precipitation-dissolution reactions are slow. Based on this assumption the distribution of a given constituent among its possible chemical forms can be described with conditional stability constants. For the simplest of descriptions we require data describing the dissolved constituent concentrations, pH, electrolytic conductivity and equations describing the mass-balance of each constituent in terms of chemical species.

The pH of the soil is the major factor determining the solution speciation of metals. In acidic soils (approx. pH 4.5–6) copper, lead and zinc are most commonly in their divalent cationic forms (Cu^{2+}, Pb^{2+} and Zn^{2+}). For example the solubility of zinc changes a hundred fold as the pH increases from pH 4.5 to pH 7.5 (Russell 1988). At higher pH values hydroxide and carbonate species become increasingly common (e.g. $Zn(OH)^+$, $Zn(HCO_3^+)$ and $ZnCO_3^0$) the ratios of which depend upon the partial pressure of carbon dioxide. Nevertheless it is the properties of the divalent cation forms and their specific interactions with the soil solids that are critical in determining the fate and mobility of copper, lead and zinc in soils.

Nevertheless inputs into the soil caused by anthropogenic activity are diverse and are subject to modifications by weathering in the soil.

When the solubility of a solid phase controls the concentration of an element in solution via precipitation-dissolution mechanisms an *activity-ratio* diagram can be constructed, which plots the changes in the solution concentration of the species, typically against changes in pH.

Oxidation-reduction (*redox*) reactions occur when electrons are transferred between species. Loss of an electron results in *oxidation*, the receipt of an electron *reduction*. A redox half-reaction describes the transfer of protons (H^+) and electrons (e^-) from the soil solution to an oxidized species. Reduction therefore causes a consumption of protons and a lower soil pH. The redox status of a soil can be described with reference to the negative logarithm of the free-electron activity, or *pE value* in a similar manner to pH. Large values of pE describe an electron-poor environment, typical of well aerated soils, having a predominance of oxidized species, just as large values of pH favor proton-poor species (alkalis). In a low pE environment, typical of waterlogged soil, reduced species are common and electron-rich species are favored.

The most commonly affected elements are C, N, O, S, Mn and Fe and in contaminated soils As, Se, Cr, Hg and Pb. In a closed system the respiration of micro-organisms leads to a depletion of oxygen, a build-up of carbon dioxide and a reduction sequence (O_2 (g), NO_3^- (aq.), Mn (s), Fe^{2+} (aq), SO_4^{2-} (aq)). As pE becomes negative sulphur reduction can take place and copper, lead and zinc (also Mn, Fe and Zn) present in high enough concentrations can react to form insoluble metal sulfides. On oxidation and consequent acidification of the soil these metals are released into solution. If they exceed the buffer capacity of the soil (see below) they may become mobile and enter the groundwater to pose a considerable risk to human health. This is a situation common when tidal flats come into agricultural production, but is far more severe for mine tailing ponds that can contain very high concentrations of the requisite constituents (metals, organics and sulfides).

B3. Solid phase constituents and complex formation

The majority of lead, copper and zinc compounds entering the soil are rapidly transformed into ionic species, in solution, which then form complexes with the solid phases of the soil. To understand how, it is necessary to consider the properties of the constituents forming the solid phase and most importantly to understand how pH acts as the master variable controlling the form and strength of the interactions. In large part it is the ability of the soil to bind strongly with heavy metal cations and the total number of functional sites within a soil, at a given pH, that determine their fate and mobility (see *Figure B1*).

The inorganic constituents in sand, silt and clay are usually crystalline and have a regularly ordered and repeating arrangement of atoms governed by lattice coordination properties. The sand and silt fractions are comprised of crystalline mineral particles from primary rocks and microcrystalline aggregates, such as calcium carbonate, gypsum, ferric or aluminum hydroxides and silica, that have formed as a result of weathering or from biotic residues. The clay fraction differs mineralogically from the sand and silt fractions by being dominantly composed of the more thermodynamically stable products of weathering (*secondary* minerals) such as mica, smectite and kaolinite (see below). Clay minerals occur as particles typically smaller than 2 μm. Their much smaller size means that they have a very large surface area relative to the sand and silt fractions regardless of their proportion in the soil.

Aluminum, oxygen and silicon are the dominant elements that form clay minerals (*aluminosilicates*). The comparatively large O^{2-} ions form a closely packed sheet with the much smaller metallic *cations* filling the gaps between the oxygen ions to balance the electric charges. Silicon ions (Si^{4+}) are very small and fit into the space between 4 oxygens arranged as a tetrahedron

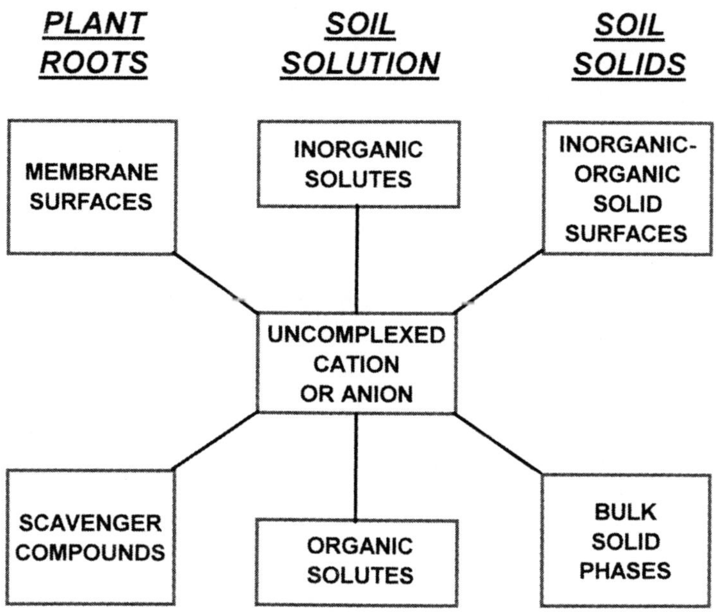

Figure B1: Schematic of competition among functional groups in soil.

(SiO^4). These can subsequently polymerize to form very stable silicate chains and sheets (*lamellae*). Aluminum ions (Al^{3+}) are larger and cannot form tetrahedra. Instead they occupy a larger space within an octahedron formed by six oxygens. These may also polymerize to form sheets. Only two-thirds of the octahedral spaces need to be filled to achieve charge balance (e.g. dioctahedral $Al(OH)_3$). However, if divalent Mg^{2+} or Ca^{2+} ions substitute for Al^{3+} all the spaces must be filled (e.g. trioctahedral $Mg(OH)_2$). The tetrahedral and di-(tri)octahedral sheets are the building blocks of clays. The 'isolated' oxygen of the SiO_4 tetrahedra bonds with the di-(tri)octahedral layer to form 1:1 lamellae from a single layer of each (e.g. kaolinite), and the 2:1 lamellae by sandwiching a di-(tri)octahedral sheet between two tetrahedral sheets (e.g. smectite).

This idealized picture must then be placed into the context of a heterogeneous and diverse chemical environment within which the lamellae (trans)form. Aqueous acidic dissolution of primary rock minerals releases a variety of metal ions depending on the geochemistry of the environment, Mg^{2+}, Fe^{2+} and Fe^{3+} being the most abundant. Any one of these ions that is sufficiently small may substitute for any other in a process called 'isomorphic substitution'. This process is so named because the valency is not important, simply the size.[4] Substitution distorts the shape of the lamellae and gives a

[4] Ca^{2+} cannot therefore be a substituting ion.

net permanent[5] negative charge to the crystal. Cations from the soil solution balance the charge by binding electrostatically to the particle surfaces. The position of the charge deficit within the lamellae will determine the strength of the complex formed.

Cations can also bind as a result of the ionization of protons from the surface of lamellae upon contact with the soil solution. Hydroxyl (OH) groups on the surface and broken edges of 1:1 clays receive protons (H^+) from the soil solution to generate a net positive charge that must be balanced by exchangeable anions. The water molecule can be exchanged with an organic or inorganic anion that can then form a more stable bond with a metal cation, in a process called ligand exchange. The amount of this 'variable charge' depends on the acidity of the soil solution. For soils dominated by 1:1 clays (predominantly tropical soils) a reduction in the pH of the soil is accompanied by a reduction in buffering capacity. Such a change is likely to cause the release of cations into the soil solution, increased uptake by plants and/or a loss from the soil to the groundwater.

Adsorption reactions in soils operate over a range of time scales, from minutes to weeks. Readily exchangeable ions reach equilibrium rapidly by film diffusion. There are many ways that specifically adsorbed ions adsorb and desorb; there are often multiple mechanisms, causing the long time to equilibrium. In general the relative affinity of a soil adsorbent for a metal cation is a function of ionic radius (for fixed valency) via the ionic potential (valence/radius).

$Cs^+ > Rb^+ > K^+ > Na^+ > Ba^{2+} > Sr^{2+} > Ca^{2+} > Mg^{2+} > Hg^{2+} > Cd^{2+} > Zn^{2+}$
(order of decreasing strength of adsorption)

Firstly the larger the ionic radius the smaller the electric field and the less likely the ion is to remain *solvated.* Secondly, the ion is more readily polarized, a necessary condition for the creation of a covalent bond. This generalization breaks down for several transition metals (e.g. Mn^{2+}, Fe^{2+} and Ni^{2+}) as a result of their specific electron configurations and the many possible ways these can interact with organic matter and clay functional groups.

$Cu^{2+} > Ni^{2+} > Co^{2+} > Fe^{2+} > Mn^{2+}$
(order of decreasing strength of adsorption)

Organic solids also contribute significantly to the chemistry of the soil. While the solids themselves are very diverse and complex, hence poorly characterized, it is possible to generalize with regard to their functional behavior. The modes of reaction between ions in solution and the organic matter functional groups are similar to those already described. However,

---

[5] In the sense that it is a function of the crystal composition.

soil humus[6] has properties that can have significant effects on soil process, despite the often small proportion of organic matter relative to other solids. *Polyfunctionality* refers to the great diversity of functional group types it can posses (carboxyl, carbonyl, amino, phenolic OH being just a few). At the macromolecular level humus has *anionic character*, which increases the reactivity of functional sites and therefore strengthens complexation. As a result they form very strong ligands where two functional groups are attached to the metal. The electron configuration of the metal will determine the favorability and strength of such complexation. Soil organic matter, either as a solid or in solution, binds very strongly to Cu^{2+}. The ligand metal complex is called a *chelate* if in solution. Complexation of copper by organic matter is therefore the major control of it's mobility and availability, to a lesser extent that of zinc and to a minimal extent that of lead. Being in solution, chelates can significantly increase the mobility and bioavailability of copper in soils.

The reactivity of these functional groups largely depends on the acidity of the solution; however their charge is always negative. This means that, in soils dominated by 1:1 (variable charge) clays, organic matter is the principal constituent able to adsorb metal cations regardless of changes in the acidity of the soil solution. Losses of soil organic matter that occur upon deforestation, cultivation or drainage can cause the loss of the majority of negative charge sites in tropical soils,[7] whereas the permanent negative charge of the 2:1 clay dominated soils, typical of temperate regions, means that the reduction upon loss of organic matter is less significant. Nevertheless, the importance of organic matter as a sink for heavy metals should not be underestimated. The greatest concentration of organic matter is in the top few centimeters of the soil that receives soil pollutants directly from the atmosphere. It is clear that any change in the total content or functional quality of the pool of organic matter can have considerable impacts on the fate of contaminants.

To summarize, clay and organic matter surfaces interact with the soil solution to define the soil chemistry.[8] We can distinguish between complexes where at least one water molecule lies between the ion and the molecule that binds it, *outer-sphere* complexes that are caused by electrostatic forces, and *inner-sphere* complexes, where no water molecule interposes, which involve either some degree of ionic or covalent bonding. Inner-sphere complexes are stronger and more typical of heavy metal ions that are small and have a particular affinity, such as Cu^{2+}. If a solvated ion does not form a surface complex, but instead neutralizes surface charge diffusely it is said to be adsorbed in the *diffuse-ion swarm* as is the case for Zn^{2+} and Pb^{2+}. This

---

6 Humus refers to all organic materials except unaltered or partially altered biomass, i.e. all materials that are not synthesized directly by the soil flora and fauna.

7 Despite the reduction of acidity concomitant with the release of base cations such as Ca^{2+} and Mg^{2+}.

8 We discuss the complex and poorly understood properties of the organic solids later.

division based on the strength of bonding reflects the mobility of an element or molecule within the soil profile. Outer-sphere complex ions are *readily exchangeable* and readily leached while covalently bonded inner-sphere complexed elements and molecules may sometimes be considered to all intents and purposes 'irreversible' under normal soil conditions. The ability of a soil to buffer pollutants is closely related therefore to the soils cation exchange capacity, which is largest for clay soils and lowest for sandy soils. Therefore regulations defining critical loads are often based on soil texture and organic matter content.

This situation is further complicated as the hydrogen ion (H^+) competes directly with metal cations for adsorption sites. So, as the soil pH falls, the strength of adsorption of heavy metals is reduced and the concentration in solution increased. This poses potential threats if the soil environment becomes acidified (see *Figure B2*).

A plot of *ionic potential* (ionic radius/valency) versus the ratio of elemental concentration in organisms to that in the soil (the *enrichment factor*, EF) shows that the uptake of different metal elements depends on whether they solvate in aqueous solution or hydrolyze[9] (*Figure B2*) and then the degree with which they form strong complexes with organic matter and clays. Higher EF signifies greater bio-availability and therefore the degree of accumulation in foodstuffs). Along this sequence we can differentiate between Ca^{2+} a solvating cation (as are Cu^{2+}, Pb^{2+}, and Zn^{2+}), Al^{3+} a hydrolyzing cation and Mo^{6+} an oxyanion-forming cation. Hydrolyzed species and oxyanions may become strongly adsorbed or precipitate under certain solution conditions and are therefore rarely amongst the 'essential' elements. Fortunately copper strongly binds to the soil through specific adsorption with organic matter and is therefore poorly available for plant uptake, and unlikely to become mobile unless as a result of severe pH changes or by soil erosion attached to dust particles. Zinc on the other hand is only weakly adsorbed as a diffuse-ion, less strongly bound to organic matter than Cu^{2+} and is therefore more readily bio-available.

Competition between the H^+ ion and metal cations means that acidification of the soil can cause complexed heavy metals to be desorbed and become mobile in the soil. Acidification can be caused by many anthropogenic activities, for example, sulphur deposition and fertilizer applications, but also inadvertently by reforestation[10] of agricultural land. A soil has an acid-neutralizing capacity with respect to additions of H^+ in the same way that it is described as having a buffering capacity with respect metal cations. This is a function of the quantity of materials able to neutralize acidity via dissolution (e.g. $CaCO^3$) or by desorption (Ca^{2+}, Mg^{2+}).[11] Once this capacity

[9] At an arbitrary pH 7.

[10] Often proposed as a means to increase terrestrial carbon sinks.

[11] Described as base cations to distinguish them from Al^{3+} a hydrolyzing cation that contributes to acidity.

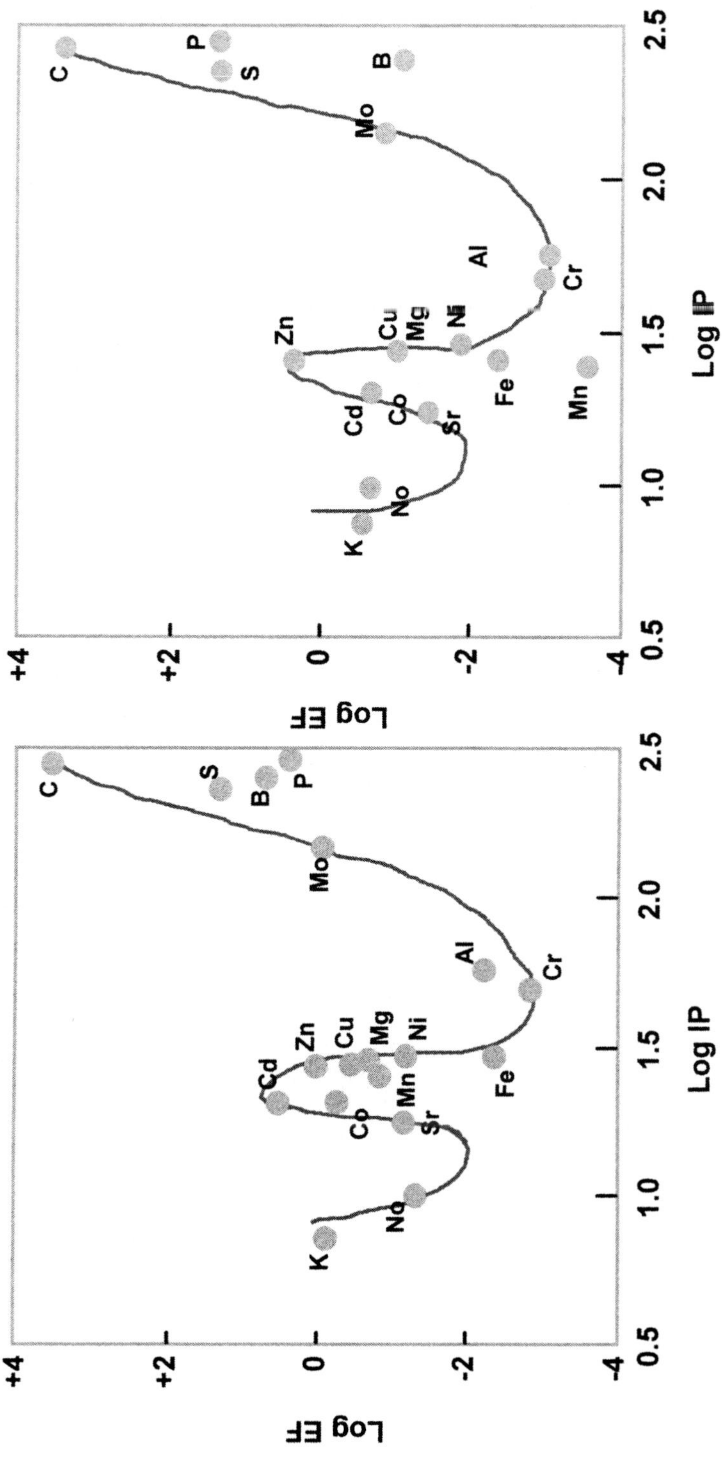

Figure B2: Banin-Navrot plots for terrestrial plants and animals.

has been exceeded dramatic changes to soil processes can occur that may lead to the release of heavy metals from their inner-sphere complexes and precipitates. Once in solution, either as free-ions or as chelates, they are increasingly bio-available and, as a result, a threat to the quality of water and food sources on a global scale.

B4. Summary

The fate of heavy metals entering the soil is a function of specific or exchange adsorption at the mineral-solution interface, the precipitation of sparingly soluble solids, and the formation of relatively stable organo-metal complexes with the organic matter in the soil.

The buffering capacity of the soil is largely determined by the negative net electrostatic charge held by the solid phase of the soil, but can be considerably modified by dynamic properties of the soil such as pH and moisture status. While the spatial variability of soil and inputs of pollution has lead to widely accepted concerns about exceeding the buffering capacity of the soil locally,[12] the diffuse nature of much of the pollution entering the soil by atmospheric deposition means that changes that may occur to soil properties and processes over time, over large areas, such as those caused by acidification, remain the subject of considerable concern for the future and have lead to the alarmingly termed concept of the *toxic time bomb*.

[12] e.g. landfill sites or other forms of 'point source pollution'.

APPENDIX C

GLOBAL COPPER MODEL

C1. Introduction

In this appendix, a model of the global copper system and the generation of scenarios of the consumption of refined copper are described. The copper consumption scenarios function as input to the copper system model. The purpose of the model is to generate scenarios of the evolution over time of the global copper system, especially of the societal stocks of copper.

The global copper system model is presented in *Section C2*. *Section C3* describes its calibration and the generation of copper consumption scenarios is presented in *Section C4*. *Section C5* describes the resulting copper system scenarios. *Section 2.9* of the main report discussed both the generated copper consumption scenarios and the copper system scenarios.

C2. A model of the global copper system

The model of the global copper system (*Figure C1*) is based on a model of the copper system in the USA by Zeltner et al. (Zeltner et al. 1999). The model consists of six processes and eight stocks. Life times and collection rates in the Zeltner model are shown in *Table C1*.

The world is divided into four world regions: OECD90, REF, ASIA and ALM. OECD90 includes the countries that were members of OECD in 1990; that is, North America, Western Europe, Australia, Japan and New Zealand. REF includes countries undergoing economic reform in Central and Eastern Europe and the former Soviet Union. ASIA includes the countries in Asia that were not members of OECD in 1990. ALM includes countries in Africa, the Middle East and Latin America. These are the same regions used by the Intergovernmental Panel on Climate Control (IPCC). For a full definition of the regions, see *Table C2*. Each region is considered separately. However, primary resources (S1), concentration processes (P2), smelting and refining processes (P3), gangue (S11) and slag (S12) are modeled on a global rather than a regional basis.

S1) *Primary resources*: Primary resources consist of copper that has not yet been mined. There are no restrictions in the model on how much primary copper can be mined. The annual mine production of copper goes into the concentration stage. Primary resources are modeled on a global basis.

P2) *Concentration*: In concentration, the copper minerals are separated from the gangue and copper concentrate is produced. The annual production of copper concentrate then goes into smelting and refining. A small but non-zero

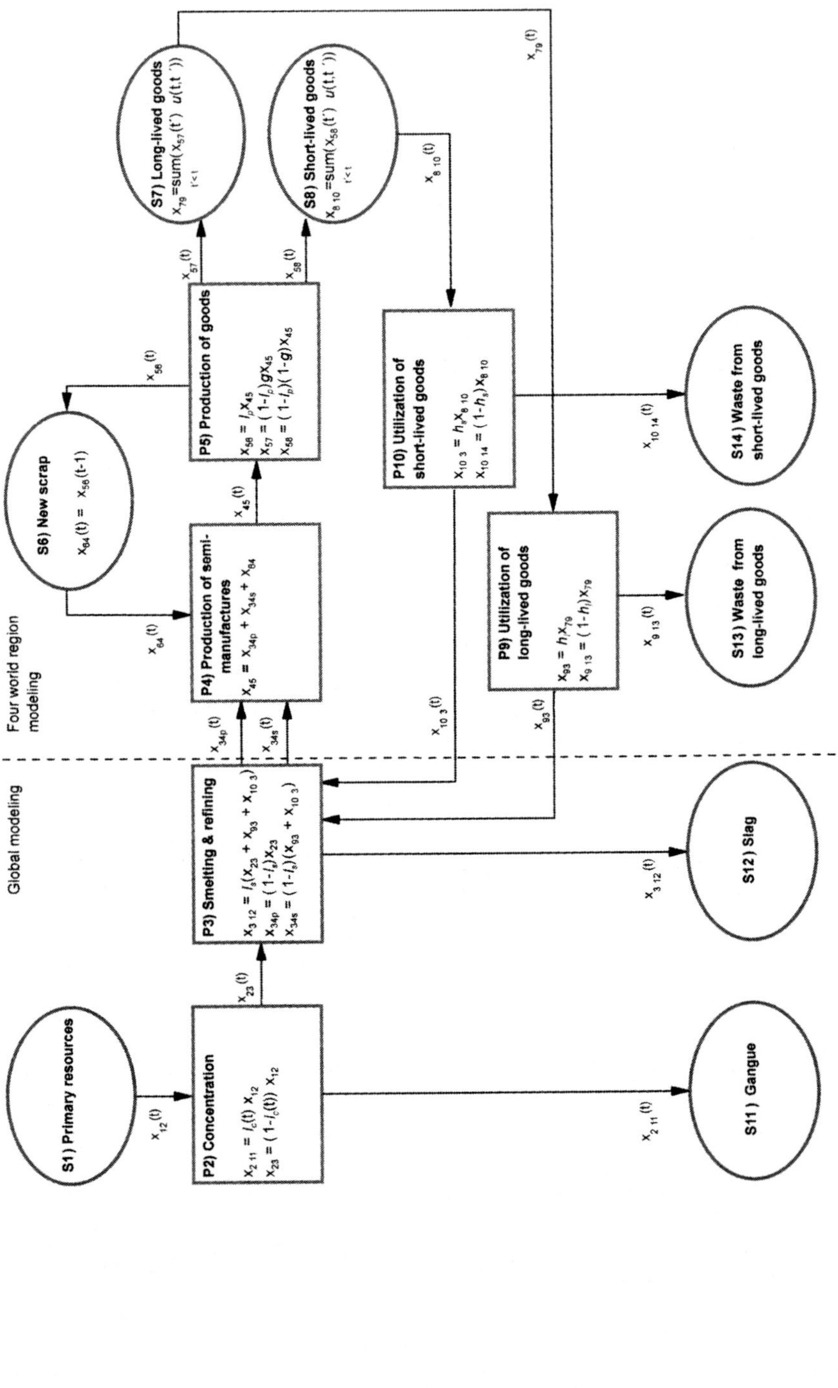

Figure C1 (3.12): Model of the global copper system.

Table C1: Lifetimes, standard deviations and collection rates

	Lifetimes and collection rates					
	t_l year	s_l year	h_l –	t_s year	s_s year	h_s –
Scenario 1	30	20	36%	5	3	18%
Scenario 2	50	30	50%	10	7	25%
Scenario 3	70	40	70%	15	10	35%
Scenario 4	90	50	96%	20	15	48%

Source: Zeltner et al. (1999).

share, $l_c(t)$, of copper in mine production is lost in gangue. Concentration (as discussed in *Sections 2.2.2*, *2.2.3* of the main report) is modeled on a global basis.

S11) *Gangue*: Gangue is waste from the concentration stage. Copper lost in concentration is accumulated in gangue, as noted above.

P3) *Smelting and refining*: In smelting and refining, both primary resources (concentrate) and secondary resources (old scrap) are treated to produce copper metal. The annual production of refined copper goes into the production of semi-manufactures such as wire, tube, sheet and castings. A share, l_s, of copper in concentrate and old scrap is lost in slag. Smelting and refining (as discussed in Sections 2.2.5–2.2.8 of the main report) is modeled on a global basis.

S12) *Slag*: Slag is waste from the smelting and refining stage. Copper lost in smelting and refining is accumulated in slag.

P4) *Production of semi-manufactures*: Copper from the smelting and refining stage as well as new scrap is used to produce semi-manufactured copper products. The driving variable, the input to the model, is the annual consumption of refined copper in each of the regions (primary and secondary) ($x_{34p}(t) + x_{34s}(t)$); that is, the annual production of semi-manufactures less the use of new scrap. For the years 1900–1997, statistics of the consumption of refined copper are used (*Table C3*). For the years 1998–2100, scenarios of the consumption of refined copper are used (*Section C4*). The annual production of semi-manufactures goes into the production of finished goods. No losses are assumed in the production of semi-manufactures. It is further assumed that the entire production of semi-manufactures in a region goes to the production of finished goods in the same region. That is, no net trade of semi-manufactures is assumed between regions.

P5) *Production of finished goods*: Two generic types of products are produced: long-term products and short-term products. A share, l_p, of semi-manufactures is lost and goes into new scrap. The share g of produced goods goes to long-term usage. It is assumed that the entire production of finished goods in a region is used in the same region. That is, no net trade of copper-containing finished goods is assumed between regions.

S6) *New scrap*: New scrap is the amount of copper lost in the production

Table C2: Definition of world regions

OECD90 Region

North America

Canada	Puerto Rico	United States of America	Virgin Islands
Guam			

Western Europe

Andorra	Denmark	Greece	Liechtenstein	Norway
Austria	Faeroe Islands	Greenland	Luxembourg	Portugal
Azores	Finland	Iceland	Madeira	Spain
Belgium	France	Ireland	Malta	Sweden
Canary Islands	Germany	Isle of Man	Monaco	Switzerland
Channel Islands	Gibraltar	Italy	Netherlands	Turkey
Cyprus				

Pacific OECD

Australia	Japan	New Zealand

REF Region (countries undergoing economic reform)

Central and Eastern Europe

Albania	Czech Republic	Poland	Yugoslavia
Bosnia & Herzegovina	The Former Yugoslav	Romania	
Bulgaria	Hungary	Slovak Republic	
Croatia	Republic of Macedonia	Slovenia	

Newly independent states of the former Soviet Union

Armenia	Estonia	Kyrgyzstan	Republic of Moldova	Turkmenistan
Azerbaijan	Georgia	Latvia	Russian Federation	Ukraine
Belarus	Kazakhstan	Lithuania	Tajikistan	Uzbekistan

ASIA Region

Centrally planned Asia and China

Cambodia	Hong Kong	Laos (PRD)	Mongolia	Viet Nam
China	Korea (DPR)			

South Asia

Afghanistan	Bhutan	Maldives	Pakistan	Sri Lanka
Bangladesh	India	Nepal		

Other Pacific Asia

American Samoa	Indonesia	Philippines	Thailand
Brunei Darussalam	Malaysia	Republic of Korea	Tonga
Fiji	Myanmar	Singapore	Vanuatu
French Polynesia	New Caledonia	Solomon Islands	Western Samoa
Gilbert-Kiribati	Papua New Guinea	Taiwan, province of China	

Table C2 *(Continued)*

ALM Region (Africa and Latin America)				
Middle East and North Africa				
Algeria	Iran	Lebanon	Qatar	Tunisia
Bahrain	Israel	Libya	Saudi Arabia	United Arab Emirates
Egypt	Jordan	Morocco	Sudan	Yemen
Iraq	Kuwait	Oman	Syria	

Latin America and Caribbean			
Antigua & Barbuda	Costa Rica	Guatemala	Peru
Argentina	Cuba	Guyana	Saint Kitts & Nevis
Bahamas	Dominica	Haiti	Santa Lucia
Barbados	Dominican Republic	Honduras	Saint Vincent & Grenadines
Belize	Ecuador	Jamaica	Suriname
Bermuda	El Salvador	Martinique	Trinidad & Tobago
Bolivia	French Guyana	Mexico	Uruguay
Brazil	Grenada	Netherlands Antilles	Venezuela
Chile	Guadeloupe	Nicaragua	
Colombia	Guatemala	Panama	

Sub-Saharan Africa			
Angola	Djibouti	Malawi	Seychelles
Benin	Equatorial Guinea	Mali	Sierra Leone
Botswana	Eritrea	Mauritania	Somalia
British India Ocean Territory	Ethiopia	Mauritius	South Africa
Burkina Faso	Gabon	Mozambique	Saint Helena
Burundi	Gambia	Namibia	Swaziland
Cameroon	Ghana	Niger	Tanzania
Cape Verde	Guinea-Bissau	Nigeria	Togo
Central African Republic	Kenya	Reunion	Uganda
Chad	Lesotho	Rwanda	Zaire
Comoros	Liberia	Sao Tome & Principe	Zambia
Cote d'Ivoire Congo	Madagascar	Senegal	Zimbabwe

Source: IPCC (2000).

of finished goods. The entire production of new scrap in a given year goes back into the production of semi-manufactures during the subsequent year. No losses of new scrap are assumed. It is assumed that the entire production of finished goods in a region is used in the same region. That is, no net trade of copper-containing new scrap is assumed between regions.

S7) *Long-term usage*: The residence times of copper products in the stocks of long-lived and short-lived goods (S7 and S8) are assumed to be normally distributed around a mean. The 'outflow' $x_{57}(t')$ from the long-lived stock at time t of products produced at time t' is

$$x_{57}(t)u(t, t') = \frac{1}{\sigma_l\sqrt{2\pi}} e^{-\frac{1}{2\sigma_l^2}(t - t' - \tau_l)^2} \qquad (t > t') \qquad \text{(C1)}$$

Table C3: Annual consumption of refined copper in 1900–1997 (1,000 metric tons)

	Europe	Czech	Hung.	Poland	USSR	Yugosl.	Asia	Iran	Japan	Saudi	Turkey	Other Asia	Africa	Amer.	Canada	US	Austr. & Ocean	Eastern Countr.	China*	N. Korea	Vietn.	Cuba	World	OECD 90	REF	ASIA	ALM
1900																							452	436	11	4	1
1901																							529	510	13	5	1
1902																							557	537	14	5	1
1903																							593	572	15	5	1
1904																							662	638	16	6	2
1905																							693	668	17	6	2
1906																							725	699	18	6	2
1907																							725	699	18	6	2
1908																							766	738	19	7	2
1909																							851	820	21	7	2
1910																							872	841	22	8	2
1911																							887	855	22	8	2
1912																							1019	982	25	9	3
1913																							982	947	24	9	2
1914																							926	893	23	8	2
1915																							1058	1020	26	9	3
1916																							1376	1326	34	12	3
1917																							1443	1391	36	13	4
1918																							1395	1345	35	12	3
1919																							1050	1012	26	9	3
1920																							967	932	24	8	2
1921																							528	509	13	5	1
1922																							863	832	21	8	2
1923																							1250	1205	31	11	3
1924																							1331	1283	33	12	3
1925																							1410	1359	35	12	4
1926	623	13			24		94		82				4	728	16	712	9						1459	1406	36	13	4
1927	734	14			44		84		76				4	661	15	646	9						1492	1423	57	8	4
1928	737	17			45		82		74				4	745	15	729	7						1574	1501	63	7	4
1929	686	14			46		72		64				4	826	19	807	8						1595	1524	60	7	4
1930	719	17			62		77		69				3	592	18	574	5						1397	1307	79	8	3
1931	631	14	5	5	56		77		70				3	427	15	409	4						1142	1050	80	7	5
1932	565	13	5	5	44		81		73				3	261	24	236	5						915	837	66	7	5
1933	649	13	5	7	41		92		83				3	342	29	308	6						1091	1008	66	9	8
1934	803	18	9	11	55		125		113				4	337	39	293	8						1277	1163	93	12	9
1935	910	22	8	12	93		150		135				4	446	41	400	10						1520	1362	134	15	9
1936	950	29	10	15	128		142		127				4	647	45	595	13						1755	1549	181	15	10
1937	1106	31	12	17	157		200		183				4	695	56	631	14						2018	1773	217	16	12
1938	1200	32	15	26	165		218		200				5	428	48	369	17						1869	1598	238	17	16

1939																		1631	230	18	20
1940																		1665	223	20	23
1941																		1699	216	21	26
1942																		1732	208	22	29
1943																		1766	201	23	32
1944																		1800	193	24	35
1945																		1833	186	25	39
1946	688	22	54	27			5	1188	73	1077	24	157					2144	1867	179	26	42
1947	831	16	64	41			13	1474	99	1327	21	173					2576	2302	190	23	61
1948	833	15	90	64			15	1428	100	1289	21	191					2577	2292	205	25	54
1949	828	17	74	45			17	1174	92	1025	20	221					2335	1993	239	29	74
1950	890	19	96	64			17	1438	97	1292	18	241					2701	2341	260	33	66
1951	918	19	129	95			23	1507	122	1286	33	279					2888	2434	298	34	122
1952	941	19	125	93			25	1516	118	1343	35	321					2964	2511	340	32	81
1953	911	18	121	98			12	1510	99	1355	14	337					2903	2459	355	23	67
1954	1321	18	117	100			18	1317	93	1139	45	352					3171	2680	370	17	103
1955	1437	21	133	112			21	1559	126	1363	45	390					3586	3062	411	22	92
1956	1436	23	182	148			23	1566	132	1380	31	443					3681	3104	465	34	77
1957	1533	27	223	176			31	1403	107	1227	37	464					3691	3053	491	47	101
1958	1684	40	222	147		17	34	1349	112	1135	60	785	75				4133	3097	750	133	154
1959	1622	38	288	219		15	29	1538	116	1327	64	835	90				4376	3312	783	144	137
1960	1931	46	384	304		18	35	1421	107	1225	72	912	110				4756	3593	847	172	143
1961	2004	41	462	372		22	37	1552	129	1327	63	952	120				5069	3853	873	189	154
1962	1928	53	402	301		23	40	1683	138	1451	77	1025	120				5155	3842	958	198	157
1963	1971	56	455	352		24	44	1838	154	1582	84	1041	120				5432	4087	977	199	190
1964	2179	56	541	458		18	48	1997	184	1656	99	1055	120				5918	4518	99	185	224
1965	2157	59	504	428	7	2	44	2220	209	1844	102	1166	120				6193	4687	1105	188	213
1966	2003	63	536	483	10	5	38	2530	248	2156	108	1227	130				6443	4945	1160	169	169
1967	1973	65	683	616	8	7	38	2113	204	1800	88	1302	140				6197	4624	1228	191	154
1968	2175	67	761	695	9	7	40	2099	232	1704	105	1346	150				6526	4853	1263	200	210
1969	2342	75	893	807	12	12	45	2359	222	1951	106	1409	180				7154	5365	1304	242	244
1970	2488	89	918	821	14	12	44	2272	229	1860	114	1486	220				7322	5437	1355	292	238
1971	2370	75	910	806	15	12	58	2278	220	1833	117	1599	260				7332	5287	1413	337	294
1972	2511	83	1085	951	15	13	66	2510	224	2030	113	1700	270				7965	5761	1513	356	335
1973	2650	85	1335	1202	14	12	84	2722	231	2221	136	1813	270	13		1	8740	6368	1614	391	368
1974	2679	109	1010	881	15	10	93	2557	248	1995	123	1846	281	14		1	8307	5832	1660	397	417
1975	2406	103	943	827	16	7	87	1891	185	1399	98	2021	300	15	1	1	7444	4828	1807	408	402
1976	2627	100	1217	1050	18	9	76	2386	206	1812	121	2111	320	15	1	1	8539	5735	1874	476	454
1977	2762	118	1326	1127	22	9	76	2603	222	1990	108	2182	330	15	1	1	9057	6113	1953	514	477
1978	2752	125	1481	1241	26	12	95	2821	250	2197	126	2253	350	16	1	1	9527	6467	2009	568	482
1979	2834	132	1596	1330	22	7	98	1868	243	2158	129	2317	360	15	1	1	9842	6585	2072	613	573
1980	2843	149	1466	1158	33	15	120	2557	209	1862	130	2274	370	15	1	1	9390	6087	2035	645	624
1981	2656	149	1633	1254	27	19	115	2689	242	2025	144	2273	370	15	1	1	9509	6200	2034	718	557

Table C3 *(Continued)*

	Europe	Czech	Hung.	Poland	USSR	Yugosl.	Asia	Iran	Japan	Saudi	Turkey	Other Asia	Africa	Amer.	Canada	US	Austr. & Ocean	Eastern Countr.	China*	N. Korea	Vietn.	Cuba	World	OECD 90	REF	ASIA	ALM
1982	2634					136	1614	6	1239	0	39	18	106	2359	149	1762	132	2245	400	17	1	2	9090	5819	1961	731	579
1983	2602					126	1680	10	1212	0	7	11	97	2533	195	2013	128	2471	600	17	1	2	9510	6070	1976	1018	445
1984	2748					130	1878	10	1364		55	8	108	2741	216	2116	125	2453	600	18	1	2	10053	6494	1963	1060	537
1985	2724					147	1766	10	1226	6	76	2	90	2796	223	2144	126	2383	550	20	1	3	9885	6372	1957	1017	540
1986	2765					123	1886	18	1211	9	77	5	103	2835	226	2138	118	2476	590	20	1	3	10183	6411	1986	1178	609
1987	2803					127	2050	20	1277	17	75	7	100	2918	232	2127	125	2433	540	22	2		10430	6513	1996	1218	704
1988	2861					119	2189	34	1331	40	71	8	104	2912	236	2206	131	2462	550	25	2		10660	6717	2004	283	656
1989	3062					119	2484	34	1447	50	101	10	101	2871	213	2206	134	2393	610	25	3		11046	7045	1874	1480	648
1990	3135					93	2673	40	1577	49	109	14	96	2709	181	2150	125	2023	512	30	3		10761	7183	1572	1428	576
1991	3173					80	2880	52	1613	54	98	20	99	2665	159	2058	104	1754	590	30	3		10674	7125	1212	1665	672
1992	3268					70	2827	61	1411	80	117	54	103	2782	156	2176	126	1696	882	27	3	1	10801	7183	854	2017	748
1993	3100					50	3004	52	1384	93	109	63	106	3007	186	2359	150	1607	985	27	3	1	10674	7238	642	2317	776
1994	3359					50	3207	44	1375	89	108	48	122	3380	199	2678	148	1346	798	15	1	1	11561	7817	581	2357	805
1995	3465					50	3370	48	1415	85	139	64	115	3228	190	2530	169	1728	1148	15	1	1	12076	7858	614	2784	821
1996	3372					44	3673	54	1480	145	160	116	112	2459	218	2621	190	1678	1161	15	1	1	12483	7996	545	2895	1047
1997	3561					43	3763	56	1441	150	188	115	118	3681	225	2770	188	1765	1241	15	1	1	13076	8330	551	3069	1126

Sources: 1990–1925 [MetallGesellschaft & WBMS 1997], 1926–1938 [McMahnon 1965, Table 38], 1939–1945 extrapolated, 1946–1957 [McMahnon 1965, Table 39], 1958–1964 [MetallGesellschaft 1968], 1965–1972 [MetallGesellschaft 1976], 1973–1981 [MetallGesellschaft 1984], 1982–1985 [MetallGesellschaft 1993], 1986–1995 [MetallGesellschaft & WBMS 1997] and 1996–1997 [MetallGesellschaft & WBMS 1998].

1900–1925 Mine production of copper. 1926–1938 Unspecified. 1958–1997 Apparent consumption of refined copper.

* In 1958–1964, China denotes 'Other Asian countries in the East block', In 196972, China includes North Korea.

1900–1925 OECD90, REF, ASIA and ALM are asumed to have thef the consumption of refined copper as in 1926, 96%, 2.5%, 0.87% and 0.25%, respectively.

1926–1938 OECD90 = (Europe – Czechoslovakia – Hungary – Poland – USSR) + Japan + (Canada + United States) + Australia & Oceania
REF = (Czechoslovakia + Hungary + Poland + USSR); ASIA = (Asia – Japan); ALM = Africa + (America – Canada – United States).

1946–1957 OECD90 = (Europe – Yugoslavia) + Japan + (Canada + United States) + Australia & Oceania; REF = Yugoslavia + Eastern Countries;
ASIA = (Asia – Japan); ALM = Africa + (America – Canada – United States).

1958–1997 OECD90 = (Europe – Yugoslavia) + (Japan +Turkey) + (Canada + United States) + Australia & Oceania
REF = Yugoslavia + (Eastern countries – C;hina – Cuba – North Korea – Vietnam);
ASIA = (Asia – Iran – Japan – Saudi Arabia – Turkey – Other Asia) + China + North Korea + Vietnam);
ALM = (Iran + Saudi Arabia + Other Asia) + Africa + (America – Canada – United States) + Cuba.

where π_l is the mean residence time and σ_l is the standard deviation. The total outflow from the long-lived stock $x_{79}(t)$ at time t is thus

$$x_{79}(t) = \sum_{t' < t} X_{57}(t)u(t, t') = \sum_{t' < t} x_{57}(t') \frac{1}{\sigma_l \sqrt{2\pi}} e^{1 - \frac{1}{2\sigma_l^2}(t - t' - \tau_l)^2} \tag{C2}$$

S8) *Short-term usage*: Short-term usage is modeled the same way as long-term usage but with a shorter mean residence time and standard deviation.

P9) *Old scrap from long-term usage*: A share η_l of long-lived goods is collected as old scrap and recycled to smelting and refining. The remainder $(1 - \eta_l)$ is not recycled and becomes waste.

P10) *Old scrap from short-term usage*: A share η_s of short-lived goods is collected as old scrap and recycled to smelting and refining. The remainder $(1 - \eta_s)$ is not recycled and becomes waste.

S13) *Waste from long-term usage*: Copper not recycled from long-lived products is accumulated in this stock.

S14) *Waste from short-term usage*: Copper not recycled from short-lived products is accumulated in this stock. The model is run in discrete time steps of one year from 1900 to 2100. All variables except primary resources are assumed zero in 1900.

C3. Calibration of the model

As mentioned in the previous section, for the years 1900–1997, statistics of the consumption of refined copper, *Table C3*, are used as inputs to the model. These data are also used to calibrate the model parameters to fit historical data. The applied parameters are summarized in *Table C4*.

Losses in smelting and refining l_s are derived from global cumulative smelter production of copper from ore and global cumulative mine production of copper as content in ore in 1900–1997. In this period, according to the data from *Table C5*, cumulative smelter production was 96.95 percent of cumulative mine production. Consequently, losses in smelting have been on average 3.05 percent. We further assume losses in refining to be 0.26 percent, which leads to total losses in smelting and refining l_s of 3.3 percent, the same value as assumed in Zeltner et al.

The share of semi-manufactures that goes to new scrap l_p is calibrated against data for OECD90 in 1958–1997. In OECD90, in this period, the share of new scrap declined from around 30 percent to 25 percent (*Figure C2*). In the model, the share of new scrap l_p is assumed to be constant and is set to 26 percent. This is a slightly higher value than the 24 percent assumed in Zeltner et al. This assumption leads to a consumption of new scrap quite in line with statistics of the recovery of new scrap in OECD90 in 1958–1997 (*Figure C3*). Since 26 percent is closer to the values in 1975–1997, a better fit of the consumption of new scrap is shown in this period than in 1958–1975.

Table C4: Parameters for copper system scenarios Sc1–8

	IPCC scenario	IU-curve					Consumption scenarios	Lifetimes and collection rates							Losses					
			a	b	c	d			t_1	s_1	t_s	s_s	h_1	h_s	l_{c1}	l_{c2}	t_1	l_s	l_p	g
		—	kg/cap	—	—	—			yr.	yr.	yr.	yr.	—	—	—	—	year	—	—	—
Sc1	**B2**	**HIU**	**10**	**55**	**1**	**1**	**ConS1**	**LLT**	**50**	**30**	**10**	**7**	**65%**	**32.5%**	**40%**	**10%**	**1940**	**3.3%**	**26%**	**80%**
Sc2	B1	HIU					ConS2	LLT												
Sc3	B2	LIU				0.9975	ConS3	LLT												
Sc4	B1	LIU					ConS4	LLT												
Sc5	B2	HIU				1	ConS1	HLT	70	40	15	10	95%	47.5%						
Sc6	B1	HIU					ConS2	HLT												
Sc7	B2	LIU				0.9975	ConS3	HLT												
Sc8	B1	LIU					ConS4	HLT												

Table C5: Annual mine production, smelter production and recovery of new and old scrap in 1900–1997 (1,000 metric tons)

	Mine production world	Smelter production world	Recovery of new scrap								Recovery of old scrap							
			Europe	Yugoslavia	Japan	Canada	US	Australia	West countr.	Brazil	OECD 90	Europe	Yugoslavia	Japan	Canada	US	Australia	OECD 90
1900	452	499																
1901	529	535																
1902	557	553																
1903	593	591																
1904	662	648																
1905	693	694																
1906	725	721																
1907	725	712																
1908	766	764																
1909	851	852																
1910	872	890																
1911	887	915																
1912	1019	1017																
1913	982	1022																
1914	926	962																
1915	1058	1092																
1916	1376	1421																
1917	1443	1457																
1918	1395	1460																
1919	1050	982																
1920	967	946																
1921	528	557																
1922	863	866																
1923	1250	1223																
1924	1331	1352																
1925	1410	1395																
1926	1468	1453																
1927	1513	1500																
1928	1716	1695																
1929	1948	1895																
1930	1596	1578																
1931	1387	1354																
1932	897	911																
1933	1008	1011																
1934	1267	1258																

Table C5 *(Continued)*

	Mine production world	Smelter production world	Recovery of new scrap								Recovery of old scrap							
			Europe	Yugoslavia	Japan	Canada	US	Australia	West countr.	Brazil	OECD 90	Europe	Yugoslavia	Japan	Canada	US	Australia	OECD 90
1935	1467	1455																
1936	1696	1673																
1937	2285	2259																
1938	2010	1939																
1939	2124	2077																
1940	2397	2413																
1941	2525	2540																
1942	2672	2680																
1943	2693	2682																
1944	2509	2520																
1945	2172	2167																
1946	1848	1845																
1947	2229	2259																
1948	2323	2327																
1949	2258	2301																
1950	2525	2519																
1951	2662	2650																
1952	2766	2731																
1953	2802	2816																
1954	2852	2829																
1955	3112	3087																
1956	3470	2443																
1957	3556	3505																
1958	3449	3416							1313		1313							528
1959	3693	3603							1495		1495							593
1960	4242	4287							1476		1476							673
1961	4394	4364							1513		1513							660
1962	4555	4574							1556		1556							681
1973	4624	4635							1679		1679							662
1964	4799	4903							1885		1885							759
1965	4963	4964							2105		2105							902
1966	5220	5097							2074	20	2054							990
1967	5060	4891							1936	17	1919							905

1968	5459	5508							2110	18	2092							1019
1969	5943	5973							2347	17	2330							1113
1970	6403	6310							2156	15	2141							1138
1971	6477	6377							2222	13	2209							917
1972	7063	7009							2298	14	2284							947
1973	7511	7291	970	31	422	36	957	43			2397	431	4	133	32	403	33	1028
1974	7576	7560	835	48	394	37	878	43			2139	508	10	127	25	439	32	1120
1975	7182	7266	652	48	290	20	663	36			1612	420	14	77	25	301	28	836
1976	7658	7607	813	49	364	20	801	37			1986	439	15	95	31	319	28	897
1977	7854	7808	745	39	363	23	848	35			1975	454	51	85	32	320	31	871
1978	7671	7722	838	43	370	18	888	34			2104	439	47	105	35	383	23	938
1979	7765	7757	848	37	406	25	985	40			2267	455	38	130	37	499	33	1116
1980	7806	7643	964	42	402	19	860	44			2248	499	40	125	40	515	34	1173
1981	8383	7995	907	54	443	21	884	49			2250	492	42	120	29	494	27	1120
1982	8062	7858	814	50	476	23	698	46			2008	527	42	127	24	468	18	1121
1983	8071	7874	814	43	518	24	784	42			2139	592	37	147	35	402	34	1173
1984	8271	8100	826	35	560	34	877	48			2311	546	47	114	35	307	26	981
1985	8398	8222	874	54	520	26	802	41			2209	618	34	134	25	372	31	1145
1986	8474	8299	880	58	555	25	835	38			2274	576	41	115	24	406	21	1101
1987	8654	8446	891	51	595	24	877	50			2386	594	40	109	31	415	36	1144
1988	8753	8460	1053	54	666	34	951	40			2690	630	10	101	47	446	27	1241
1989	9019	8793	1921	52	649	36	935	41			2630	702	7	107	38	480	35	1355
1990	8986	8306	954	45	585	36	944	40			2514	657	8	115	47	441	24	1276
1991	9116	8227	1031	56	614	36	845	34			2504	706	3	109	27	418	35	1291
1992	9338	8776	918	49	595	36	907	34			2441	745	48	115	37	433	32	1314
1993	9455	8705	879	35	578	36	878	38			2374	797	13	90	44	460	24	1402
1994	9418	8784	895	35	628	38	975	36			2537	781	17	94	45	405	24	1332
1995	10066	8828	1017	35	643	42	1003	36			2706	916	17	148	91	352	24	1514
1996	11098	9533	976	36	594	39	991	36			2600	963	44	117	83	339	24	1482
1997	11421	9974	1057	36	639	39	948	36			2683	921	17	87	97	285	24	1397

Sources: Mine production and smelter production [MetallGesellschaft & WBMS 1997]; Recovery of new and old scrap, 1958–1964 [MetallGesellschaft 19€8], 1965–1972 [MetallGesellschaft 1976], 1973–1981 [MetallGesellschaft 1984], 1982–1985 [MetallGesellschaft 1993], 1986–1995 [MetallGesellschaft & WBMS 1997] and 1996–1997 [MetallGese■schaft & WBMS 1998].

Recovery of new scrap
1965–1972 OECD90 = Western countries – Brazil
1973–1997 OECD90 = Europe – Yugoslavia + Japan + Canada + USA + Australia

Recovery of old scrap
1973–1997 OECD90 = Europe – Yugoslavia + Japan + Canada + USA + Australia

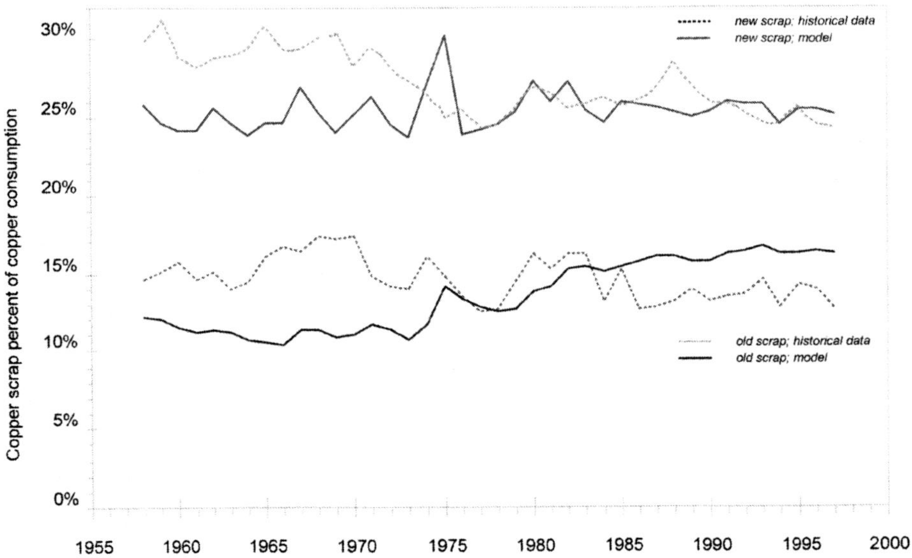

Figure C2: Recovery of copper in new and old scrap in OECD90, 1958–1997; scrap as a percentage of consumption.

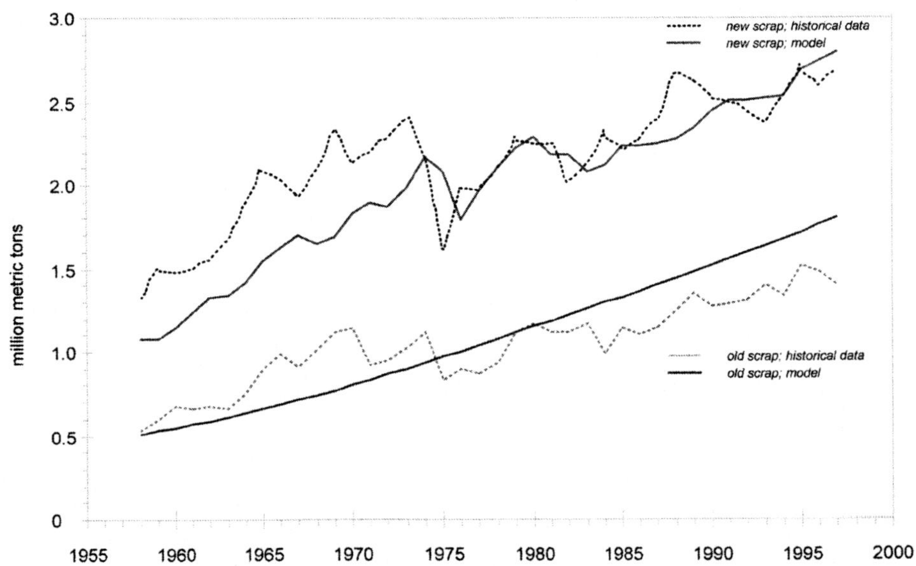

Figure C3: Recovery of copper in new and old scrap in OECD90, 1958–1997.

The share of produced goods that goes to long-term usage g is set to 80 percent based on the use of copper in applications in the US (*Figure C4*). Zeltner et al assume a slightly higher value of 82 percent.

For assumed mean residence times (τ_l, τ_s) and standard deviations (σ_l, σ_s) the collection rates (η_l, η_s) are calibrated against the global cumulative smelter production of copper from ore in 1900–1997. The global cumulative production of refined primary copper in the model $x_{34p}(t)$ in 1900–1997 should be 99.74 percent of the global cumulative smelter production in the same period. The 99.74 percent is in order to account for losses in refining. It is further assumed that the collection of goods in long term usage is twice as efficient as in short-term usage; that is, the same assumption as in Zeltner et al.

Zeltner et al. present scenarios based on four different assumptions on mean residence times and standard deviations (*Table C1*). For the mean residence time of long-term products τ_l they choose four equidistantly distributed values between two extreme values of 30 and 50 years. They then assume 'reasonable' values for the three other parameters σ_l, τ_s, η_s. Longer residence time leads to higher collection rates and shorter residence time to smaller collection rates. In this report, we use the values of σ_l, τ_l, τ_s, η_s. of Scenario 2 and 3 in Zeltner et al.

In our calibration, we get collection rates 30–36 percent higher than in Zeltner et al. With the parameters of Scenario 2 in Zeltner et al., the calibration leads to a collection rate of long-term products η_l of 65 percent, and

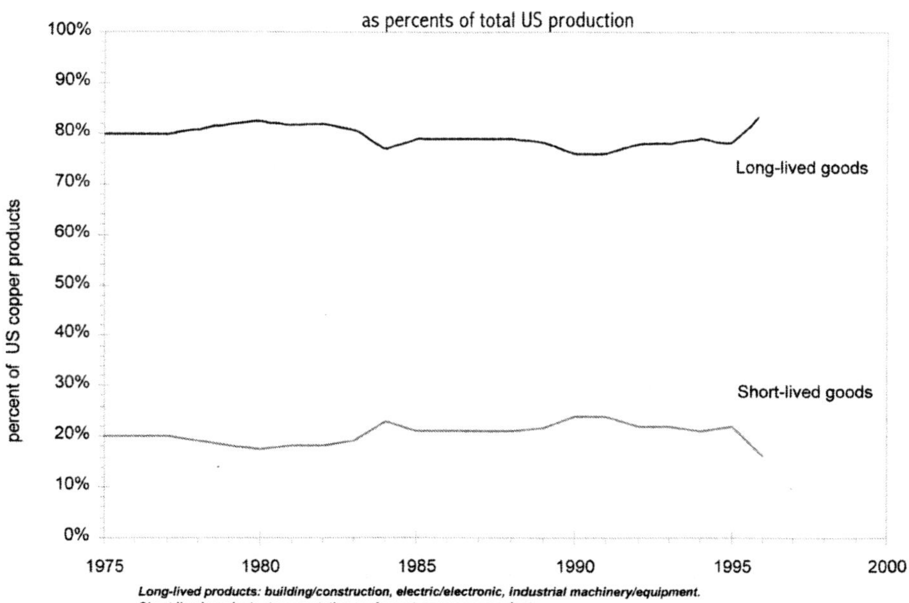

Figure C4: US production of long- and short-lived copper products; 1975–1996 as percents of total US production.

consequently a collection rate of short-term products η_s of 32.5 percent. The modeled global cumulative production of refined primary copper is 99.87 percent of the global cumulative smelter production. Zeltner et al. get collection rates of 50 percent and 25 percent, respectively. With the parameters of Scenario 3 in Zeltner et al., the calibration leads to a collection rate of long-term products η_l of 95 percent, and consequently a collection rate of short-term products η_s of 47.5 percent. The modeled global cumulative production of refined primary copper is 99.80 percent of the global cumulative smelter production. Zeltner et al. get collection rates of 70 percent and 35 percent, respectively. This is not due to the slightly different setting of parameters in the production of goods g and l_p, but rather to different geographical scope and calibration method. They match the appearance of old scrap in the USA very well.

However, compared to data of the global smelter production, the modeled global annual smelter production is generally too low in 1900–1955 and generally too high thereafter (*Figures C5 and C6*). That implies that the collection rates in the model are too high before 1955 and too low thereafter; that is, the collection rates seem to have increased with time.

On the other hand, the modeled recovery of old scrap in OECD90 is higher than data from 1975 and lower before that (*Figure C2*). This implies that the collection rates in the model are too high in 1975–1997 and too low in 1958–1974 and that the collection rates have decreased or that export and import account for the differences.

Losses in concentration $l_c(t)$ are assumed to be time dependent. Losses in

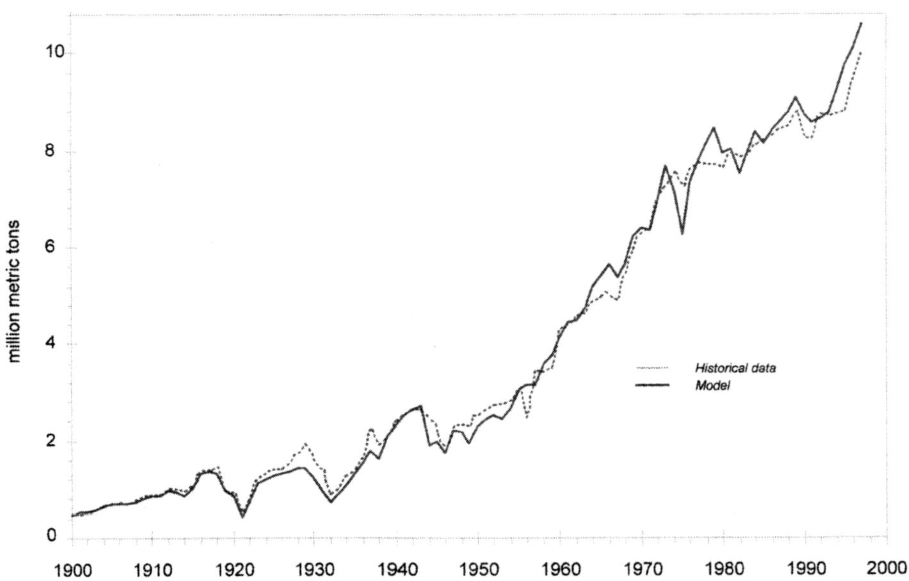

Figure C5: Global smelter production of copper from ore; 1900–1998.

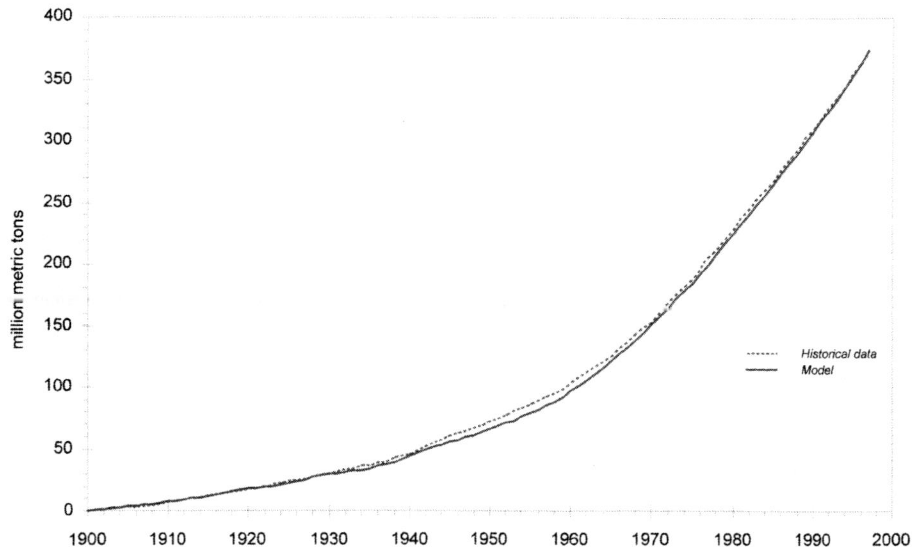

Figure C6: Global cumulative smelter production of copper from ore; 1990–1998.

concentration in 1900 l_{c1} decline linearly to l_{c2} in t_1, and stay at that level thereafter. l_{c1} is set to 40 percent, l_{c2} is set to 10 percent and t_1 is set to 1940.

C4. Copper consumption scenarios

As mentioned, the driving variable, the input to the model, is the annual consumption of refined copper (primary and secondary) ($x_{34_p}(t) + x_{34_s}(t)$) in the four regions, respectively.

For the years 1900–1997, statistics of the consumption of refined copper is used (*Table C3*). For the years 1998–2100, scenarios have been generated. Empirical research in resource economics has found that metal intensity of use (metal demand per unit GDP) often can be described as a function of per capita income, and that its general shape is an inverse U-shaped curve. This phenomenon is called the 'Intensity of Use' hypothesis (see for example Malembaum 1978; Tilton 1990; Roberts 1996; Cleveland & Ruth 1997). The consumption of refined copper is assumed to have the shape of such an inverse U-shaped curve. The intensity of use IU is modeled as in (Van Vuuren et al. 2000); that is,

$$\mathrm{IU}(y)t = \frac{a}{y(t) + \dfrac{b}{y(t)_c}} d(t - t_0) \tag{C3}$$

where $y(t)$ is GDP/capita (thousand US\$/capita) and a (kg/capita), b (thousand US\$/capita) and the exponent c are parameters, d is a factor that scales down the intensity of use with time and t_0 is the first year. If c and d are equal to 1.0, then for low values of $y(t)$, IU grows linearly with the slope a/b and for high values of $y(t)$, IU declines as $a/y(t)$. The parameter a thus represent the asymptotic per capita level of consumption of refined copper. d is a factor that scales down the intensity of use with time and t_0 is the first year. IU maximum is $a/(2\sqrt{b})$, reached at $y(t) = \sqrt{b}$.

The intensity of use of the consumption of refined copper in OECD90, REF, ASIA and ALM in 1960–1997 is shown in *Figure C7*. During this period, in the poorest region ASIA, the intensity of use has risen from 0.1 kg/k\$US (kilograms per 1,000 US dollars) to 0.4 kg/k\$US. In the next poorest region, ALM, the intensity of use has been lower. It has risen from 0.1 to 0.25 kg/k\$US. The intensity of use in REF has declined drastically, from above 1.4 kg/k\$US to just above 0.2 kg/k\$US. In OECD90, the intensity of use has declined from approximately 0.75 kg/k\$US to just below 0.45 kg/k\$US.

Two intensity of use curves are employed to generate copper consumption scenarios, one high and one low. For both curves, a is set to 10 kg/capita, b is set to 55 and c is set to 1.0. The parameter a is set at 10 kg/capita because that is roughly the per capita consumption of refined copper in the USA in the late 1990s. The parameter b is set to give a good fit with the intensity of use in ASIA. These parameters lead to a slope of 0.18 in the linear region and a maximum intensity of use of 0.67 kg/k\$US, reached at 7,400 \$US/(capita, year).

For the high intensity of use curve, d is set to 1.0. For the low intensity of

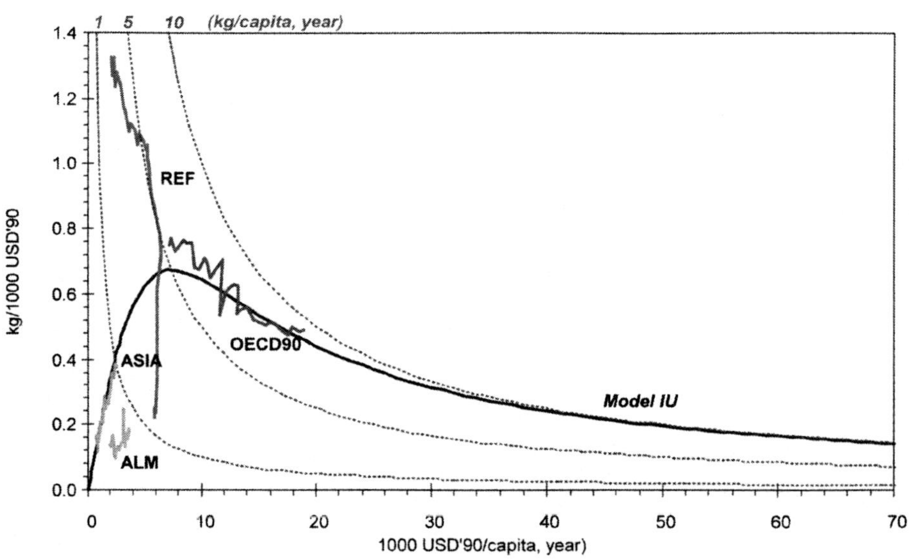

Figure C7 (3.11): Historical and modeled Intensity of Use (consumption of refined copper) as a function of GDP/capita in 1960–1997.

use curve, d is set to 0.9975; that is, the intensity of use curve is scaled down by 0.25 percent each year (starting in 1999; that is, t_0 is 1998). Consequently, the intensity of use curve in 2010 is 97 percent, in 2025 93 percent, in 2050 88 percent and in 2100 77 percent of the intensity of use curve in 1998. The rate of decline per annum, 0.25 percent, is the same value as that assumed in the 'egalitarian scenario' in van Vuuren et al. The high intensity of use curve was shown in *Figure C7*.

The consumption of refined copper $z(t)$ is then

$$z(t) = \mathrm{IU}(y(t)) \times w(t) \qquad \text{(C4)}$$

where $w(t)$ is Gross Regional Product (GRP) or Gross World Product (GWP) in billions of 1990 US dollars/year at purchasing parity price (PPP) exchange rates.

Four scenarios of the consumption of refined copper are derived, ConSc1–ConSc4. They are based on the high and low intensity of use curves described above and Scenario B2 and B1 of the evolution of the population and economy in (IPCC 2000). Historical values for the regions for GDP and population are shown in *Table C6*; IPCC projections of these variables to 2100 are shown for Scenarios B1 (*Table C7*) and B2 (*Table C8*). The evolutions of the global population and GDP in the Scenarios B2 and B1 are shown in *Figure C8*. In Scenario B2, the world population in 1990 of 5.4 billion capita grows to 9.4 billion capita in 2050 and doubles to the year 2100. In Scenario B1, the world population peaks at 8.7 billion capita in 2050 and then decreases

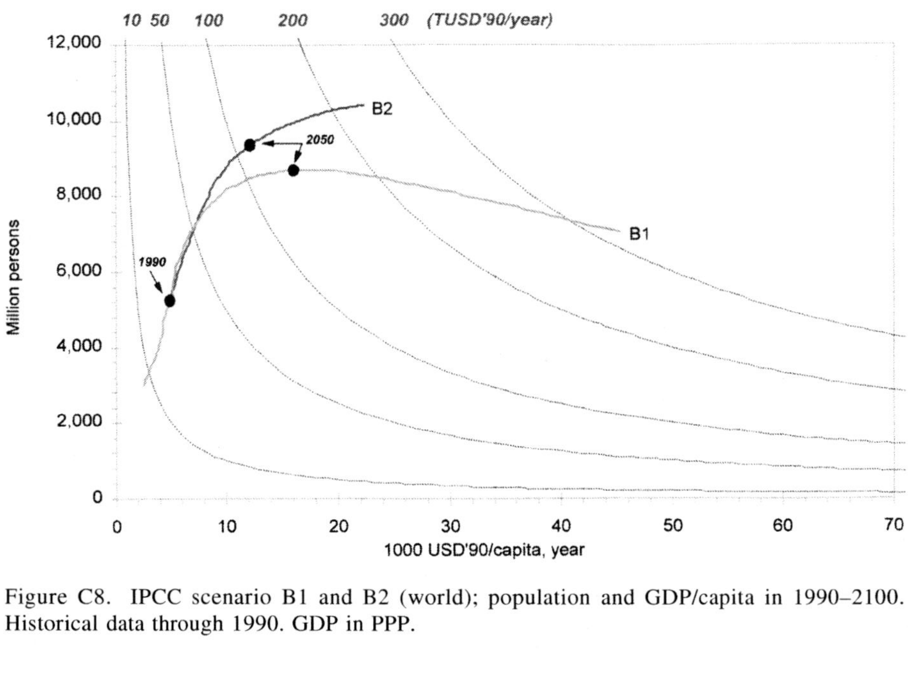

Figure C8. IPCC scenario B1 and B2 (world); population and GDP/capita in 1990–2100. Historical data through 1990. GDP in PPP.

Table C6: Global and Regional Population and GDP, 1960–1990

Year	Population (Million capita)					GDP in PPP (TUSD '90)				
	OECD 90	REF	ASIA	ALM	World	OECD 90	REF	ASIA	ALM	World
1960	668	308	1503	544	3022	4.8	0.6	1.2	1.1	7.7
1961	676	312	1537	559	3083	5.0	0.7	1.2	1.1	8.0
1962	684	316	1571	574	3145	5.3	0.7	1.2	1.2	8.4
1963	692	320	1607	589	3208	5.5	0.7	1.3	1.2	8.8
1964	700	324	1643	605	3273	5.9	0.8	1.4	1.4	9.4
1965	709	329	1680	622	3339	6.2	0.9	1.4	1.4	9.9
1966	715	332	1723	638	3408	6.5	0.9	1.5	1.5	10.5
1967	722	335	1766	656	3479	6.8	1.0	1.5	1.6	10.9
1968	729	338	1811	673	3551	7.2	1.1	1.5	1.7	11.5
1969	736	341	1856	691	3625	7.5	1.1	1.7	1.8	12.2
1970	743	344	1903	710	3701	7.9	1.2	1.9	2.0	12.9
1971	750	348	1947	729	3774	8.1	1.3	1.9	2.1	13.5
1972	757	351	1992	748	3849	8.5	1.3	2.0	2.3	14.1
1973	764	354	2038	768	3925	9.0	1.5	2.1	2.4	15.0
1974	771	357	2085	789	4003	9.1	1.6	2.2	2.6	15.4
1975	779	360	2133	810	4082	9.0	1.7	2.3	2.7	15.7
1976	785	363	2173	832	4153	9.5	1.7	2.3	2.9	16.4
1977	791	366	2213	855	4226	9.8	1.8	2.5	3.0	17.2
1978	797	370	2255	879	4300	10.2	1.9	2.7	3.1	18.0
1979	803	373	2297	903	4375	10.6	2.0	2.9	3.3	18.7
1980	809	376	2339	927	4452	10.7	2.0	3.1	3.4	19.1
1981	815	379	2384	953	4530	10.8	2.0	3.2	3.4	19.5
1982	820	382	2429	980	4610	10.8	2.1	3.3	3.4	19.6
1983	826	385	2475	1007	4692	11.1	2.2	3.6	3.3	20.1
1984	831	388	2521	1035	4775	11.6	2.3	3.9	3.4	21.1
1985	837	391	2569	1064	4860	12.0	2.3	4.2	3.5	21.9
1986	843	394	2617	1092	4945	12.3	2.4	4.3	3.5	22.5
1987	849	396	2665	1121	5031	12.7	2.4	4.5	3.6	23.3
1988	854	399	2715	1151	5119	13.3	2.5	4.9	3.7	24.4
1989	861	402	2765	1182	5210	13.8	2.6	5.1	3.7	25.2
1990	859	413	2798	1192	5262	14.1	2.6	5.3	3.8	25.7

The GDP is originally in USD = 85 (PPP). The GDP is transformed to USD = 90 (PPP) by multiplying the GDP in OECD90, REF, ASIA and ALM by 1.16, 0.97, 1.19 and 1.22, respectively.

Table C7: IPCC Scenario B1

Year	Population (Million capita)					GDP in PPP (TUSD '90)				
	OECD 90	REF	ASIA	ALM	World	OECD 90	REF	ASIA	ALM	World
1990	859	413	2798	1192	5262	14.1	2.6	5.3	3.8	25.7
2000	919	419	3261	1519	6118	17.7	2.2	8.2	5.1	33.3
2010	965	427	3620	1875	6887	22.4	2.6	12.0	7.6	44.6
2020	1007	433	3937	2241	7618	28.1	3.3	17.3	12.8	61.6
2030	1043	435	4147	2557	8182	33.3	4.3	24.6	20.1	82.2
2040	1069	433	4238	2791	8531	38.3	5.3	34.1	30.6	108.4
2050	1081	423	4220	2980	8704	43.6	6.4	46.1	44.0	140.0
2060	1084	409	4085	3089	8667	48.5	8.2	57.4	57.8	171.8
2070	1089	392	3867	3115	8463	52.5	10.3	67.7	73.6	204.1
2080	1098	374	3589	3064	8125	58.4	12.8	79.3	92.0	242.5
2090	1108	357	3258	2934	7657	65.1	15.3	91.7	109.1	281.3
2100	1110	339	2882	2727	7058	72.7	18.1	103.1	124.8	318.8

Source: IPCC (2000).

Table C8: IPCC Scenario B2

Year	Population (Million capita)					GDP in PPP (TUSD '90)				
	OECD 90	REF	ASIA	ALM	World	OECD 90	REF	ASIA	ALM	World
1990	859	413	2798	1192	5262	14.1	2.6	5.3	3.8	25.7
2000	916	415	3248	1511	6090	18.3	2.4	9.3	4.9	34.8
2010	953	417	3649	1872	6891	23.0	2.7	15.1	6.2	46.9
2020	982	418	4008	2263	7671	26.3	3.3	22.4	8.2	60.2
2030	994	416	4312	2649	8371	28.8	4.3	30.7	11.7	75.5
2040	988	411	4538	2992	8929	31.3	5.6	39.3	17.0	93.2
2050	976	406	4696	3289	9367	33.5	7.2	49.3	23.9	113.9
2060	965	396	4790	3554	9705	35.9	9.5	59.0	32.4	136.8
2070	951	389	4856	3764	9960	39.2	11.6	68.7	41.2	160.7
2080	941	384	4902	3931	10158	42.4	13.3	78.5	49.6	183.8
2090	934	381	4938	4053	10306	46.1	14.8	89.2	57.2	207.4
2100	928	379	4968	4139	10414	50.4	16.2	100.4	64.9	231.8

Source: IPCC (2000).

to 7.1 billion capita in 2100. The economic growth is much higher in scenario B1 than in Scenario B2. In Scenario B2, the global GDP/capita in 1990 of 4,900 $US/capita increases with a factor of 2.5 to 12,200 $US/capita in 2050 and with a factor of 4.5 to 22,300 $US/capita in 2100. In Scenario B1, the global GDP/capita in 1990 increases with a factor of 3.3 to 16,100 $US/capita in 2050 and with a factor of 9.2 to 45,200 $US/capita in 2100. The evolution of the population and GDP in the four regions in Scenario B1 and B2 are shown in *Figures C9 and C10*, respectively.

In order to avoid discontinuities in the transition from statistics to modeled data, the consumption of refined copper in 1998–2009 in each region is a linear extrapolation of the values in 1997 and 2010.

The global consumption of refined copper in ConSc1–ConSc4 is shown in *Figure C11*, and the regional consumption is shown in *Figures C12–C15*. The global per capita consumption of refined copper in ConSc1–ConSc4 is shown in *Figure C16*, and the regional per capita consumption are shown in *Figures C17–C20*. Global consumption and per capita consumption for the four consumption scenarios for the years 2010, 2025, 2050 and 2100 are shown in *Table C9*. The scenarios were discussed in *Section 2.9*.

C5. Copper system scenarios

In total, eight copper system scenarios are generated, Sc1–Sc8. These are based on the two different assumptions on lifetimes and collection rates of copper products discussed in *Section C3*, and the four different copper consumption scenarios, ConSc1–ConSc4, discussed in *Section C4*. The characteristics of the eight system scenarios are shown in *Table C4*. Sc1 is the base case: IPCC scenario B2, high intensity of use curve and low residence times and hence high collection rates. Comparative results are shown in *Figures C21–C33*. The system scenario results (direct and per capita) are summarized in *Tables C10–C11* and are discussed in *Section 2.9*.

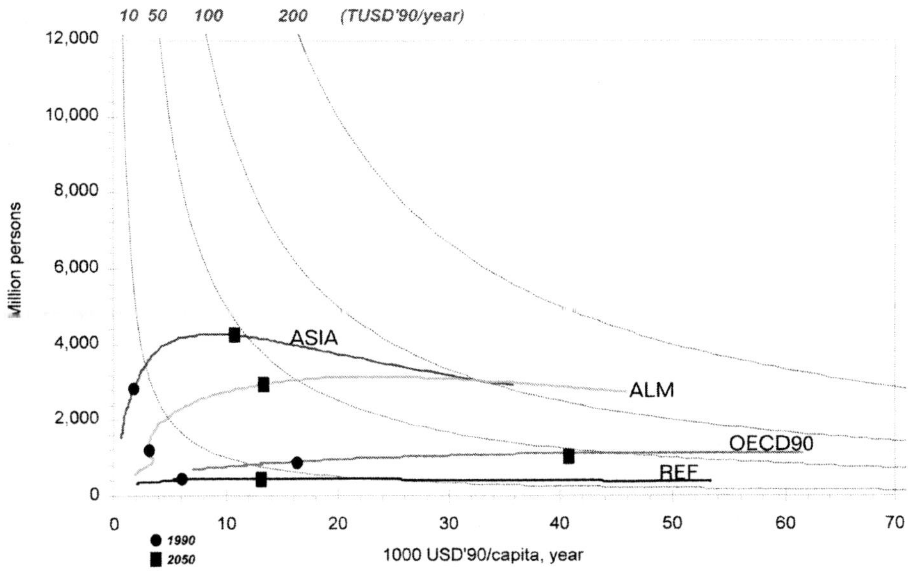

Figure C9. IPCC scenario B1 (regional); population and GDP/capita in 1990–2100. Historical data through 1990. GDP in PPP.

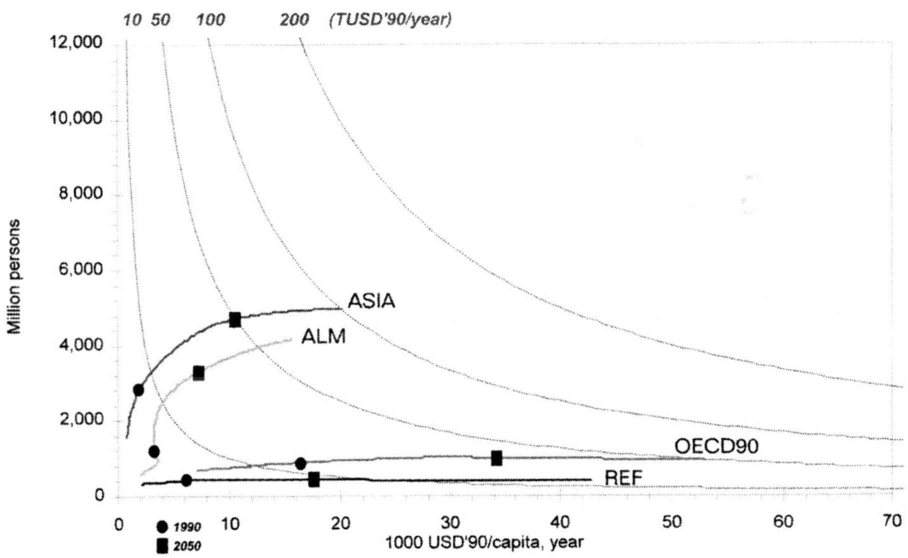

Figure C10. IPCC scenario B2 (regional); population and GDP/capita in 1990–2100. Historical data through 1990. GDP in PPP.

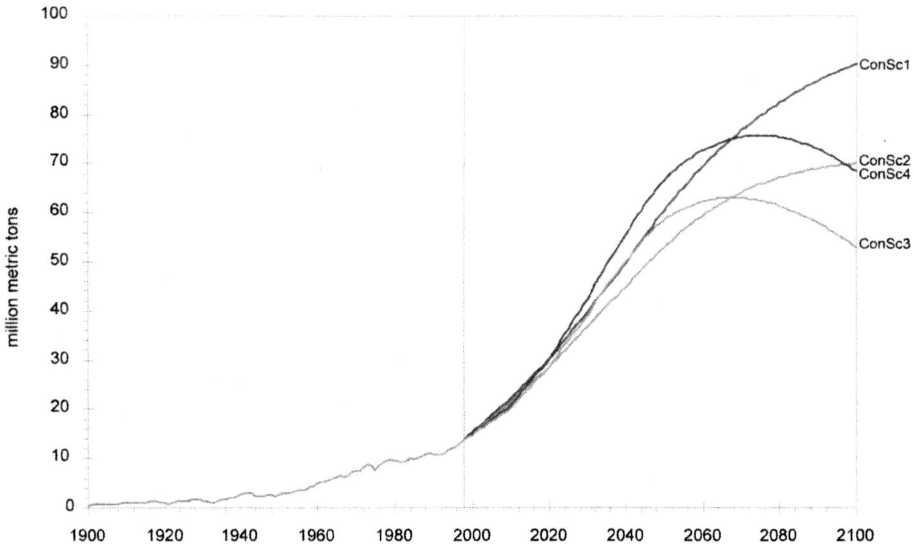

Figure C11 (3.15): Global consumption of refined copper, scenarios ConSc1 through ConSc4.

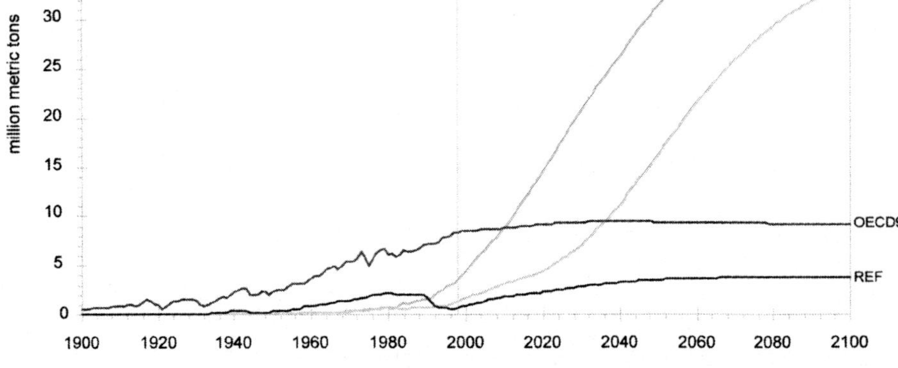

Figure C12 (3.16): Regional consumption of refined copper, scenarios Sc1 and Sc4.

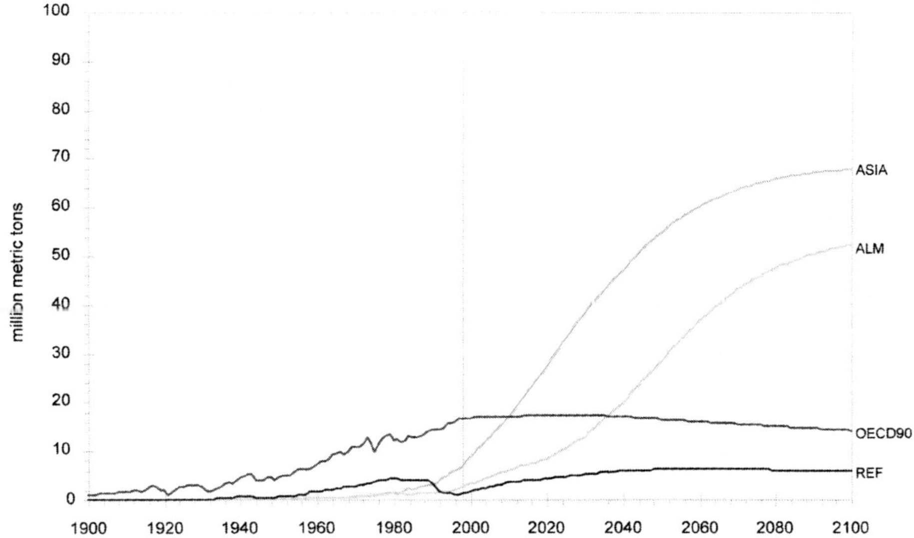

Figure C13 (3.17): Regional consumption of refined copper, scenarios Sc2 and Sc6.

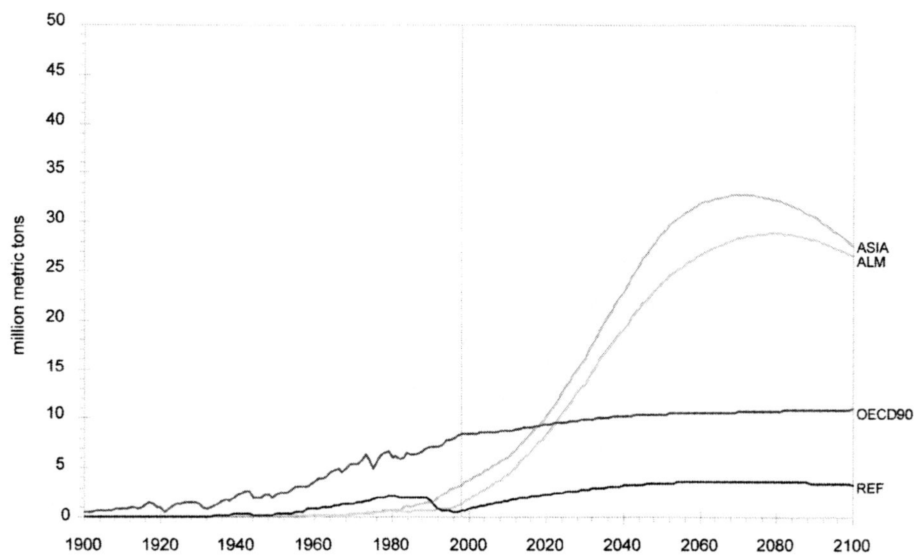

Figure C14 (3.18): Regional consumption of refined copper, scenarios Sc3 and Sc7.

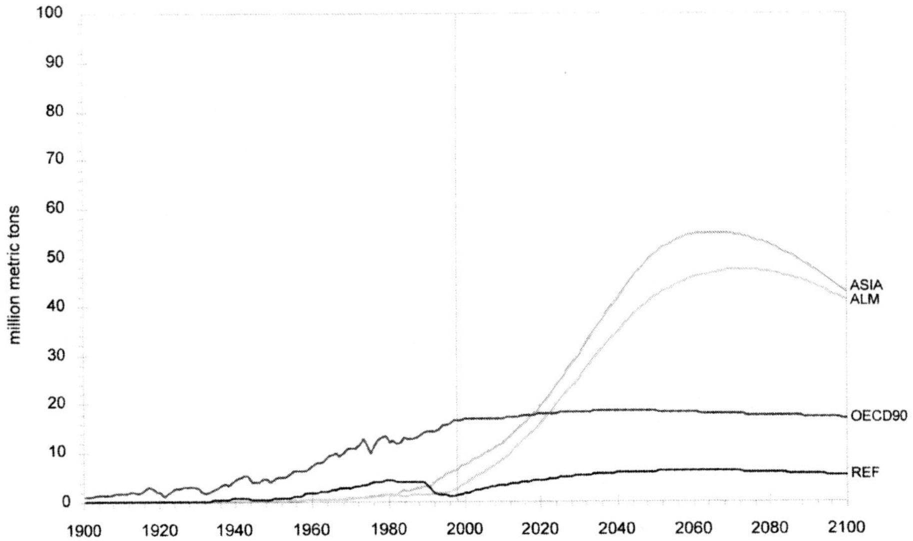

Figure C15 (3.19): Regional consumption of refined copper, scenarios Sc4 and Sc8.

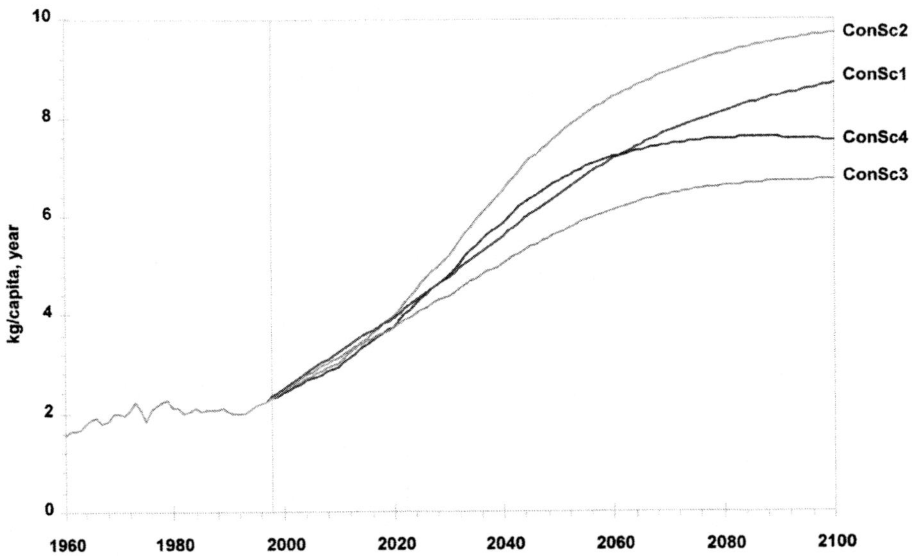

Figure C16: Global per capita consumption of refined copper, ConSc1–ConSc4.

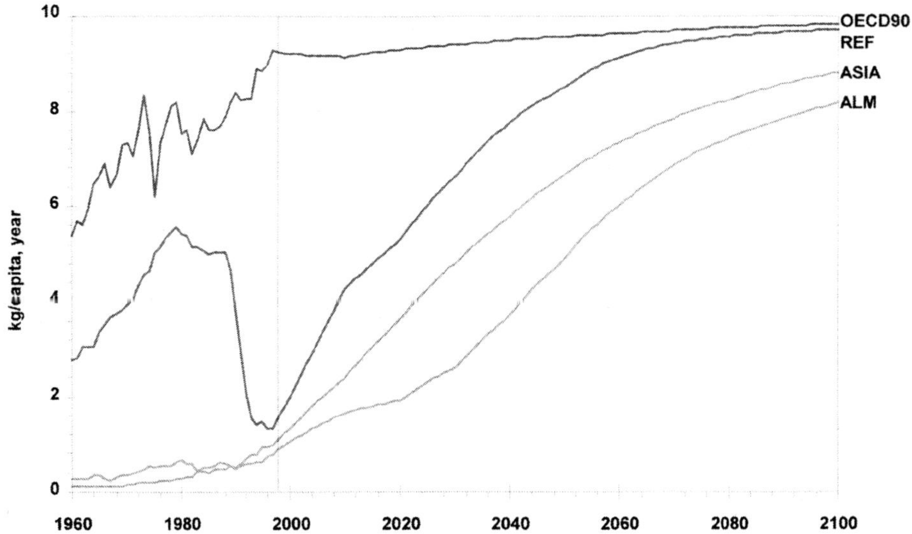

Figure C17: Regional per capita consumption of refined copper with IPCC scenario B2 and high IU, ConSc1. Input to Sc1 to Sc5.

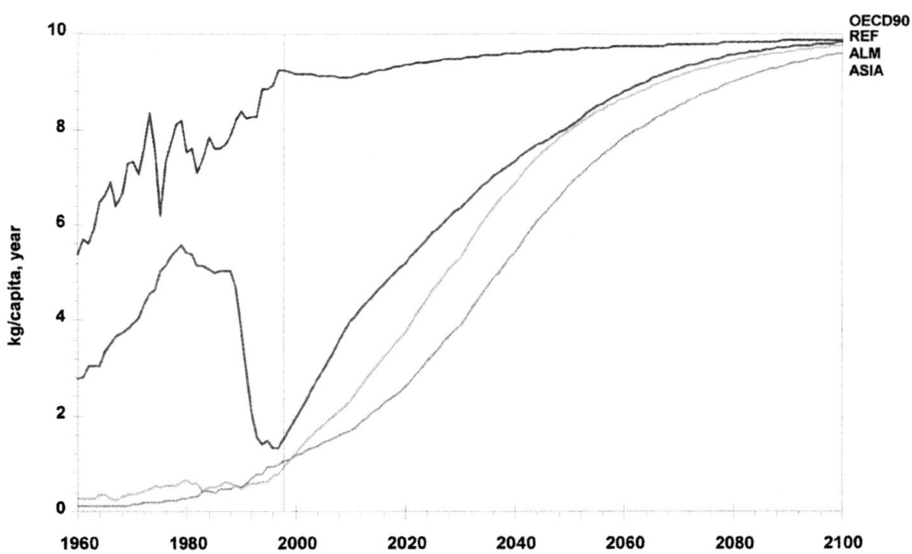

Figure C18: Regional per capita consumption of refined copper with IPCC scenario B1 and high IU, ConSc2. Input to Sc2 and Sc6.

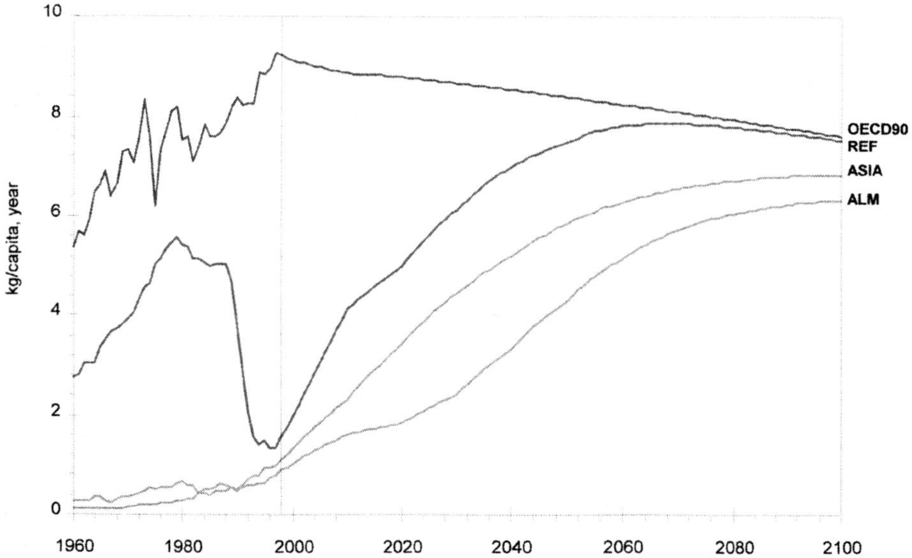

Figure C19: Regional per capita consumption of refined copper with IPCC scenario B2 and low IU, ConSc3. Input to Sc3 and Sc7.

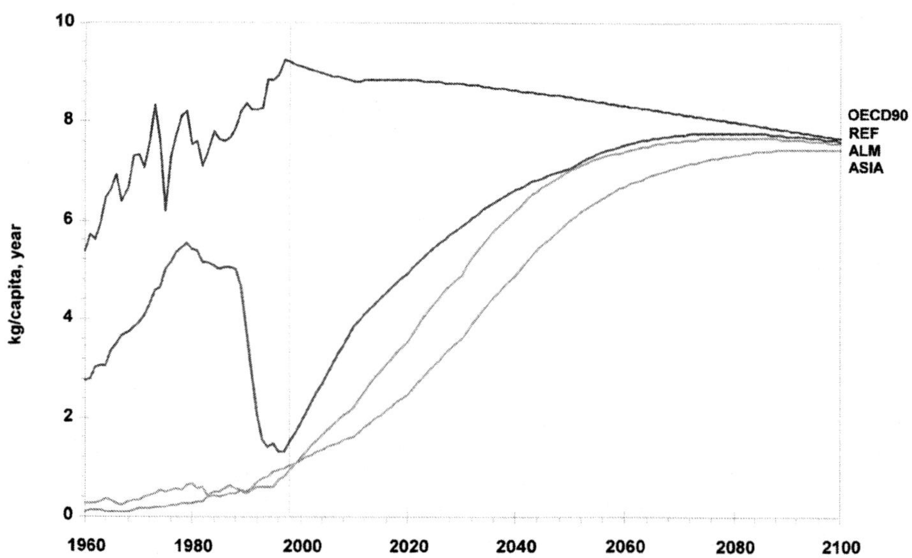

Figure C20: Regional per capita consumption of refined copper with IPCC scenario B1 and low IU, ConSc4. Input to Sc4 and Sc8.

Table C9 (= Table 3.5): Copper consumption scenario results in 2010, 2025, 2050 and 2100; consumption of refined copper

	World				OECD90				REF			
	2010	2025	2050	2100	2010	2025	2050	2100	2010	2025	2050	2100
	(1000 metric tons)											
ConSc1	22232	35061	60230	90394	8706	9228	9326	9110	1767	2508	3452	3679
ConSc2	20794	36510	66520	68491	8759	9657	10456	10960	1694	2528	3405	3326
ConSc3	21574	32770	52879	70026	8448	8625	8188	7057	1715	2344	3030	2850
ConSc4	20178	34124	58401	53058	8500	9026	9180	8490	1644	2363	2990	2577
	(kg/capita)											
ConSc1	3.2	4.4	6.4	8.7	9.1	9.3	9.6	9.8	4.2	6.0	8.5	9.7
ConSc2	3.0	4.6	7.6	9.7	9.1	9.4	9.7	9.9	4.0	5.8	8.1	9.8
ConSc3	3.1	4.1	5.6	6.7	8.9	8.7	8.4	7.6	4.1	5.6	7.5	7.5
ConSc4	2.9	4.3	6.7	7.5	8.8	8.8	8.5	7.6	3.9	5.4	7.1	7.6

	ASIA				ALM			
	2010	2025	2050	2100	2010	2025	2050	2100
	(1000 metric tons)							
ConSc1	8687	17698	31331	43784	3072	5626	16122	33821
ConSc2	6066	13244	28869	27632	4274	11081	23789	26573
ConSc3	8430	16542	27507	33918	2981	5258	14155	26200
ConSc4	5886	12378	25346	21406	4148	10357	20886	20585
	(1000 metric tons)							
ConSc1	2.4	4.3	6.7	8.8	1.6	2.3	4.9	8.2
ConSc2	1.7	3.3	6.8	9.6	2.3	4.6	8.0	9.7
ConSc3	2.3	4.0	5.9	6.8	1.6	2.1	4.3	6.3
ConSc4	1.6	3.1	6.0	7.4	2.2	4.3	7.0	7.5

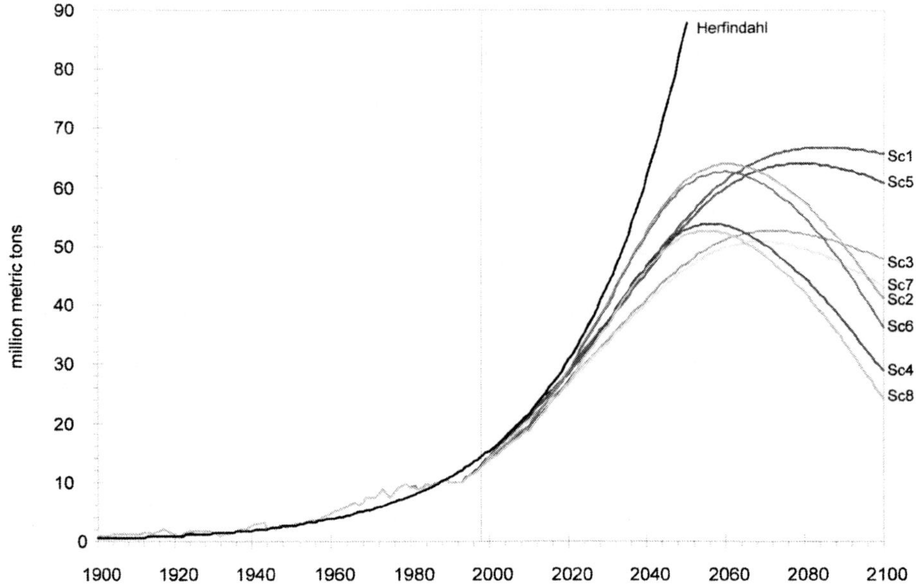

Figure C21 (3.20): Global mine production of copper: 1990–1998; comparison of copper system scenarios Sc1–Sc8.

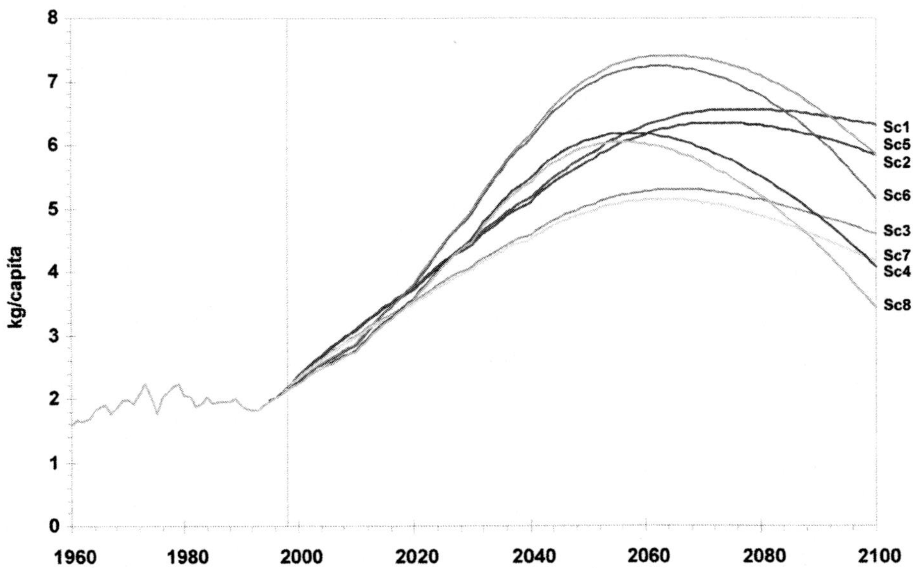

Figure C22: Global per capita mine production of copper: 1990–1998; comparison of copper system scenarios Sc1–Sc8.

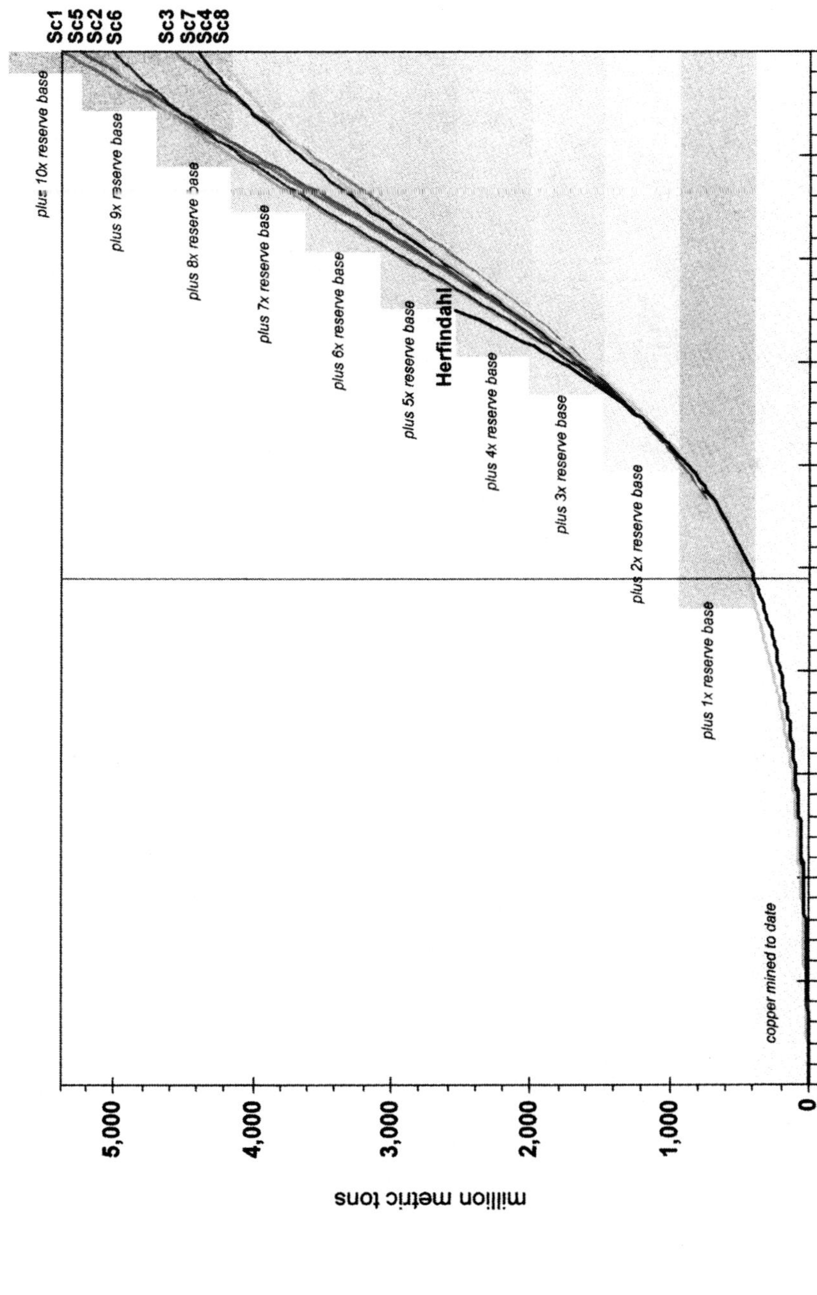

Figure C23 (3.21): Global cumulative mine production of copper: 1990–1998; comparison of copper system scenarios Sc1–Sc8.

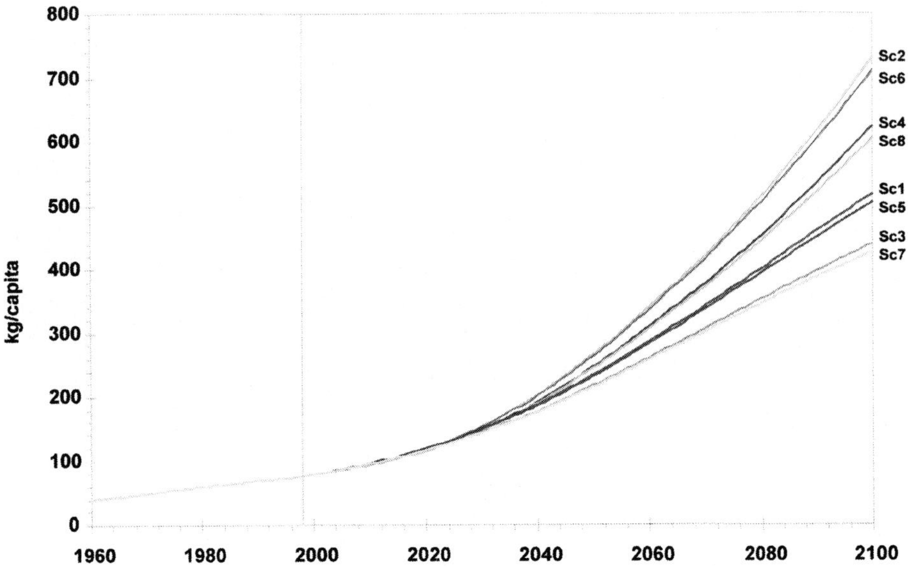

Figure C24: Global cumulative per capita mine production of copper: 1990–1998; comparison of copper system scenarios Sc1–Sc8.

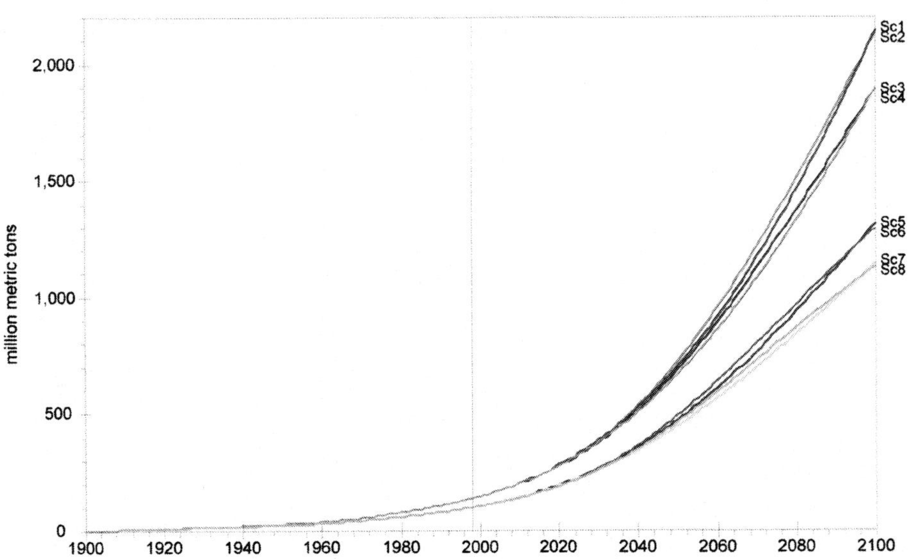

Figure C25 (3.22): Global stock of waste copper: 1990–1998; comparison of copper system scenarios Sc1–Sc8.

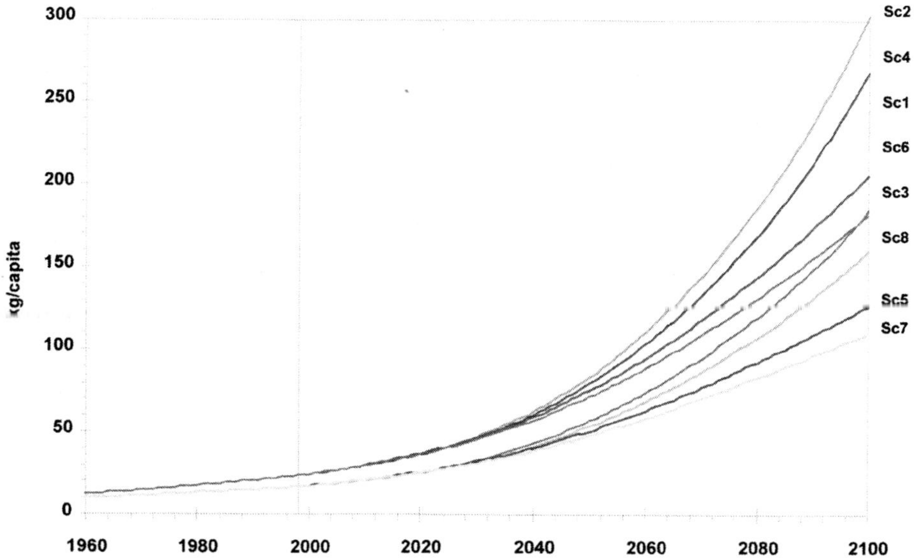

Figure C26: Global per capita stock of waste copper: 1990–1998; comparison of copper system scenarios Sc1–Sc8.

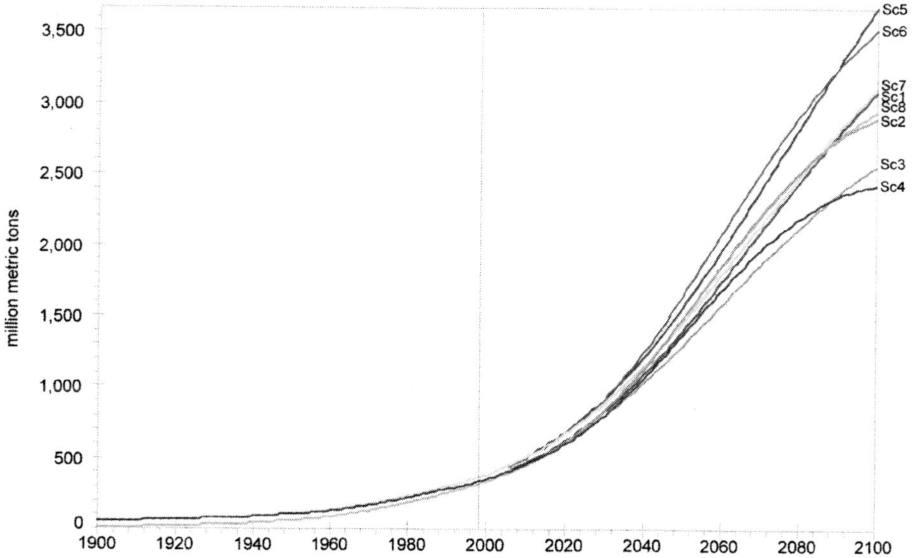

Figure C27 (3.23): Global stock of long-lived copper products: 1990–1998; comparison of copper system scenarios Sc1–Sc8.

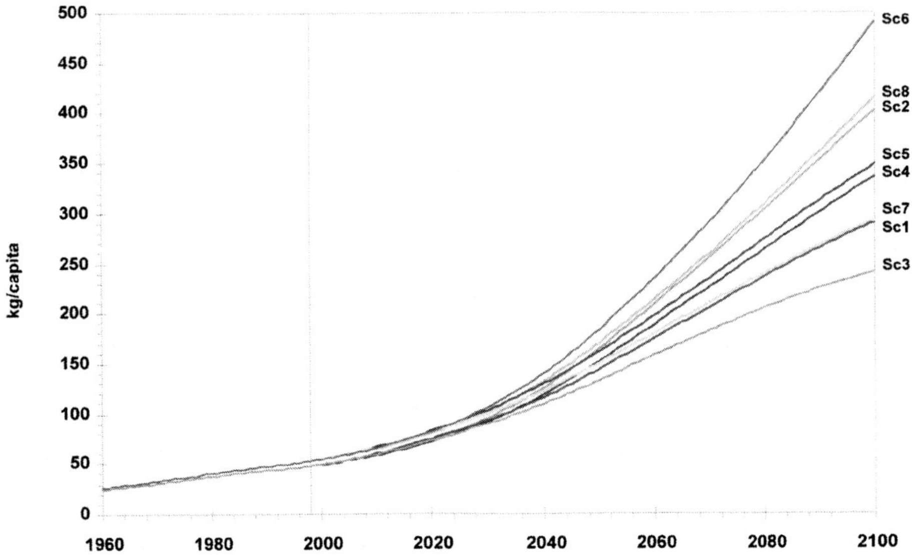

Figure C28: Comparison of copper system scenarios Sc1–8. Global per capita stock of long-lived copper products: 1990–1998; comparison of copper system scenarios Sc1–Sc8.

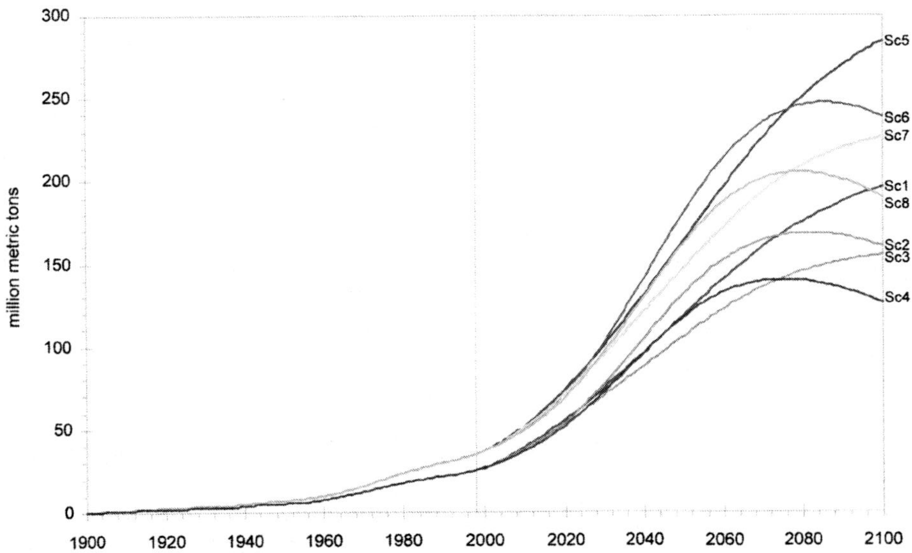

Figure C29 (3.24): Global stock of short-lived copper products: 1990–1998; comparison of copper system scenarios Sc1–Sc8.

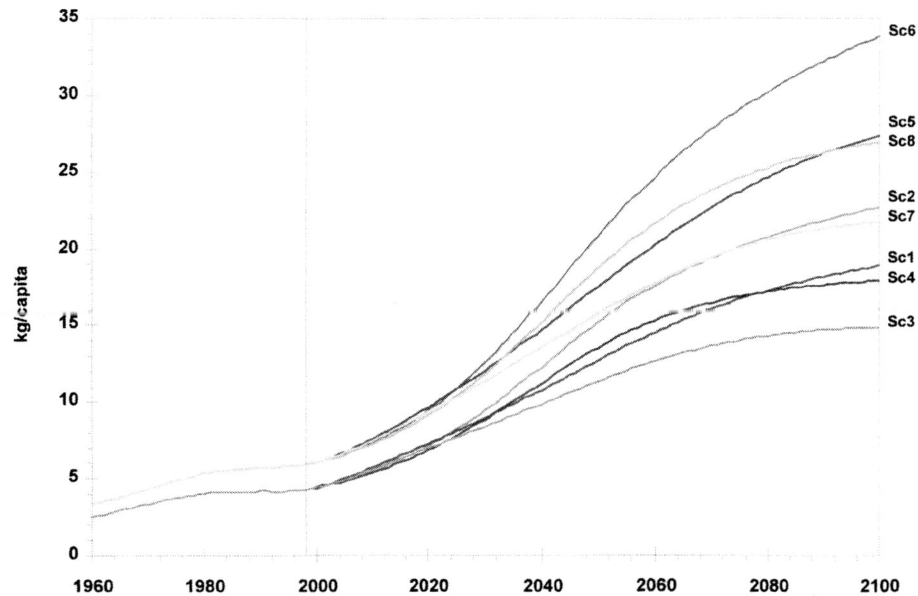

Figure C30: Comparison of copper system scenarios Sc1–8. Global per capita stock of short-lived copper products: 1990–1998; comparison of copper system scenarios Sc1–Sc8.

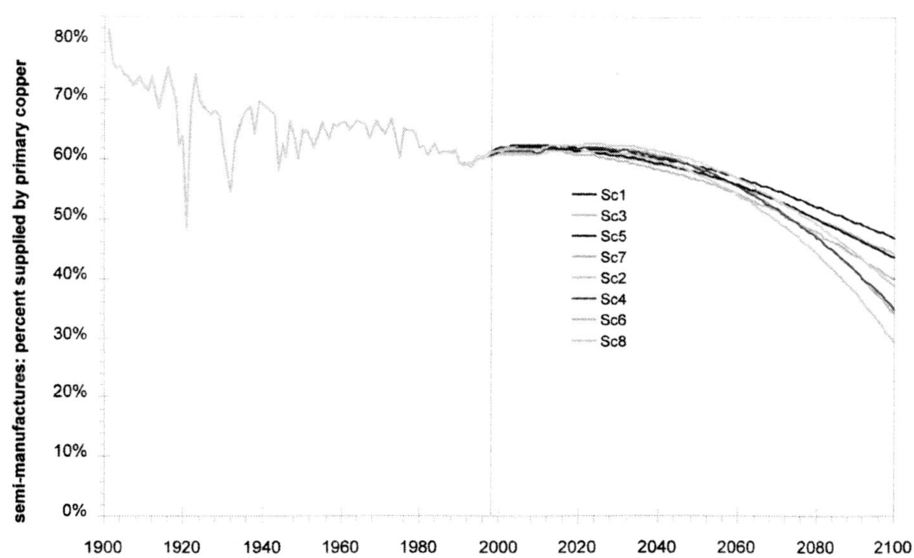

Figure C31: Global primary copper share of copper supplied to semi-manufactures: 1990–1998; comparison of copper system scenarios Sc1–Sc8.

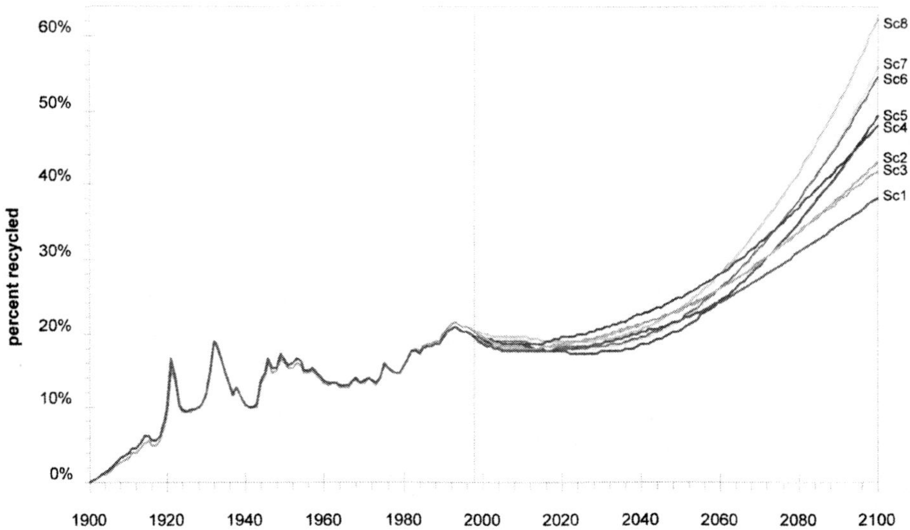

Figure C32 (3.14): Global copper recycling rate: 1990–1998; comparison of copper system scenarios Sc1–Sc8.

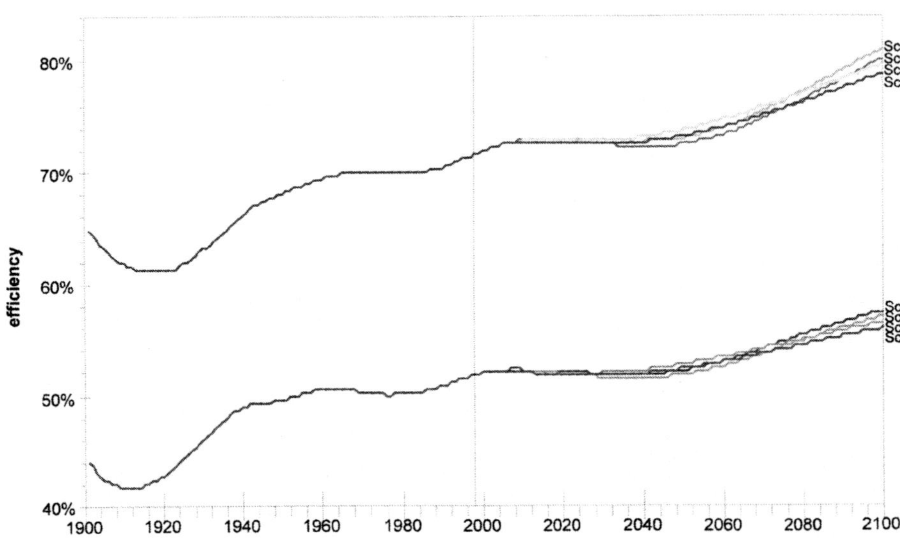

Figure C33 (3.13): Global copper recycling (separation) efficiency: 1990–1998; comparison of copper system scenarios Sc1–Sc8.

Table C10: Global results of the copper system scenarios Sc1–Sc8, 1997, 1910, 1925, 2050 and 2100

			Sc1	Sc2	Sc3	Sc4	Sc5	Sc6	Sc7	Sc8
–S1	1997	(Tg)	–438	–438	–438	–438	–438	–438	–438	–438
Primary	2010		–661	–650	–656	–645	–659	–648	–654	–643
resources	2025		–1073	–1061	–1044	–1032	–1068	–1056	–1039	–1027
	2050		–2187	–2307	–2035	–2142	–2167	–2289	–2015	–2123
	2100		–5393	–5159	–4580	–4405	–5253	–5019	–4440	–4265
x12	1997	(Gg/yr)	12183	12183	12183	12183	12080	12080	12080	12080
Mining	2010		21262	19657	20528	18970	21126	19514	20388	18824
	2025		33280	35048	30793	32451	32994	34770	30499	32163
	2050		54696	61277	47135	52904	53786	60384	46205	51986
	2100		65706	41203	47927	28844	60898	36315	43260	24117
S11	1997	(Tg)	52	52	52	52	52	52	52	52
Gangue	2010		74	73	73	72	74	73	73	72
	2025		115	114	112	111	115	113	112	111
	2050		226	239	211	222	225	237	209	220
	2100		547	524	466	448	533	510	452	434
x211	1997	(Gg/yr)	1218	1218	1218	1218	1208	1208	1208	1208
Gangue	2010		2126	1966	2053	1897	2113	1951	2039	1882
	2025		3328	3505	3079	3245	3299	3477	3050	3216
	2050		5470	6128	4714	5290	5379	6038	4620	5199
	2100		6571	4120	4793	2884	6090	3631	4326	2412
S12	1997	(Tg)	15	15	15	15	15	15	15	15
Slag	2010		23	23	23	23	23	23	23	23
	2025		38	37	37	36	38	37	37	36
	2050		79	82	74	77	79	82	74	77
	2100		213	207	184	180	213	207	184	180
x312	1997	(Gg/yr)	446	446	446	446	446	446	446	446
Slag	2010		759	710	736	689	759	710	736	689
	2025		1197	1246	1118	1165	1197	1246	1118	1165
	2050		2055	2270	1805	1993	2055	2270	1805	1993
	2100		3085	2337	2390	1811	3085	2337	2390	1811
S13	1997	(Tg)	26	26	26	26	2	2	2	2
Waste	2010		43	43	43	43	4	4	4	4
long-term	2025		73	73	73	73	7	7	7	7
	2050		169	170	166	166	16	16	16	16
	2100		679	698	624	641	67	69	62	64
x913	1997	(Gg/yr)	1023	1023	1023	1023	94	94	94	94
Waste	2010		1560	1550	1555	1546	145	144	145	144
long-term	2025		2520	2490	2485	2456	239	236	236	233
	2050		5336	5472	5073	5192	514	525	492	502
	2100		15481	15612	13366	13525	1618	1638	1425	1446
S14	1997	(Tg)	42	42	42	42	28	28	28	28
Waste	2010		64	63	63	63	43	43	43	42
short-term	2025		104	102	103	101	69	68	69	68
	2050		228	234	218	223	154	156	148	150
	2100		705	708	619	623	502	509	444	451
Sx10 14	1997	(Gg/yr)	1368	1368	1368	1368	964	964	964	964
Waste	2010		1987	1938	1965	1918	1344	1322	1334	1313
short-term	2025		3375	3297	3237	3161	2240	2180	2166	2107
	2050		6542	7226	5902	6518	4547	4937	4147	4500
	2100		11616	9868	9260	7873	8765	7767	7076	6280

Table C10 *(Continued)*

			Sc1	Sc2	Sc3	Sc4	Sc5	Sc6	Sc7	Sc8
S11+S12+	1997	(Tg)	135	135	135	135	97	97	97	97
S13+S14	2010		203	201	202	201	143	142	143	141
Waste	2025		330	327	325	321	228	226	224	222
	2050		703	724	669	688	473	491	447	463
	2100		2144	2136	1893	1892	1315	1294	1142	1129
x211+x312+	1997	(Gg/yr)	4055	4055	4055	4055	2712	2712	2712	2712
x913+x1014	2010		6432	6164	6309	6049	4360	4128	4254	4028
Waste	2025		10420	10538	9919	10026	6975	7139	6570	6722
	2050		19403	21095	17493	18993	12495	13771	11064	12194
	2100		36753	31938	29808	26093	19557	15374	15217	11949
S6	1997	(Tg)	5	5	5	5	5	5	5	5
Newscrap	2010		8	7	7	7	8	7	7	7
	2025		12	13	11	12	12	13	11	12
	2050		21	23	18	20	21	23	18	20
	2100		32	24	25	19	32	24	25	19
S7	1997	(Tg)	274	274	274	274	302	302	302	302
Long-term	2010		411	403	407	400	456	448	453	445
usage	2025		666	657	646	638	740	731	721	712
	2050		1344	1428	1241	1316	1509	1593	1403	1478
	2100		3021	2839	2507	2368	3622	3462	3047	2927
S8	1997	(Tg)	25	25	25	25	34	34	34	34
Short-term	2010		39	38	38	37	52	50	51	50
usage	2025		65	65	61	61	87	86	83	82
	2050		119	132	106	118	164	181	147	162
	2100		197	160	155	126	285	239	226	190
x23	1997	(Gg/yr)	10964	10964	10964	10964	10872	10872	10872	10872
Concentrate	2010		19136	17691	18475	17073	19013	17562	18350	16942
	2025		29952	31543	27715	29206	29695	31293	27449	28947
	2050		49226	55149	42422	47613	48408	54346	41584	46788
	2100		59136	37083	43135	25960	54809	32683	38934	21705
x34p	1997	(Gg/yr)	10603	10603	10603	10603	10513	10513	10513	10513
Refined	2010		18505	17108	17866	16510	18386	16983	17744	16382
primary	2025		28963	30502	26801	28242	28715	30260	26544	27992
	2050		47602	53329	41022	46042	46810	52552	40212	45244
	2100		57184	35859	41711	25103	53000	31605	37649	20989
x34s	1997	(Gg/yr)	2473	2473	2473	2473	2562	2562	2562	2562
Refined	2010		3727	3686	3708	3668	3846	3811	3830	3796
secondary	2025		6098	6008	5969	5881	6347	6249	6226	6132
	2050		12628	13191	11858	12359	13420	13967	12667	13157
	2100		33210	32631	28314	27954	37394	36886	32376	32069
x45	1997	(Gg/yr)	17408	17408	17408	17408	17408	17408	17408	17408
Semis	2010		29709	27817	28844	27008	29709	27817	28844	27008
	2025		46924	48756	43895	45608	46924	48756	43895	45608
	2050		80934	89456	71118	78606	80934	89456	71118	78606
	2100		121999	92785	94591	71942	121999	92785	94591	71942
x56	1997	(Gg/yr)	4526	4526	4526	4526	4526	4526	4526	4526
Newscrap	2010		7724	7233	7499	7022	7724	7233	7499	7022
	2025		12200	12676	11413	11858	12200	12676	11413	11858
	2050		21043	23259	18491	20438	21043	23259	18491	20438
	2100		31720	24124	24594	18705	31720	24124	24594	18705

Table C10 *(Continued)*

			Sc1	Sc2	Sc3	Sc4	Sc5	Sc6	Sc7	Sc8
x64	1997	(Gg/yr)	4332	4332	4332	4332	4332	4332	4332	4332
Newscrap	2010		7477	7024	7270	6830	7477	7024	7270	6830
	2025		11863	12246	11125	11484	11863	12246	11125	11484
	2050		20704	22937	18239	20205	20704	22937	18239	20205
	2100		31604	24295	24566	18884	31604	24295	24566	18884
x57	1997	(Gg/yr)	10306	10306	10306	10306	10306	10306	10306	10306
Long-term	2010		17587	16468	17075	15989	17587	16468	17075	15989
products,	2025		27779	28863	25986	27000	27779	28863	25986	27000
inflow	2050		47913	52958	42102	46535	47913	52958	42102	46535
	2100		72223	54929	55998	42590	72223	54929	55998	42590
x58	1997	(Gg/yr)	2576	2576	2576	2576	2576	2576	2576	2576
Short-term	2010		4397	4117	4269	3997	4397	4117	4269	3997
products,	2025		6945	7216	6496	6750	6945	7216	6496	6750
inflow	2050		11978	13240	10525	11634	11978	13240	10525	11634
	2100		18056	13732	14000	10647	18056	13732	14000	10647
x79	1997	(Gg/yr)	2921	2921	2921	2921	1871	1871	1871	1871
Long-term	2010		4458	4428	4444	4416	2907	2889	2899	2881
products,	2025		7201	7115	7099	7016	4775	4727	4715	4668
outflow	2050		15245	15633	14493	14835	10278	10502	9840	10037
	2100		44231	44605	38188	38642	32358	32755	28504	28927
x810	1997	(Gg/yr)	2026	2026	2026	2026	1836	1836	1836	1836
Short-term	2010		2944	2872	2911	2841	2559	2519	2541	2501
products,	2025		5000	4885	4795	4683	4267	4152	4126	4014
outflow	2050		9692	10705	8744	9656	8660	9404	7899	8571
	2100		17209	14620	13719	11664	16695	14795	13478	11962
x93	1997	(Gg/yr)	1899	1899	1899	1899	1778	1778	1778	1778
Long-term	2010		2897	2878	2889	2870	2762	2745	2754	2737
products to	2025		4681	4625	4614	4560	4536	4490	4479	4434
smelting	2050		9909	10161	9421	9643	9764	9977	9348	9535
	2100		28750	28993	24822	25117	30740	31117	27079	27481
x103	1997	(Gg/yr)	658	658	658	658	872	872	872	872
Short-term	2010		957	933	946	923	1216	1196	1207	1188
products to	2025		1625	1588	1558	1522	2027	1972	1960	1907
smelting	2050		3150	3479	2842	3138	4114	4467	3752	4071
	2100		5593	4751	4459	3791	7930	7027	6402	5682
x34p (x34p+	1997	(%)	61%	61%	61%	61%	60%	60%	60%	60%
x34s+x64)	2010		62%	61%	62%	61%	62%	61%	62%	61%
Shareprimary	2025		62%	63%	61%	62%	61%	62%	60%	61%
of supply to	2050		59%	60%	58%	59%	58%	59%	57%	58%
semis/yr	2100		47%	39%	44%	35%	43%	34%	40%	29%
(x103+x93)	1997	(%)	20%	20%	20%	20%	21%	21%	21%	21%
/(x57+x58)	2010		18%	19%	18%	19%	18%	19%	19%	20%
/yr	2025		18%	17%	19%	18%	19%	18%	20%	19%
Recycling	2050		22%	21%	23%	22%	23%	22%	25%	23%
rate	2100		38%	49%	42%	54%	43%	56%	48%	62%
(x103+x93)	1997	(%)	52%	52%	52%	52%	71%	71%	71%	71%
/(x810+x79)	2010		52%	52%	52%	52%	73%	73%	73%	73%
/yr	2025		52%	52%	52%	52%	73%	73%	73%	73%
Separation	2050		52%	52%	53%	52%	73%	73%	74%	73%
efficiency	2100		56%	57%	56%	57%	79%	80%	80%	81%

Table C11: Global per capita results of the copper system scenarios Sc1BSc8, 1997, 1910, 1925, 2050 and 2100.

		Sc1	Sc2	Sc3	Sc4	Sc5	Sc6	Sc7	Sc8
–S1	1997 (kg/capita)	–75	–75	–75	–75	–75	–75	–75	–75
Primary	2010	–96	–94	–95	–94	–96	–94	–95	–93
resources	2025	–134	–134	–130	–131	–133	–134	–130	–130
	2050	–233	–265	–217	–246	–231	–263	–215	–244
	2100	–518	–731	–440	–624	–504	–711	–426	–604
x12	1997 (kg/cap/yr)	2.1	2.1	2.1	2.1	2.1	2.1	2.1	2.1
Mining	2010	3.1	2.9	3.0	2.8	3.1	2.8	3.0	2.7
	2025	4.1	4.4	3.8	4.1	4.1	4.4	3.8	4.1
	2050	5.8	7.0	5.0	6.1	5.7	6.9	4.9	6.0
	2100	6.3	5.8	4.6	4.1	5.8	5.1	4.2	3.4
S11	1997 (kg/capita)	8.8	8.8	8.8	8.8	8.8	8.8	8.8	8.8
Gangue	2010	10.7	10.6	10.7	10.5	10.7	10.5	10.6	10.5
	2025	14.4	14.4	14.0	14.1	14.3	14.4	13.9	14.0
	2050	24.2	27.4	22.6	25.5	24.0	27.2	22.4	25.3
	2100	52.5	74.2	44.7	63.5	51.2	72.2	43.4	61.5
x2 11	1997 (kg/cap/yr)	0.2	0.2	0.2	0.2	0.2	0.2	0.2	0.2
Gangue	2010	0.3	0.3	0.3	0.3	0.3	0.3	0.3	0.3
	2025	0.4	0.4	0.4	0.4	0.4	0.4	0.4	0.4
	2050	0.6	0.7	0.5	0.6	0.6	0.7	0.5	0.6
	2100	0.6	0.6	0.5	0.4	0.6	0.5	0.4	0.3
S12	1997 (kg/capita)	2.6	2.6	2.6	2.6	2.6	2.6	2.6	2.6
Slag	2010	3.3	3.3	3.3	3.3	3.3	3.3	3.3	3.3
	2025	4.7	4.7	4.6	4.6	4.7	4.7	4.6	4.6
	2050	8.4	9.5	7.9	8.9	8.4	9.5	7.9	8.9
	2100	20.4	29.3	17.7	25.5	20.4	29.3	17.7	25.5
x3 12	1997 (kg/cap/yr)	0.1	0.1	0.1	0.1	0.1	0.1	0.1	0.1
Slag	2010	0.1	0.1	0.1	0.1	0.1	0.1	0.1	0.1
	2025	0.1	0.2	0.1	0.1	0.1	0.2	0.1	0.1
	2050	0.2	0.3	0.2	0.2	0.2	0.3	0.2	0.2
	2100	0.3	0.3	0.2	0.3	0.3	0.3	0.2	0.3
S13	1997 (kg/capita)	4.4	4.4	4.4	4.4	0.4	0.4	0.4	0.4
Waste	2010	6.2	6.2	6.2	6.2	0.6	0.6	0.6	0.6
long-term	2025	9.1	9.2	9.1	9.2	0.8	0.8	0.8	0.8
	2050	18.1	19.5	17.7	19.1	1.7	1.8	1.7	1.8
	2100	65.2	98.9	59.9	90.8	6.4	9.7	6.0	9.0
x9 13	1997 (kg/cap/yr)	0.2	0.2	0.2	0.2	0.0	0.0	0.0	0.0
Waste	2010	0.2	0.2	0.2	0.2	0.0	0.0	0.0	0.0
long-term	2025	0.3	0.3	0.3	0.3	0.0	0.0	0.0	0.0
	2050	0.6	0.6	0.5	0.6	0.1	0.1	0.1	0.1
	2100	1.5	2.2	1.3	1.9	0.2	0.2	0.1	0.2
S14	1997 (kg/capita)	7.2	7.2	7.2	7.2	4.7	4.7	4.7	4.7
Waste	2010	9.2	9.2	9.2	9.2	6.2	6.2	6.2	6.2
short-term	2025	13.0	13.0	12.8	12.8	8.6	8.7	8.6	8.6
	2050	24.4	26.9	23.3	25.6	16.4	17.9	15.8	17.2
	2100	67.7	100.2	59.5	88.3	48.2	72.1	42.7	64.0

Table C11 *(Continued)*

			Sc1	Sc2	Sc3	Sc4	Sc5	Sc6	Sc7	Sc8
x10 14	1997	(kg/cap/yr)	0.2	0.2	0.2	0.2	0.2	0.2	0.2	0.2
Waste	2010		0.3	0.3	0.3	0.3	0.2	0.2	0.2	0.2
short-term	2025		0.4	0.4	0.4	0.4	0.3	0.3	0.3	0.3
	2050		0.7	0.8	0.6	0.7	0.5	0.6	0.4	0.5
	2100		1.1	1.4	0.9	1.1	0.8	1.1	0.7	0.9
S11+S12+	1997	(kg/capita)	23	23	23	23	17	16	17	16
S13+S14	2010		29	29	29	29	21	21	21	20
Waste	2025		41	41	41	41	28	29	28	28
	2050		75	83	71	79	51	56	48	53
	2100		206	303	182	268	126	183	110	160
x2 11+	1997	(kg/cap/yr)	0.7	0.7	0.7	0.7	0.5	0.5	0.5	0.5
x3 12+	2010		0.9	0.9	0.9	0.9	0.6	0.6	0.6	0.6
+x9 13	2025		1.3	1.3	1.2	1.3	0.9	0.9	0.8	0.9
+x10 14	2050		2.1	2.4	1.9	2.2	1.3	1.6	1.2	1.4
Waste	2100		3.5	4.5	2.9	3.7	1.9	2.2	1.5	1.7
S6	1997	(kg/capita)	0.8	0.8	0.8	0.8	0.8	0.8	0.8	0.8
New	2010		1.1	1.1	1.1	1.0	1.1	1.1	1.1	1.0
scrap	2025		1.5	1.6	1.4	1.5	1.5	1.6	1.4	1.5
	2050		2.2	2.7	2.0	2.3	2.2	2.7	2.0	2.3
	2100		3.0	3.4	2.4	2.7	3.0	3.4	2.4	2.7
S7	1997	(kg/capita)	46.9	46.8	46.9	46.8	51.8	51.6	51.8	51.6
Long-term	2010		59.6	58.6	59.1	58.1	66.2	65.1	65.7	64.6
usage	2025		83.0	83.2	80.5	80.7	92.3	92.6	89.8	90.1
	2050		143.5	164.0	132.5	151.2	161.1	183.1	149.7	169.8
	2100		290.1	402.2	240.7	335.5	347.8	490.4	292.6	414.8
S8	1997	(kg/capita)	4.2	4.2	4.2	4.2	5.9	5.8	5.9	5.8
Short-term	2010		5.7	5.5	5.6	5.4	7.6	7.3	7.4	7.2
usage	2025		8.1	8.2	7.7	7.8	10.8	10.9	10.3	10.4
	2050		12.7	15.1	11.3	13.5	17.5	20.8	15.7	18.6
	2100		18.9	22.7	14.9	17.9	27.3	33.8	21.7	26.9
x23	1997	(kg/cap/yr)	1.9	1.9	1.9	1.9	1.9	1.9	1.9	1.9
Concen-	2010	2.8	2.6	2.7	2.5	2.8	2.6	2.7	2.5	
trate	2025		3.7	4.0	3.5	3.7	3.7	4.0	3.4	3.7
	2050		5.3	6.3	4.5	5.5	5.2	6.2	4.4	5.4
	2100		5.7	5.3	4.1	3.7	5.3	4.6	3.7	3.1
x34p	1997	(kg/cap/yr)	1.8	1.8	1.8	1.8	1.8	1.8	1.8	1.8
Refined	2010		2.7	2.5	2.6	2.4	2.7	2.5	2.6	2.4
primary	2025		3.6	3.9	3.3	3.6	3.6	3.8	3.3	3.5
	2050		5.1	6.1	4.4	5.3	5.0	6.0	4.3	5.2
	2100		5.5	5.1	4.0	3.6	5.1	4.5	3.6	3.0
x34s	1997	(kg/cap/yr)	0.4	0.4	0.4	0.4	0.4	0.4	0.4	0.4
Refined	2010		0.5	0.5	0.5	0.5	0.6	0.6	0.6	0.6
secondary	2025		0.8	0.8	0.7	0.7	0.8	0.8	0.8	0.8
	2050		1.3	1.5	1.3	1.4	1.4	1.6	1.4	1.5
	2100		3.2	4.6	2.7	4.0	3.6	5.2	3.1	4.5

Table C11 *(Continued)*

		Sc1	Sc2	Sc3	Sc4	Sc5	Sc6	Sc7	Sc8
x45	1997 (kg/cap/yr)	3.0	3.0	3.0	3.0	3.0	3.0	3.0	3.0
Semis	2010	4.3	4.0	4.2	3.9	4.3	4.0	4.2	3.9
	2025	5.9	6.2	5.5	5.8	5.9	6.2	5.5	5.8
	2050	8.6	10.3	7.6	9.0	8.6	10.3	7.6	9.0
	2100	11.7	13.1	9.1	10.2	11.7	13.1	9.1	10.2
x56	1997 (kg/cap/yr)	0.8	0.8	0.8	0.8	0.8	0.8	0.8	0.8
New scrap	2010	1.1	1.1	1.1	1.0	1.1	1.1	1.1	1.0
	2025	1.5	1.6	1.4	1.5	1.5	1.6	1.4	1.5
	2050	2.2	2.7	2.0	2.3	2.2	2.7	2.0	2.3
	2100	3.0	3.4	2.4	2.7	3.0	3.4	2.4	2.7
x64	1997 (kg/cap/yr)	0.7	0.7	0.7	0.7	0.7	0.7	0.7	0.7
New scrap	2010	1.1	1.0	1.1	1.0	1.1	1.0	1.1	1.0
	2025	1.5	1.6	1.4	1.5	1.5	1.6	1.4	1.5
	2050	2.2	2.6	1.9	2.3	2.2	2.6	1.9	2.3
	2100	3.0	3.4	2.4	2.7	3.0	3.4	2.4	2.7
x57	1997 (kg/cap/yr)	1.8	1.8	1.8	1.8	1.8	1.8	1.8	1.8
Long-term	2010	2.6	2.4	2.5	2.3	2.6	2.4	2.5	2.3
products,	2025	3.5	3.7	3.2	3.4	3.5	3.7	3.2	3.4
inflow	2050	5.1	6.1	4.5	5.3	5.1	6.1	4.5	5.3
	2100	6.9	7.8	5.4	6.0	6.9	7.8	5.4	6.0
x58	1997 (kg/cap/yr)	0.4	0.4	0.4	0.4	0.4	0.4	0.4	0.4
Short-term	2010	0.6	0.6	0.6	0.6	0.6	0.6	0.6	0.6
products,	2025	0.9	0.9	0.8	0.9	0.9	0.9	0.8	0.9
inflow	2050	1.3	1.5	1.1	1.3	1.3	1.5	1.1	1.3
	2100	1.7	1.9	1.3	1.5	1.7	1.9	1.3	1.5
x79	1997 (kg/cap/yr)	0.5	0.5	0.5	0.5	0.3	0.3	0.3	0.3
Long-term	2010	0.6	0.6	0.6	0.6	0.4	0.4	0.4	0.4
products,	2025	0.9	0.9	0.9	0.9	0.6	0.6	0.6	0.6
outflow	2050	1.6	1.8	1.5	1.7	1.1	1.2	1.1	1.2
	2100	4.2	6.3	3.7	5.5	3.1	4.6	2.7	4.1
x8 10	1997 (kg/cap/yr)	0.3	0.3	0.3	0.3	0.3	0.3	0.3	0.3
Short-term	2010	0.4	0.4	0.4	0.4	0.4	0.4	0.4	0.4
products,	2025	0.6	0.6	0.6	0.6	0.5	0.5	0.5	0.5
outflow	2050	1.0	1.2	0.9	1.1	0.9	1.1	0.8	1.0
	2100	1.7	2.1	1.3	1.7	1.6	2.1	1.3	1.7
x93	1997 (kg/cap/yr)	0.3	0.3	0.3	0.3	0.3	0.3	0.3	0.3
Long-term	2010	0.4	0.4	0.4	0.4	0.4	0.4	0.4	0.4
products	2025	0.6	0.6	0.6	0.6	0.6	0.6	0.6	0.6
to smelting	2050	1.1	1.2	1.0	1.1	1.0	1.1	1.0	1.1
	2100	2.8	4.1	2.4	3.6	3.0	4.4	2.6	3.9
x10 3	1997 (kg/cap/yr)	0.1	0.1	0.1	0.1	0.1	0.1	0.1	0.1
Short-term	2010	0.1	0.1	0.1	0.1	0.2	0.2	0.2	0.2
products	2025	0.2	0.2	0.2	0.2	0.3	0.2	0.2	0.2
to smelting	2050	0.3	0.4	0.3	0.4	0.4	0.5	0.4	0.5
	2100	0.5	0.7	0.4	0.5	0.8	1.0	0.6	0.8

APPENDIX D

GLOSSARY

aerobic refers to an environment in which oxygen is present
acid mine drainage acidic water, usually from the oxidation of pyrite, which drains from areas disturbed by mining
adsorption the process by which liquids, gases, or dissolved substances attach to the surface of solids; compare to ion exchange
anaerobic referring to an environment in which oxygen is lacking
anion a negatively charged ion
atomic weight weight of an atom (usually on a scale of oxygen = 16), determined largely by the number of protons and neutrons
basic oxygen furnace steelmaking furnace with a jet of pure oxygen
beneficiation physical separation of an ore mineral from its ore by crushing, grinding, froth flotation, and other methods
bioaccumulation accumulation of an element or compound in living organisms; extreme bioaccumulation is biomagnification
biosphere all living and dead organic matter on Earth
blast furnace furnace used to convert iron ore to pig iron
blister copper copper produced by smelting; commonly contains gold, silver, and other metals that must be removed in a refinery
brass an alloy of copper and zinc
bronze an alloy of copper and tin
calcine to heat and drive off a gas that is part of a compound; often used for driving off CO_2 from limestone ($CaCO_3$) to form lime (CaO)
catalyst a substance that enhances the rate of a chemical reaction without participating in it
cation a positively charged ion
chelate a compound with a ring-like structure, in which a central cation is connected to each surround anion by at least two bonds; commonly seen in organometallic molecules
compound a substance containing more than one element and having properties different from its constituent elements
concentrate a product in which the proportion of ore mineral has been increased significantly above its concentration in ore; usually by crushing, grinding, or mineral separation by flotation
continuous smelting smelting process that includes roasting, matte production, and conversion to metal in one continuous process
dross waste product taken off molten metal, essentially metallic in character
electric arc furnace steelmaking furnace that is used largely with scrap iron or sponge iron
electrowinning concentration of one or more dissolved ions onto electrodes

placed in a solution; basis for hydrometallurgy; commonly referred to as solvent extraction-electrowinning (SX-EW)

flash furnace smelter in which injections of concentrate, preheated air, and flux are heated rapidly to high temperatures

flocculation electrostatic attraction of colloidal particles to one another to form larger grains

flue-gas desulfurization (FGD) removal of sulfur from gases emitted by power plants, smelters, refineries and other industrial installations

flux substance added to lower the melting temperature of a material

fly ash small ash particles that go up the stack during the combustion of coal and other materials

froth flotation method used to make a concentrate from a slurry of pulverized ore by coating one or more minerals with an organic liquid that causes them to attach to bubbles and float to the surface

fulvic acid humic material that remains in solution when alkaline extract (humic acid) is acidified

galvanized steel steel coated with a thin film of zinc to prevent corrosion

gangue mineral waste minerals in an ore deposit, commonly quartz and calcite

greenhouse gas a gas (especially H_2O, CO_2, CH_4) that prevents incoming solar energy from being radiated back into space, thus warming the Earth

humic acid humic material that is soluble in dilute alkali solutions but precipitates when the solution is acidified

hydrometallurgy the separation of a desired metal from an ore or concentrate by dissolution and later precipitation or electrowinning

ion atom or group of atoms that has lost (cation) or gained (anion) one or more electrons

kiln long horizontal cylindrical furnace that is tilted slightly to allow material to progress through it; widely used in the production of cement, iron or pellets and other mineral products

lime CaO; produced by calcining limestone or calcite

lithosphere (1) entire solid Earth (environmental geochemistry); (2) the rigid, outer 100 km of Earth including crust and upper mantle (geology) marginal demand extra demand for a commodity; usually related to short-term market changes

matte molten metal sulfide, formed by melting sulfide minerals in the early stages of smelting, usually involving loss of some S as SO_2 mineral a naturally occurring, inorganic, crystalline solid with a regular chemical composition

mineral deposit any unusual mineral concentration, regardless of whether it can be extracted at a profit (see ore deposit)

mobile (mobility) in geochemistry the term refers to the degree to which an element or compound can be dispersed from it rock or mineral source, often by dissolution in water

ore deposit mineral deposit that can be extracted at a profit

organometallic compound compound in which metals are connected to complex molecules consisting largely of carbon and hydrogen
overburden rock or regolith covering a mineral or ore deposit
pyrolysis chemical decomposition by heat
pyrometallurgy smelting processes based on thermal decomposition of ore minerals
reduction chemical process in which valance electrons are added to elements; commonly characterized by the scarcity of free oxygen
refining removal of trace amounts of impurity elements from a smelted metal
regolith unconsolidated rock and soil material underlying bedrock
resource concentration of natural material that can be extracted now or in the future; includes reserves and is divided into demonstrated resources that have been measured to some degree and inferred resources that are only thought to be present
reverberatory furnace a type of furnace used in copper and other types of smelting
roast to heat a rock, mineral or ore to drive away gases
rotary kiln a type of kiln that rotates, causing mineral material to flow from one end to the other
scrubber device to recover SO_2 and other acid-forming gases produced by combustion of coal and other fossil fuels and by smelters
slag calcium-rich silicate waste material from smelting of metal ore concentrates
slimes very fine-grained rock and mineral material produced by excessive grinding during beneficiation
smelt to separate metal from its ore mineral by pyrometallurgy
solvent extraction (SX) concentration of a desired element (such as gold) from a primary solution by dissolving it in a smaller volume of a second solution that is mixed with the primary solution but separates from it because it is immiscible; first step in solvent extraction-electrowinning (SX-EW) process
steel iron-based alloy containing up to 2 percent C
strip mining form of open pit mining in which flat-lying ore bodies are mined in linear zones (strips) and the overburden is replaced after mining
tailings waste material remaining after processing of pulverized ore to make a concentrate of the desired ore mineral(s)
tetraethyl lead organometallic compound $Pb(C_2H_5)_4$ added to some gasoline to prevent premature combustion

Subject index

A
abiotic oxidation 47
accumulation
in agricultural soils 177
of byproduct 98
of copper in the anthroposphere 70- 80
acid
mine drainage 47
rock drainage (ARD) 47, 175, 177, 180
wastes 50, 51
acidification 102, 174, 175, 204, 208
adenosine-triphosphate (ATP) 102
agitation leaching 34
agricultural soils
accumulation of copper and zinc 177
emissions of copper and zinc 177
airborne emissions
global estimates 54
aluminum 6, 61, 172
applications 61
exergy content 40
sulfate 146
annual consumption of copper 74
uncertainty in the estimates 74
anthropogenic
activities 195
emissions 79, 81, 202
sources 201
stocks 170
anthroposphere
accumulation of copper stocks 70-80
antimony 130, 165, 174
aqueous
acidic dissolution 205
phase speciation 203, 203
ARD, see acid rock drainage
arsenic 14, 131-135, 161, 165, 174, 177-179
benign use 179
emissions 132-134
historical uses 131, 178
atmospheric emissions 120-127, 132, 201
atomic substitution 16
ATP (adenosine-triphosphate) 102
automotive brake linings 64
availability
of copper 169-182
of recoverable metal 151

B
baghouse dust 164
Banin Navrot plots 209
behavior of copper in soil 201-210
beneficiation 23, 30-33, 103
wastes 48-50
bimodal distribution 16
bio-availability 102
biological activity 22
bismuth 135, 174
box model 4
brass 6, 60, 118, 154, 155
applications 60
recycling 118
scrap 154
bronze 154
Bronze Age 6, 7
byproducts 98, 101-147, 165-167, 173
availability 173, 174
relationships in metals 129
uses 173, 174

C
cadmium 135-140, 165, 166, 174, 177-179
consumption patterns 137
emissions 133, 138, 139
pollution 179
recovery rate 165
recycling 166
zinc ratio 179
calcine 114
calibration of the copper model 219-227
carcinogens 4, 5
cations 204, 206
cementation 34, 35
ceramic superconductors 69, 172
chalcopyrite ores 34
chelate 207
chemical
exergy 194, 195

mechanical polishing 159
uses of copper 64, 81
chemistry of
copper 21-23
lead 101
zinc 101
Chile
copper smelters 33
chimney-manto deposits 19
chloride leaching 34
clay minerals 204
collection rate of copper 232
complex formation 204-210
concentration 211
from sulfide ores 30, 31
in sewage water 82
in the open ocean 52
techniques 152, 174
conductivity
of copper 59
consumption
annual 74, 216
definition 130
of copper 12, 74, 211, 239
in electrical equipment 59
patterns 59-70
scenarios 97, 227-239
trends 59-70
contaminants 159, 160, 181
contingent valuation 2
Contop cyclone smelting 37
converters 37, 38
continuous 37
copper compounds 64
copper cycle 85
copper extraction
variations in rate 10
copper mixtures 198
copper nickel alloys 63
copper smelts 35
copper system scenarios 220, 244-252
corrosion
rate 82
resistance 21, 59, 81, 82
cradle to cradle approach 6
versus cradle to grave approach 6, 176
cumulative copper in use 70

D
demands 59-100, 171-173
for copper wire 62
deposition 18, 29, 69, 102-104, 122-124, 133, 140, 159, 178
acid 98
atmospheric 138, 139, 177
desulfurization 23
dilute aqueous waste streams 159
dioxide emissions 146
Disability Adjusted Life Years 5
discovery rate 170
disposition 59-100
dissipative losses 79, 80-83
via chemicals 80
via copper-containing products 80
via corrosion 80
via food 80
distribution in fresh water 22
drossing 106
double peak distribution 19
ductility 59
dump leaching 33

E
economic flows 75
eco-toxicity 3, 4
efficiency measures 30
electric furnaces 35, 36
electrical
applications 60, 68
conductivity 21, 59
power usage 83
electrification 172
electrolytic refining 38, 39
electronic product recycling 158
electrowinning 34, 38, 50, 51
El Teniente smelter 36
embodiment of copper
in landfills 70
in long-lived goods 70
in mine waste dumps 70
in short-lived goods 70
in soil 70
emissions 50, 54, 120-127, 162
airborne 54
arsenic 132-134
atmospheric 120-127
cadmium 133, 138, 139, 140

dioxide 146
in agricultural soils 177
environmental 154
global 55-57
from metal consumption 130
from metallurgical operations 125
in the ocean 57
lead 120-127
metallic 126
monetary damage estimates 127
natural 79
non-ferrous 127
production related 46-56
recycling 161, 162
wastes 161, 162
zinc 120-127, 177
end uses 155, 161
energy
concept 193-199
economizing on 61
efficiency measure 43
flows 40-44, 106, 114-117
recovery 168
versus exergy 40
enrichment 102
soil 103
environmental 208
damage approach 3
implications 98
themes approach 3
exergy
accounting
advantages 195, 196
calculations 193-195
definition and description 193-195
chemical 194-196
concept 193
environmental indicator 195-197
flows 40-45, 106, 114-117
intensiveness 17
kinetic 193
potential 193
resource accounting 195-197
versus energy 40
waste accounting 195-195
exhaustible resources 169
exploitable reserves 18
Europe
cadmium deposition 140
copper balance 68
lead deposition 124

F
facing 64
final electrolytic refining 23, 106
finished goods 213
finishing operations
wastes from 54
fire refining 38, 39
flash
converting 37, 39
smelters 35, 36, 174
flotation
agents 31, 157
in non-ferrous metal mills 32
fluidized bed roasters 35
forecasting assumptions
economic growth 89
intensity of use 89
models 83
prices 100
recycling efficiency 89
foundry production 24
French copper production 24
froth flotation 30
fuels composition 197-200
future demands 83-100, 149-168

G
galvanizing 117
gangue 48, 49, 80, 175, 213
geochemical mineralization 16
geology of
copper 14-18
lead 18, 19
zinc 18, 19
germanium 140, 141, 166
German copper production 25
by products 50
emission 49, 50
energy consumption 50
waste 49, 50
glass fiber 62, 68
global
consumption of refined copper 91, 232, 234, 236
copper model 86, 211-252

copper production 12, 22, 23, 30, 93, 94, 171, 240-242
copper recycling efficiency 90, 246
copper recycling rate 90, 246
demand 180
electrification 175
energy system 171-173
estimates of airborne emissions 54
peak production 152
population 230-232
primary copper share 245
resources 181
stock of long-lived copper products 95, 243
stock of short-lived copper products 96, 244
stock of waste copper 95, 242, 243
resources of copper 181
Global Warming Potentials 3
gold 11, 141, 142, 151, 166, 173, 177
gravity separation 30, 157

H
heap leaching 33, 34, 39, 50
heat exchangers 63
heavy metals
buildup 177, 180
in soil 201-210
high temperature incineration 168
high temperature superconductors (HST) 69, 70, 172
historical background
of copper production 6
of mining 6-13
home scrap 153
household wiring 172
HST, see high temperature superconductors
Hubbert-type depletion model 96
human toxicity 3, 5, 128
monetary damage estimates 128
hydro-metallurgical methods 33, 164, 174
hydrothermal deposits 18, 19
categories 18, 10

I
income pattern 83
indicators for elements 80
indium 142, 166, 173
industry transformation 176
inner-sphere complexes 207
in situ leaching 33
intensity of use 83-87, 228, 229
historical 84, 228, 229
modeled 85, 86, 228, 229
Isasmelt lance process 37
isomorphic substitution 205

J
Japan
copper balance 67
copper production 26

K
kinetic energy 193

L
lamellae 205
lance converting 37
last mile connection 62
LCA, see life cycle assessment
leaching
of copper 33, 34
wastes 50, 51
leaching-cementation process 34
lead 6, 11, 101-147, 169, 173, 201
availability 173, 174
behavior in soil 201-210
biological effect 102
chemical uses 109, 110
chemistry 101-103
crystal 110
compounds 110
concentration levels in Greenland ice 13
demand 111, 112
deposition 124
emissions 120-127
exergy content 40
historical production 13
mining 109, 110
exergy flows 107, 108
mass flows 104
mixtures 198
peak production 173
physical properties 101-103
primary unit mass 108

process technology 103-108
recycling 151, 175, 176
secondary 164
solder 111
sources 109-112
uses 109-112, 173, 174
wastes 120-127
life cycle
analogy 1
assessment 1, 3-6
characterization 2, 3
inventory analysis 1
methodology 1
results 4
weighting 2, 3
perspective 1
life time of copper 212, 232
liquid process wastes 168
long-term
forecasting 83
economic growth 89
intensity of use 89
recycling efficiency 89
prospect 180-182
usage 215
long-lived copper products 225
losses 80-83, 167, 168
types 85
low grade scrap 156

M
magnetic separation process 156
malleability 59
market for copper
future trends 65
marmatite 112
mass
balance 178
flows of copper production 28
metals
accumulated use 19
future availability 19
in scrap 157
in soils 201
mill tailings 48
mineralogical barrier 17, 19, 44
mini-mills 149
mining 27-30
environmental implications 98
history 6-13, 169, 170
production 221
wastes 46-48, 76
Mississippi Valley Type 18
Mitsubishi lance smelter 36
mixtures composition 197-200
molybdenum 151
multimedia modeling 4
multiple hearths 35

N
natural emissions 79
net energy analysis 3
new scrap 153
nickel-cadmium batteries 138, 139
nickel-zinc batteries 119
non-electrical uses of copper 62, 63, 107
non-ferrous metal recycling system 176
non-recoverable losses 81
no observable effect level 4
Noranda smelters 35, 36, 159, 174, 176

O
old scrap
domestic 71
recovery efficiency 70
open pit mining 27
optimal
control theory 57
extraction 56, 57
production 46-56
ore
lower grade 154
minerals 8
mining 23, 103, 104, 112
outer-sphere complexes 207
overburden 46-49, 175
oxidation rate 47, 203
treatment strategies 47
oxidation-reduction (redox) 203, 204
oxide ores 18, 33

P
physical properties of
copper 21-23
lead 101
zinc 101
platinum 173

pollution problems 177-180
polyfunctionality 207
post consumer copper scrap 154
potential energy 193
precipitation 22
Predicted No-Effect Concentration 4
prices 71, 72, 98, 175, 181
  declining 150, 171
  forecasting 100
primary processing
  resources 211
  future trends 39, 40
process technology 23-27
  of lead 103
  of the copper cycle 85
pyrolysis 156
pyro-metallurgical operations 23, 114, 174
pyro-metallurgy 174

Q
quantity-grade distribution function 169

R
recoverable
  copper deposits 17
    exploitable reserves 18
  metal 71, 151, 153, 167, 168
recovery efficiency 73, 151, 156
  from old scrap 153-157, 221-224
  of
    byproduct metals 165
    electronic scrap 157-161
    lead 151, 162, 163
    zinc 151, 164
  rate 27
recycling
  byproduct metals 165
  efficiency 70, 73, 150, 151, 176, 181
  electronic scrap 157-161
  emissions 161, 162
  future 149-168
  lead 162, 163
  nature of use 167
  old scrap 117
  performance 150
  rate 70, 150, 151
  steel 159-161
  technology 149
  wastes 54, 161, 162
  zinc 164
redox (oxidation-reduction) 203
reduction technology 174
reference system 194
refinery output 23
refining 114, 213
  final 106
  fire 38, 39
  technology 174
regional
  consumption of refined copper 91-93, 234-238
  population 230-232
reserves 30, 170, 171
  official 170
resource
  accounting 195
  exhaustion 98
reverberatory furnaces 35, 39
rhenium 142, 173
roasting 35-37, 112-114
roofing 64
runoff ration 81

S
scarce elements 14
  chemical characteristics 15
  distributional patterns 14-16
  loss reducing 167
scrap 153-155
  electronic 157-161
  high grade 154
  low grade 154, 156
  metals content 157
  new 213
  old 117, 156, 219
  recycling 150, 153-157
  recovery 153-157, 221-224
  reprocessors 149
secondary
  copper 150, 153
  lead 162, 163
  metallurgical operations 150
  minerals 204
  steel 159, 160. 162
    quality 181
  zinc 164
sedimentary exhalative deposits 18, 19

selenium 143, 167, 173
semi-manufactures 213
separation efficiency 70
sewage water 82
SFA, see Substance Flow Analysis
shadow prices 2
short-lived copper products 225
short-term usage 219
silver 11, 143, 167, 173, 177
sintering 105, 112
skarn deposits 19
slag 213
smelters
  types 35
smelting 35-37, 105, 106, 114, 213
  production 221, 226, 227
  wastes 51-54
soil
  adsorption reactions 206
  acidic 203, 208
  behavior of heavy metals 201-210
  chemistry 206, 207
  functional groups 205
  organic material 207
  polluted 201
  solution 203
solid phase constituents 204-210
solvent extraction 34, 35, 38, 50
  electro-winning technology (SX-EW) 23, 38, 39, 50, 171, 174
sorption 22
sources of copper 21-58
sphalerite 112
steel
  copper contamination 159
stocks 70, 85
  categories 70
  changes 77, 78
submerged tuyere converting 37
Substance Flow Analysis (SFA) 5
  in a life cycle perspective 5
sulfide 112
  galena 103
  ores 32, 33, 101
    concentration 30, 31
    reduction 174
sulfur 11, 143-147, 161
  balance 175
  recovery 44-46, 144, 145, 174-177
    efficiency 45
    improvements 46
sulfuric acid
  uses 145, 146
supergene sulfide ores 33
supply of copper 21-58
Sweden
  annual change in the copper store 78
  copper balance 66
  copper production 26
  cumulative evolution of the copper reservoir 78
  use of copper compounds 64, 65
swell ratio 47, 49
SX-EW, see solvent extraction – electrowinning
system scenarios 232-252

T
tailings 47, 49, 120
target system 194
telecommunications channel capacity 62, 63
tellurium 147, 167, 173
terrestrial environment 177
thallium 147, 174
thermal conductivity 21, 59
thermodynamic distance 194, 196
topsoil
  , buildup of heavy metals in 177
toxic
  byproducts 177-181
  heavy metals 180
  Release Inventory (TRI) 3
  releases 54
  time bombs 180
  trace
  elements, essential versus non-essential 102
  levels 22
traditional copper markets 68
TRI, see Toxic Release Inventory

U
Uniform System for the Evaluation of Substances 4
United Kingdom
  copper production 25
United States

accumulation of copper-in-use 74, 75
copper ore grade percent 29
copper prices 72
copper production 24, 99
electronic product recycling 158
unit risks 4

V
vacuum distillation 160
vat leaching 34, 38
Vanyukov smelters 37
vein deposits 19
volcanogenic massive sulfide deposits 19

W
waste
acid 52, 54
cumulative 76
electrical 156
electronic 156
from
finishing operations 54
industry 76
long-term usage 219
municipalities 76
short-term usage 219
mine-related 76, 80
production-related 46-56, 162
recycling 54, 161, 162
separation 167
sources 152
water pipes 63
willingness to accept (WTA) 2
willingness to pay (WTP) 2
limitations 3
wire 62
world
copper production 30, 84
regions 83, 84, 211
definition 214

Z
zinc 6, 11, 101-147, 169, 173
availability 173, 174
behavior in soil 201-210
biological effect 101
chemical uses 119
chemistry 101-103
emissions 120-127, 177
exergy content 40
exergy flows 115, 116
losses 119
mining 115-116
mass flows 114
mixtures 198
physical properties 101-103
processing 112-117
recycling 118, 175, 176
recovery efficiency 118, 151
scrap recycling 117
secondary 164
sources 117-120
uptake 101
uses 117-120, 173, 174
wastes 120-127

Eco-Efficiency in Industry and Science

1. J.E.M. Klostermann and A. Tukker (eds.): *Product Innovation and Eco-efficiency.* Twenty-three Industry Efforts to Reach the Factor 4. 1997 ISBN 0-7923-4761-7
2. K. van Dijken, Y. Prince, T. Wolters, M. Frey, G. Mussati, P. Kalff, O. Hansen, S. Kerndrup, B. Søndergård, E. Lopes Rodrigues and S. Meredith (eds.): *Adoption of Environmental Innovations.* The Dynamics of Innovation as Interplay Between Business Competence, Environmental Orientation and Network Involvement. 1999 ISBN 0-7923-5561-X
3. M. Bartolomeo, M. Bennett, J.J. Bouma, P. Heydkamp, P. James, F. de Walle and T. Wolters: *Eco-Management Accounting.* 1999 ISBN 0-7923-5562-8
4. P.P.A.A.H. Kandelaars: *Economic Models of Material-Product Chains for Environmental Policy Analysis.* 1999 ISBN 0-7923-5794-9
5. J. de Beer: *Potential for Industrial Energy-Efficiency Improvement in the Long Term.* 2000 ISBN 0-7923-6282-9
6. K. Green, P. Groenewegen and P.S. Hofman (eds.): *Ahead of the Curve.* Cases of Innovation in Environmental Management. 2001 ISBN 0-7923-6804-5
7. J.B. Guinée (ed.): *Handbook on Life Cycle Assessment.* Operational Guide to the ISO Standards. 2002 ISBN 1-4020-0228-9
8. T.J.N.M. de Bruijn and A. Tukker (eds.): *Partnership and Leadership.* Building Alliances for a Sustainable Future. 2002 ISBN 1-4020-0431-1
9. M. Bennett, J.-J. Bouma and T. Wolters (eds.): *Environmental Management Accounting.* Informal and Institutional Developments. 2002 ISBN 1-4020-0552-0; Pb: ISBN 1-4020-0553-9
10. N. Wrisberg and H.A. Udo de Haas (eds.): *Analytical Tools for Environmental Design and Management in a Systems Perspective.* 2002 ISBN 1-4020-0626-8
11. R. Heijungs and S. Suh: *The Computational Structure of Life Cycle Assessment.* 2002 ISBN 1-4020-0672-1
12. M. Bennett, P.M. Rikhardsson and S. Schaltegger (eds.): *Environmental Management Accounting - Purpose and Progress.* 2003 ISBN 1-4020-1365-5; Pb: ISBN 1-4020-1366-3

KLUWER ACADEMIC PUBLISHERS – DORDRECHT / BOSTON / LONDON